Lecture Notes in Computer Scie

Commenced Publication in 1973
Founding and Former Series Editors:
Gerhard Goos, Juris Hartmanis, and Jan van Leeuwen

Editorial Board

Alan P. Sexton (Ed.)

Dataspace: The Final Frontier

26th British National Conference on Databases, BNCOD 26
Birmingham, UK, July 7-9, 2009
Proceedings

Volume Editor

Alan P. Sexton
University of Birmingham
School of Computer Science
Edgbaston, Birmingham B15 2TT, UK
E-mail: a.p.sexton@cs.bham.ac.uk

Library of Congress Control Number: 2009929446

CR Subject Classification (1998): H.4, H.3, H.2, H.2.8, I.2.4, I.2.6

LNCS Sublibrary: SL 3 – Information Systems and Application, incl. Internet/Web and HCI

ISSN 0302-9743
ISBN-10 3-642-02842-X Springer Berlin Heidelberg New York
ISBN-13 978-3-642-02842-7 Springer Berlin Heidelberg New York

springer.com

Printed in Germany

Typesetting: Camera-ready by author, data conversion by Scientific Publishing Services, Chennai, India
Printed on acid-free paper SPIN: 12708833 06/3180 5 4 3 2 1 0

Preface

This volume contains the proceedings of the 26th British National Conference on Databases (BNCOD 2009) held in July 2009 at the University of Birmingham. The theme of the conference was "Dataspace: The Final Frontier," a topic that only recently rose to attention and shows promise of crystallizing many disparate research threads into a new unified vision for the future of databases.

We were very grateful to have Michael Franklin of the University of California, Berkeley, USA, and Domenico Laforenza of the Italian National Research Council (CNR) give the keynote speeches of the conference. Michael is one of the originators of the dataspace concept, and brought us up to date with current developments and future directions in the field in his opening keynote address. Domenico is one of the leading figures in the world on the topic of grid computing, an area that is tackling some of the major issues of management and manipulation of enormous amounts of data. In his closing keynote talk, he presented XtreemOS, a European project on developing an operating system platform for grid computing.

In a new departure for BNCOD, this year we held two tutorials. The first, continuing the theme of dataspaces, was presented by Jens Dittrich of Saarland University, Saarbrücken, Germany, in which he presented iMeMex, one of the first dataspace management systems. System design concepts, dataspace modelling, dataspace indexing, dataspace query processing, and pay-as-you-go information integration were all discussed and important lessons learnt in the project as well as open research challenges were presented. The second tutorial was given by Wenfei Fan, Floris Geerts and Xibei Jia of the University of Edinburgh, UK. Their presentation was on a practical, effective and theoretically well-founded approach, based on conditional dependencies, for detecting and repairing errors in real-life data.

Three satellite workshops were co-located with BNCOD 2009: the 7th International Workshop on Teaching, Learning and Assessment of Databases (TLAD 2009), is a regular associate of BNCOD. New friends this year were the 4th International Workshop on Ubiquitous Computing (iUBICOM 2009) and the First International Workshop on Database Architectures for the Internet of Things (DAIT 2009). A summary/review paper for each of the three workshops is included in these proceedings.

The PhD Forum is a regular event at BNCOD and this year was no exception. PhD students presented the current state of their work to an expert panel who were able to offer helpful advice and guidance. Invited talks at the forum presented different views of how to survive the PhD process and make a career in academia.

There were 33 submissions to BNCOD 2009 from 12 countries. Each submission was blind reviewed by at least three Programme Committee members.

The Programme Committee decided to accept 12 full, 2 short and 5 poster papers. All accepted papers are reproduced in these proceedings, with the full papers divided into four sections: "Data Integration, Warehousing and Privacy," "Alternative Data Models," "Querying," and "Path Queries and XML."

I would particularly like to thank the keynote and tutorial speakers for their contributions, the authors for submitting their papers, the Programme Committee for their reviews, the Organizing Committee for the great effort they contributed, the BNCOD Steering Committee, who were always on hand with helpful advice, the EasyChair conference management system, which hugely simplified much of the work involved in organizing the conference, and, last, but by no means least, all the attendees who form the community that makes this conference such an enjoyable and worthwhile event each year.

May 2009 Alan P. Sexton

Organization

Programme Chair

Alan P. Sexton	University of Birmingham, UK

Programme Committee

David Bell	Queen's University Belfast, UK
Sharma Chakravarthy	University of Texas at Arlington, USA
Richard Cooper	University of Glasgow, UK
Alfredo Cuzzocrea	Universita della Calabria, Italy
Barry Eaglestone	University of Sheffield, UK
Suzanne Embury	University of Manchester, UK
Alvaro Fernandes	University of Manchester, UK
Mary Garvey	University of Wolverhampton, UK
Alex Gray	University of Wales, Cardiff, UK
Georg Gottlob	University of Oxford, UK
Jun Hong	Queen's University Belfast, UK
Ela Hunt	University of Strathclyde, UK
Mike Jackson	Birmingham City University, UK
Anne James	Coventry University, UK
Keith Jeffery	CLRC Rutherford Appleton, UK
Graham Kemp	Chalmers, Sweden
Jessie Kennedy	Napier University, UK
Lachlan M. MacKinnon	University of Abertay Dundee, UK
Nigel Martin	Birkbeck College, University of London, UK
Ken Moody	University of Cambridge, UK
David Nelson	University of Sunderland, UK
Moira Norrie	ETH Zurich, Switzerland
Werner Nutt	Free University of Bozen-Bolzano, Italy
Norman Paton	University of Manchester, UK
Alex Poulovassilis	Birkbeck College, University of London, UK
Mark Roantree	Dublin City University, Ireland
Alan P. Sexton	University of Birmingham, UK
Jianhua Shao	Cardiff University, UK
Richard Swinbank	University of Birmingham, UK
Stratis Viglas	University of Edinburgh, UK
Richard White	Cardiff University, UK
John Wilson	University of Strathclyde, UK

Extra Reviewers

Peter Tiňo	University of Birmingham, UK

Local Organizing Committee

Mary Garvey	University of Wolverhampton, UK	(PhD Forum)
Mike Jackson	Birmingham City University, UK	(Exhibitions)
Anne James	Coventry University, UK	(Workshops)
Alan P. Sexton	University of Birmingham, UK	(Local)

BNCOD Steering Committee

Lachlan M. MacKinnon	University of Abertay Dundee, UK (Chair)
Richard Cooper	University of Glasgow, UK
Barry Eaglestone	University of Sheffield, UK
Jun Hong	Queen's University Belfast, UK
Anne James	Coventry University, UK
Keith Jeffery	CLRC Rutherford Appleton, UK
David Nelson	University of Sunderland, UK
Alan P. Sexton	University of Birmingham, UK
Jianhua Shao	Cardiff University, UK

Table of Contents

Keynote Talks

Tutorials

Data Integration, Warehousing and Privacy

Alternative Data Models

Querying

Path Queries and XML

Short Papers

Posters

Co-located Workshops

Dataspaces: Progress and Prospects

Michael J. Franklin

Computer Science Division, EECS
University of California, Berkeley
franklin@cs.berkeley.edu

Abstract. The concept of Dataspaces was proposed in a 2005 paper by Franklin, Halevy and Maier as a new data management paradigm to help unify the diverse efforts on flexible data models, data cleaning, and data integration being pursued by the database community. The key idea is to expand the scope of database technology by moving away from the schema-first nature of traditional data integration techniques and instead, support a spectrum of degrees of structure, schema conformance, entity resolution, quality and so on. Dataspaces offer an initial, lightweight set of useful services over a group of data sources, allowing users, administrators, or programs to improve the semantic relationships between the data in an incremental and iterative manner. This "pay-as-you-go" philosophy is at the heart of the Dataspaces approach. In this talk, I will revisit the motivation behind Dataspaces, outline the evolution of thinking on the topic since the 2005 proposal, and describe our current work focused on developing a metrics-driven framework for Dataspace systems.

1 Introduction

The traditional "Schema First" approach for integrating multiple data sources requires a significant up front analysis and design effort and has an "all or nothing" behavior in which an integrated system is non-functional until a certain critical mass is reached. In many practical scenarios, these significant up-front costs can doom the integration effort to failure due to a lack of perceived progress. In contrast, with Dataspaces, data sources can be made immediately accessible, albeit with perhaps limited functionality. Functionality and quality can then be improved over time based on a "pay-as-you-go" approach.

For example, in a Dataspace system, keyword search can be provided as base functionality for most types of data or metadata. A limitation of the basic search paradigm, however, is that it identifies relevant data but does not directly support the manipulation, summarization, and analysis of such data. In a Dataspace system, structure can be added incrementally to the data along several dimensions. Schema mapping and integration techniques can be used to create metadata to for query processing and data manipulation. Information extraction can be used to collect key subsets of data for use by structured-data applications. Inconsistency among data items can be resolved through both automated and human-assisted techniques. Taking a cue from Web 2.0 technologies, collaborative techniques can be used to allow users to pool their efforts to continually improve the quality of the Dataspace.

A.P. Sexton (Ed.): BNCOD 2009, LNCS 5588, pp. 1–3, 2009.

Pay-as-you-go thinking has gained currency in a number of areas that involve the use of data from many disparate sources. These include Personal Information Management, Health Information, eScience, and extraction of information from the Web. A key observation is that as the scale of these systems increases, automated integration techniques become increasingly important. A pay-as-you-go approach can be used to tolerate errors that arise in such processes as well as to direct the computational and human resources required to address such errors. Given the tremendous growth in data volume and diversity being driven by web-based and web-enabled applications, it is likely that pay-as-you-go integration will remain the focus of significant research effort going forward.

2 Recent Projects

In the talk, I will discuss two of our recent efforts related to Dataspaces:

Directed User-Feedback for Reference Reconciliation: A primary challenge in large-scale Dataspace integration is creating semantic equivalences between elements from different data sources that correspond to the same real-world entity or concept. While automated techniques for such matching are continually being improved, significant gains in accuracy can be gained through the judicious application of user resources for confirming proposed matches. We developed a decision-theoretic framework for ordering candidate matches for user confirmation using the concept of the value of perfect information (VPI). At the core of this concept is a utility function that quantifies the desirability of a given state; thus, we devise a utility function for dataspaces based on query result quality. It is then possible to apply VPI in concert with this utility function to order user confirmations.

Functional Dependencies for Schema Discovery: Functional dependencies (FDs) were developed in Relational Database theory for improving the quality of schemas. An FD "$X \longrightarrow Y$" indicates a relationship between attribute sets X and Y, such that any two entities that share a value for X must also share a value for Y. FDs generalize the notion of a "key" in relational databases, and as such, provide important semantic clues about data consistency. In a pay-as-you-go data integration system, functional dependencies can play an expanded role, helping to improve the fidelity of automatically generated mediated schemas, pinpointing low quality data sources, and cleaning data from multiple, disparate data sources. In order to be used in this manner, however, the FDs must be learned by observing the data, and must be extended to be probabilistic rather than absolute. The degree of conformance to an FD can then be used as a metric for directing further processing.

3 On-Going Directions

Our ongoing work is currently focused on three areas:

A Metrics-driven Framework: A key requirement of pay-as-you-go is the need for a measurement framework for gauging progress in terms of the quality and accuracy of information integration. The starting point for such a framework is the development of a set of metrics for judging the "goodness" of information integration across a number of information types and use cases. These metrics will then be analyzed and where possible, unified, so that a more general measurement framework can be developed. Such a framework will serve as a key component for future Dataspace management systems, and could provide a grounding for other information integration solutions. Fortunately, there has been significant work on data quality metrics that can serve as a starting point for this effort.

Collaborative Data Cleaning: An area of emerging interest is that of Web 2.0-inspired approaches to data sharing and collaborative data analysis, such as those developed by companies such as ManyEyes and Swivel. An important thrust in this area is the development of systems and abstractions for identifying, understanding, and perhaps resolving "interesting" aspects of a data set such as outliers, errors, or unexpected behaviors. At the user level, such mechanisms are typically attached to a visualization of the data, but underlying such visualizations there is a data model and query model that describes how the user representation maps to the underlying data. Such systems lend themselves nicely to voting schemes and other mechanisms where the "wisdom of crowds" can be brought to bear on the understanding and integration of large sets of information. A key research question here is how to leverage such techniques to better understand information and data integration quality.

Probabilistic Information Extraction: Information Extraction (IE) is the task of automatically extracting structured information from unstructured, machine-readable documents, such as the extraction of instances of structured records about corporate mergers, e.g., MergerBetween(company1,company2, date), from an online news sentence such as: "Yesterday, New York-based Foo Inc. announced the acquisition of Bar Corp." Probabilistic graphical models such as Conditional Random Fields (CRFs) are a promising technology for such extraction, achieving fairly high accuracy for some tasks. In on-going work, we are integrating CRF models and related inferencing techniques into a database query processor, to enable Information Extraction to be performed on-the-fly, effectively performing pay-as-you-go extraction. As part of this work, we are integrating a notion of probability (and more importantly, probability distributions) into the database storage and processing engine. An important task is to examine how to leverage the probabilities generated during the extraction process to help guide further extraction and identify portions of a large Dataspace where additional information or more effort is required.

Acknowledgements

The research described here is joint work with many others, including: Anish Das Sharma, Luna Dong, Minos Garofalakis, Alon Halevy, Joe Hellerstein, Shawn Jeffery, Dave Maier, Eirinaios Michelakis, and Daisy Wang. This work is supported in part by NSF grant IIS-0415175 and by a Google Research Award.

XtreemOS: Towards a Grid-Enabled Linux-Based Operating System

Domenico Laforenza

Director of the Institute for Informatics and Telematics (IIT-CNR)
and associate member at the
High Performance Computing Laboratory
Italian National Research Council (CNR)
ISTI-CNR, Via Giuseppe Moruzzi 1, 56126, Pisa, Italy
domenico.laforenza@iit.cnr.it, domenico.laforenza@isti.cnr.it

Extended Abstract

The term *"Grid Computing"* was introduced at the end of 90s by Foster and Kesselman; it was envisioned as *"an important new field, distinguished from conventional distributed computing by its focus on large-scale resource sharing, innovative applications, and, in some cases, high-performance orientation"* [1].

Defining Grids has always been difficult, but nowadays there is general agreement that Grids are distributed systems enabling the creation of Virtual Organizations (VOs) [2] in which users can share, select, and aggregate a wide variety of geographically distributed resources, owned by different organizations, for solving large-scale computational and data intensive problems in science, engineering, and commerce. These platforms may include any kind of computational resources like supercomputers, storage systems, data sources, sensors, and specialized devices.

Since the end of the 90s, a lot of water has passed under the bridge, and several researchers have proposed to revise the initial definition of Grid. More recently, researchers at the European Network of Excellence, CoreGrid [3], reached an agreement on the following definition: a Grid is *"a fully distributed, dynamically reconfigurable, scalable and autonomous infrastructure to provide location independent, pervasive, reliable, secure and efficient access to a coordinated set of services encapsulating and virtualizing resources (computing power, storage, instruments, data, etc.) in order to generate knowledge"*.

This is a more modern service-oriented vision of the Grid that stems from the conviction that, in the mid to long term, the great majority of complex software applications will be dynamically built by composing services, which will be available in an open market of services and resources. In this sense, the Grid will be conceived as a "world-wide cyber-utility" populated by cooperating services interacting as in a complex and gigantic software ecosystem.

In order to manage Grid platforms several approaches have been proposed. Some of these, which, for simplicity, we might call *"à la Globus"* [4], are based on middleware layers that link, in a loosely-coupled way, the user applications and the underlying distributed multi-domain heterogeneous resources. The use

A.P. Sexton (Ed.): BNCOD 2009, LNCS 5588, pp. 4–6, 2009.

of Grid middleware is one of the most widely adopted approaches. These middleware layers are used to address the complexity of Grid platforms and to help the user to use Grid resources in an integrated way. In some of the current Grid middleware systems (e.g., Globus [4], EGEE gLite [5] and UNICORE [6]), operating system support for Grid computing is quite minimal or non-existent because they have been developed with different goals in mind. The à la Globus approaches are designed as "sum of services" infrastructures, in which tools are developed independently in response to current needs of users. In particular, Globus started out with the bottom-up premise that a Grid must be constructed as a set of tools developed from user requirements, and consequently its versions (GT2, GT3, GT4) are based on the combination of working components into a composite Grid toolkit that fully exposes the Grid to the programmer.

While Grid Computing has gained much popularity over the past few years in scientific contexts, it is still cumbersome to use effectively in business and industrial environments. In order to help remedy this situation, several researchers proposed building a true Grid Operating System (GOS) [7,8,9,10,11]. A GOS is a distributed operating system targeted for a large-scale dynamic distributed architecture, with a variable amount of heterogeneous resources (resource may join, leave, churn). A GOS should be aware of virtual organizations (VO), spanning multiple administrative domains with no central management of users and resources (multiple administrators, resource owners, VO managers). This GOS should be composed of a consistent set of integrated system services, providing a stable Posix-like interface for application programmers. Moreover, abstractions or jobs (set of processes), files, events, etc. should be provided by a GOS.

This talk will present XtreemOS [12], a first European step towards the creation of a true open source *operating system* for Grid platforms. The XtreemOS project aims to address this challenge by designing, implementing, experimenting with and promoting an operating system that will support the management of *very large* and *dynamic* ensembles of resources, capabilities and information composing virtual organizations. Recognizing that Linux is today the prevailing operating system, XtreemOS started an effort to extend Linux towards Grid, including native support for the VOs management, and providing appropriate interfaces to the GOS services. As will be explained in this talk, the result is neither a "true" Grid operating system nor a Grid middleware environment, but rather a Linux operating systems with tightly integrated mechanisms for the quick and user-friendly creation of distributed collaborations, which share their resources in a secure and user-friendly way.

References

1. Foster, I., Kesselman, C. (eds.): The Grid: Blueprint for a new computing infrastructure. Morgan Kaufmann, CA (1999)
2. Foster, I., Kesselman, C., Tuecke, S.: The anatomy of the grid - enabling scalable virtual organizations. International Journal of Supercomputer Applications 15, 200–222 (2001)
3. CoreGrid, `http://www.coregrid.net/`

4. Globus, http://www.globus.org/ and http://www.globus.org/
5. EGEE gLite, http://glite.web.cern.ch/glite/
6. Unicore, http://unicore.sourceforge.net/
7. Vahdat, A., Anderson, T., Dahlin, M., Belani, E., Culler, D., Eastham, P., Yoshikawa, C.: WebOS: Operating system services for wide area applications. In: Proceedings of the Seventh Symposium on High Performance Distributed Computing (1998)
8. Krauter, K., Maheswaran, M.: Architecture for a grid operating system. In: Buyya, R., Baker, M. (eds.) GRID 2000. LNCS, vol. 1971, pp. 65–76. Springer, Heidelberg (2000)
9. Pike, R., Presotto, D., Dorward, S., Flandrena, B., Thompson, K., Trickey, H., Winterbottom, P.: Plan 9 from Bell Labs. Computing Systems 8(3), 221–245 (1995), http://plan9.belllabs.com/plan9dist/
10. Padala, P., Wilson, J.N.: GridOS: Operating system services for grid architectures. In: Pinkston, T.M., Prasanna, V.K. (eds.) HiPC 2003. LNCS, vol. 2913, pp. 353–362. Springer, Heidelberg (2003)
11. Mirtchovski, A.: Plan 9 — an integrated approach to grid computing. In: Proceedings of the International Parallel and Distributed Processing Symposium (IPDPS 2004) - Workshop 17, p. 273 (2004)
12. The XtreemOS European Project, http://www.xtreemos.eu/

The iMeMex Dataspace Management System: Architecture, Concepts, and Lessons Learned

Jens Dittrich

Saarland University

Abstract. The iMeMex Project was one of the first systems trying to build a so-called dataspace management system. This tutorial presents the core concepts of iMeMex. We discuss system design concepts, dataspace modelling, dataspace indexing, dataspace query processing, and pay-as-you-go information integration. We will present some important lessons learned from this project and also discuss ongoing and open research challenges.

CV

Jens Dittrich is an Associate Professor of Computer Science at Saarland University (Germany). He received his Diploma and PhD from the University of Marburg (Germany). He held positions at SAP AG (Germany) and ETH Zurich (Switzerland). His research interests are in the area of information systems and databases, and, in particular, new system architectures for information management, indexing, data warehousing, and main memory databases.

Web-site: `http://infosys.cs.uni-sb.de`

Slides and Other Material

Slides will be made available electronically[1] after the talk. An accompanying article will appear in the Data Engineering Bulletin, special issue on "New Avenues in Search" in June 2009.

[1] `http://infosys.cs.uni-sb.de/jensdittrich.php`

A.P. Sexton (Ed.): BNCOD 2009, LNCS 5588, p. 7, 2009.

Conditional Dependencies: A Principled Approach to Improving Data Quality

Wenfei Fan[1], Floris Geerts[2], and Xibei Jia[2,⋆]

[1] University of Edinburgh and Bell Laboratories
[2] University of Edinburgh

Abstract. Real-life data is often dirty and costs billions of pounds to businesses worldwide each year. This paper presents a promising approach to improving data quality. It effectively detects and fixes inconsistencies in real-life data based on conditional dependencies, an extension of database dependencies by enforcing bindings of semantically related data values. It accurately identifies records from unreliable data sources by leveraging relative candidate keys, an extension of keys for relations by supporting similarity and matching operators across relations. In contrast to traditional dependencies that were developed for improving the quality of schema, the revised constraints are proposed to improve the quality of data. These constraints yield practical techniques for data repairing and record matching in a uniform framework.

1 Introduction

Real-world data is often dirty: inconsistent, inaccurate, incomplete and/or stale. Dirty data may have disastrous consequences for everyone. Indeed, the following are real-life examples in the US, taken from [23]: (a) 800 houses in Montgomery County, Maryland, were put on auction block in 2005 due to mistakes in the tax payment data of Washington Mutual Mortgage. (b) Errors in a database of a bank led thousands of completed tax forms to be sent to wrong addresses in 2005, effectively helping identity thieves get hold of the names and bank account numbers of various people (c) The Internal Revenue Service (IRS) accused people for overdue tax caused by errors in the IRS database. There is no reason to believe that the scale of the problem is any different in the UK, or in any society that is dependent on information technology.

The costs and risks of dirty data are being increasingly recognized by all industries worldwide. Recent statistics reveal that enterprises typically find data error rates of approximately 1%–5%, and for some companies it is above 30% [28]. It is reported that dirty data costs US businesses 600 billion dollars annually [10], and that erroneously priced data in retail databases alone costs US consumers $2.5 billion each year [12]. It is also estimated that data cleaning accounts for 30%-80% of the development time and budget in most data warehouse projects [29].

⋆ Wenfei Fan and Floris Geerts are supported in part by EPSRC EP/E029213/1. Fan is also supported in part by a Yangtze River Scholar Award. Jia is supported by an RSE Fellowship.

A.P. Sexton (Ed.): BNCOD 2009, LNCS 5588, pp. 8–20, 2009.

While the prevalent use of the Web has made it possible to extract and integrate data from diverse sources, it has also increased the risks, on an unprecedented scale, of creating and propagating dirty data.

There has been increasing demand for data quality tools, to add accuracy and value to business processes. A variety of approaches have been put forward: probabilistic, empirical, rule-based, and logic-based methods. There have been a number of commercial tools for improving data quality, most notably ETL tools (extraction, transformation, loading), as well as research prototype systems, *e.g.*, Ajax, Potter's Wheel, Artkos and Telcordia (see [2,23] for a survey). Most data quality tools, however, are developed for a specific domain (*e.g.*, address data, customer records). Worse still, these tools often heavily rely on manual effort and low-level programs that are difficult to write and maintain [27].

To this end, integrity constraints yield a principled approach to improving data quality. Integrity constraints, *a.k.a.* data dependencies, are almost as old as relational databases themselves. Since Codd introduced functional dependencies in 1972, a variety of constraint formalisms have been proposed and widely used to improve *the quality of schema*, via normalization (see [13] for a survey). Recently, constraints have enjoyed a revival, for improving *the quality of data.*

Constraint-based methods specify data quality rules in terms of integrity constraints such that errors and inconsistencies in a database emerge as violations of the constraints. Compared to other approaches, constraint-based methods enjoy several salient advantages such as being declarative in nature and providing the ability to conduct inference and reasoning. Above all, constraints specify a fundamental part of the semantics of the data, and are capable of capturing semantic errors in the data. These methods have shown promise as a systematic method for reasoning about the semantics of the data, and for deducing and discovering data quality rules, among other things (see [7,14] for recent surveys).

This paper presents recent advances in constraint-based data cleaning. We focus on two problems central to data quality: (a) data repairing, to detect and fix inconsistencies in a database [1], and (b) record matching [11], to identify tuples that refer to the same real-world entities. These are undoubted top priority for every data quality tool. As an example [23], a company that operates drug stores successfully prevents at least one lawsuit per year that may result in at least a million dollars award, by investing on a tool that ensures the consistency between the medication histories of its customers and the data about prescription drugs. As another example, a recent effort to match records on licensed airplane pilots with records on individuals receiving disability benefits from the US Social Security Administration revealed forty pilots whose records turned up on both databases [23]. Constraints have proved useful in both data repairing and record matching.

For constraints to be effective for capturing errors and matching records in real-life data, however, it is necessary to revise or extend traditional database dependencies. Most work on constraint-based methods is based on traditional constraints such as functional dependencies and inclusion dependencies. These constraints were developed for schema design, rather than for data cleaning. They are not capable of detecting errors commonly found in real-life data.

In light of this, we introduce two extensions of traditional constraints, namely, conditional dependencies for capturing consistencies in real-life data (Section 2), and relative candidate keys for record matching (Section 3).

While constraint-based functionality is not yet available in commercial tools, practical methods have been developed for data cleaning, by using the revised constraints as data quality rules. We present a prototype data-quality system, based on conditional dependencies and relative candidate keys (Section 4).

The area of constraint-based data cleaning is a rich source of questions and vitality. We conclude the paper by addressing open research issues (Section 5).

2 Adding Conditions to Constraints

One of the central technical questions associated with data quality is how to characterize the consistency of data, *i.e.*, how to tell whether the data is clean or dirty. Traditional dependencies, such as functional dependencies (FDs) and inclusion dependencies (INDs), are required to hold on entire relation(s), and often fail to capture errors and inconsistencies commonly found in real-life data.

We circumvent these limitations by extending FDs and INDs through enforcing patterns of semantically related values; these patterns impose *conditions* on what part of the data the dependencies are to hold and which combinations of values should occur together. We refer to these extensions as *conditional functional dependencies* (CFDs) and *conditional inclusion dependencies* (CINDs), respectively.

2.1 Conditional Functional Dependencies

Consider the following relational schema for customer data:

customer (CC, AC, phn, name, street, city, zip)

where each tuple specifies a customer's phone number (country code CC, area code AC, phone phn), name and address (street, city, zip code). An instance D_0 of the customer schema is shown in Fig. 1.

Functional dependencies (FDs) on customer relations include:

$$f_1\text{: [CC, AC, phn]} \rightarrow \text{[street, city, zip]}, \qquad f_2\text{: [CC, AC]} \rightarrow \text{[city]}.$$

That is, a customer's phone uniquely determines her address (f_1), and the country code and area code determine the city (f_2). The instance D_0 of Fig. 1 satisfies f_1 and f_2. In other words, if we use f_1 and f_2 to specify the consistency of customer data, *i.e.*, to characterize errors as violations of these dependencies, then no errors or inconsistencies are found in D_0, and D_0 is regarded clean.

A closer examination of D_0, however, reveals that none of its tuples is error-free. Indeed, the inconsistencies become obvious when the following constraints are considered, which intend to capture the semantics of customer data:

cfd_1: ([CC = 44, zip] → [street])
cfd_2: ([CC = 44, AC = 131, phn] → [street, city = 'EDI', zip])
cfd_3: ([CC = 01, AC = 908, phn] → [street, city = 'MH', zip])

	CC	AC	phn	name	street	city	zip
t_1:	44	131	1234567	Mike	Mayfield	NYC	EH4 8LE
t_2:	44	131	3456789	Rick	Crichton	NYC	EH4 8LE
t_3:	01	908	3456789	Joe	Mtn Ave	NYC	07974

Fig. 1. An instance of customer relation

Here cfd_1 asserts that for customers in the UK ($\mathsf{CC} = 44$), zip code uniquely determines street. In other words, cfd_1 is an "FD" that is to hold on the subset of tuples that satisfies the pattern "$\mathsf{CC} = 44$", *e.g.*, $\{t_1, t_2\}$ in D_0. It is not a traditional FD since it is defined with constants, and it is not required to hold on the entire customer relation D_0 (in the US, for example, zip code does not determine street). The last two constraints refine the FD f_1 given earlier: cfd_2 states that for any two UK customer tuples, if they have area code 131 and have the same phn, then they must have the same street and zip, and moreover, the city *must* be EDI; similarly for cfd_3.

Observe that tuples t_1 and t_2 in D_0 violate cfd_1: they refer to customers in the UK and have identical zip, but they differ in street. Further, while D_0 satisfies f_1, each of t_1 and t_2 in D_0 violates cfd_2: $\mathsf{CC} = 44$ and $\mathsf{AC} = 131$, but $\mathsf{city} \neq$ EDI. Similarly, t_3 violates cfd_3.

These constraints extend FDs by incorporating conditions, and are referred to as *conditional functional dependencies* (CFDs). CFDs are introduced in [16] to capture inconsistencies in a single relation. As shown by the example above, CFDs are capable of detecting errors and inconsistencies commonly found in real-life data sources that their traditional counterparts are not able to catch.

2.2 Conditional Inclusion Dependencies

We next incorporate conditions into inclusion dependencies (INDs). Consider the following two schemas, referred to as *source* and *target*:

Source: order (asin, title, type, price)

Target: book (isbn, title, price, format) CD (id, album, price, genre)

The source database contains a single relation order, specifying various types such as books, CDs, DVDs, ordered by customers. The target database has two relations, book and CD, specifying customer orders of books and CDs, respectively. Example source and target instances D_1 are shown in Fig. 2.

To detect errors across these databases, one might be tempted to use INDs:

$$\mathsf{order}\ (\mathsf{title}, \mathsf{price}) \subseteq \mathsf{book}\ (\mathsf{title}, \mathsf{price}), \qquad \mathsf{order}\ (\mathsf{title}, \mathsf{price}) \subseteq \mathsf{CD}\ (\mathsf{album}, \mathsf{price}).$$

One cannot expect, however, to find for each book item in the order table a corresponding CD item in the CD table; this might only hold *provided that* the book is an audio book. That is, there are certain inclusion dependencies from the source to the target, but only under certain *conditions*. The following *conditional inclusion dependencies* (CINDs) correctly reflect the situation:

cind$_1$: (order (title, price, type ='BOOK') $\subseteq$ book (title, price)),
cind$_2$: (order (title, price, type ='CD') $\subseteq$ CD (album, price)),
cind$_3$: (CD (album, price, genre ='a-book') $\subseteq$ book (title, price, format ='audio')).

Here cind$_1$ states that for each order tuple t, if its type is 'book', then there must exist a book tuple t' such that t and t' agree on their title and price attributes; similarly for cind$_2$. Constraint cind$_3$ asserts that for each CD tuple t, if its genre is 'a-book' (audio book), then there must be a book tuple t' such that the title and price of t' are identical to the album and price of t, and moreover, the format of t' must be 'audio'.

	asin	title	type	price
t_4:	a23	Snow White	CD	7.99
t_5:	a12	Harry Potter	book	17.99

(a) Example order data

	isbn	title	price	format
t_6:	b32	Harry Potter	17.99	hard-cover
t_7:	b65	Snow White	7.99	paper-cover

(b) Example book data

	id	album	price	genre
t_8:	c12	J. Denver	7.94	country
t_9:	c58	Snow White	7.99	a-book

(c) Example CD data

Fig. 2. Example order, book and CD data

While D_1 of Fig 2 satisfies cind$_1$ and cind$_2$, it violates cind$_3$. Indeed, tuple t_9 in the CD table has an 'a-book' genre, but it cannot find a match in the book table with 'audio' format. Note that the book tuple t_7 is not a match for t_9: although t_9 and t_7 agree on their album (title) and price attributes, the format of t_7 is 'paper-cover' rather than 'audio' as required by cind$_3$.

Along the same lines as CFDs, conditional inclusion dependencies (CINDs) are introduced [5], by extending INDs with conditions. Like CFDs, CINDs are required to hold only on a subset of tuples satisfying certain patterns. They are specified with constants, and cannot be expressed as standard INDs. CINDs are capable of capturing errors across different relations that traditional INDs cannot detect.

2.3 Extensions of Conditional Dependencies

CFDs and CINDs can be naturally extended by supporting disjunction and inequality. Consider, for example, customers in New York State, in which most cities (CT) have a *unique* area code, except NYC and LI (Long Island). Further, NYC area codes consist of 212, 718, 646, 347 and 917. One can express these as:

ecfd$_1$: CT $\notin$ {NYC, LI} $\rightarrow$ AC
ecfd$_2$: CT $\in$ {NYC} $\rightarrow$ AC $\in$ {212, 718, 646, 347, 917}

where ecfd$_1$ asserts that the FD CT $\rightarrow$ AC holds if CT is *not in* the set {NYC, LI}; and ecfd$_2$ is defined with disjunction: it states that when CT is NYC, AC must be one of 212, 718, 646, 347, *or* 917.

An extension of CFDs by supporting disjunction and inequality has been defined in [4], referred to as eCFDs. This extension is strictly more expressive than CFDs. Better still, as will be seen shortly, the increased expressive power does not make our lives harder when it comes to reasoning about these dependencies.

2.4 Reasoning about Conditional Dependencies

To use CFDs and CINDs to detect and repair errors and inconsistencies, a number of fundamental questions associated with these conditional dependencies have to be settled. Below we address three most important technical problems, namely, the consistency, implication and axiomatizability of these constraints.

Consistency. Given a set Σ of CFDs (resp. CINDs), can one tell whether the constraints in Σ are dirty themselves? If the input set Σ is found inconsistent, then there is *no need* to check the data quality rules against the data at all. Further, the analysis helps the user discover errors in the rules.

This can be stated as the consistency problem for conditional dependencies. For a set Σ of CFDs (CINDs) and a database D, we write $D \models \Sigma$ if $D \models \varphi$ for *all* $\varphi \in \Sigma$. The *consistency problem* is to determine, given Σ defined on a relational schema $\mathcal{R}$, whether there exists a nonempty instance D of $\mathcal{R}$ such that $D \models \Sigma$.

One can specify arbitrary FDs and INDs without worrying about consistency. This is no longer the case for CFDs.

Example 1. Consider a set Σ_0 of CFDs of the form $([A = x \rightarrow [B] = \bar{x})$, where the domain of A is bool, x ranges over true and false, and $\bar{x}$ indicates the negation of x. Then there exists no nonempty instance D such that $D \models \Sigma_0$. Indeed, for any tuple t in D, no matter what $t[A]$ takes, CFDs in Σ_0 force $t[A]$ to take the other value from the finite domain bool. □

Table 1 compares the complexity bounds for the static analyses of CFDs, eCFDs and CINDs with their traditional counterparts. It turns out that while for CINDs the consistency problem is not an issue, for CFDs it is nontrivial. Worse, when CFDs and CINDs are put together, the problem becomes undecidable, as opposed to their trivial traditional counterpart. That is, the expressive power of CFDs and CINDs comes at a price of higher complexity for reasoning about them.

Implication. Another central technical problem is the *implication problem*: given a set Σ of CFDs (resp. CINDs) and a single CFD (resp. CIND) φ defined on a relational schema $\mathcal{R}$, it is to determine whether or not Σ entails φ, denoted by $\Sigma \models \varphi$, *i.e.*, whether for all instances D of $\mathcal{R}$, if $D \models \Sigma$ then $D \models \varphi$. Effective implication analysis allows us to deduce new cleaning rules and to remove redundancies from a given set of rules, among other things.

As shown in Table 1, the implication problem also becomes more intriguing for CFDs and CINDs than their counterparts for FDs and INDs.

In certain practical cases the consistency and implication analyses for CFDs and CINDs have complexity comparable to their traditional counterparts. As an example, for data cleaning in practice, the relational schema is often fixed, and only dependencies vary and are treated as the input. In this setting, the

Table 1. Complexity and finite axiomatizability

Dependencies	Consistency	Implication	Fin. Axiom
CFDs	NP-complete	coNP-complete	Yes
eCFDs	NP-complete	coNP-complete	Yes
FDs	$O(1)$	$O(n)$	Yes
CINDs	$O(1)$	EXPTIME-complete	Yes
INDs	$O(1)$	PSPACE-complete	Yes
CFDs + CINDs	undecidable	undecidable	No
FDs + INDs	$O(1)$	undecidable	No
in the absence of finite-domain attributes			
CFDs	$O(n^2)$	$O(n^2)$	Yes
CINDs	$O(1)$	PSPACE-complete	Yes
eCFDs	NP-complete	coNP-complete	Yes
CFDs + CINDs	undecidable	undecidable	No

consistency and implication problems for CFDs (resp. CINDs) have complexity similar to (resp. the same as) their traditional counterparts; similarly when no constraints are defined with attributes with a finite domain (*e.g.*, bool).

Axiomatizability. Armstrong's Axioms for FDs can be found in almost every database textbook, and are essential to the implication analysis of FDs. A finite set of inference rules for INDs is also in place. For conditional dependencies the finite axiomatizability is also important, as it reveals insight of the implication analysis and helps us understand how data quality rules interact with each other.

This motivates us to find a finite set $\mathcal{I}$ of inference rules that are *sound and complete* for implication analysis, *i.e.*, for any set Σ of CFDs (resp. CIND) and a single CFD (resp. CIND) φ, $\Sigma \models \varphi$ iff φ is provable from Σ using $\mathcal{I}$.

The good news is that when CFDs and CINDs are taken separately, they are finitely axiomatizable [16,5]. However, just like their traditional counterparts, when CFDs and CINDs are taken together, they are not finitely axiomatizable.

3 Extending Constraints with Similarity

Another central technical problem for data quality is record matching, *a.k.a.* record linkage, merge-purge, data deduplication and object identification. It is to identify tuples from (unreliable) relations that refer to the same real-world object. This is essential to data cleaning, data integration and credit-card fraud detection, among other things. Indeed, it is often necessary to correlate information about an object from multiple data sources, while the data sources may not be error free or may have different representations for the same object.

A key issue for record matching concerns how to determine matching keys [2,11], *i.e.*, what attributes should be selected and how they should be compared in order to identify tuples. While there has been a host of work on the topic, record matching tools often require substantial manual effort from human experts, or rely on probabilistic or learning heuristics (see [2,23,11] for surveys).

Constraints can help in automatically deriving matching keys from matching rules, and thus improve match quality and increase the degree of automation. To illustrate this, consider two data sources, specified by the following schemas:

card (c#, SSN, FN, LN, addr, tel, email, type),
billing (c#, FN, SN, post, phn, email, item, price).

Here a card tuple specifies a credit card (number c# and type) issued to a card holder identified by SSN, FN (first name), LN (last name), addr (address), tel (phone) and email. A billing tuple indicates that the price of a purchased item is paid by a credit card of number c#, issued to a holder that is specified in terms of forename FN, surname SN, postal address post, phone phn and email.

Given an instance (D_c, D_b) of (card,billing), for *fraud detection*, one has to ensure that for any tuple $t \in D_c$ and $t' \in D_b$, if $t[\text{c\#}] = t'[\text{c\#}]$, then $t[Y_c]$ and $t'[Y_b]$ refer to the same holder, where $Y_c = [\text{FN, LN, addr, tel, email}]$, and $Y_b = [\text{FN, SN, post, phn, email}]$. Due to errors in the data sources, however, one may not be able to match $t[Y_c]$ and $t'[Y_b]$ via pairwise comparison of their attributes. Further, it is not feasible to manually select what attributes to compare. Indeed, to match tuples of arity n, there are 2^n possible comparison configurations.

One can leverage constraints and their reasoning techniques to derive "best" matching keys. Below are constraints expressing matching keys, which are an extension of relational keys and are referred to *relative candidate keys* (RCKs):

$$\text{rck}_1\text{: card[LN]} = \text{billing[SN]} \wedge \text{card[addr]} = \text{billing[post]} \wedge \text{card[FN]} \approx \text{billing[FN]} \rightarrow \text{card}[Y_c] \rightleftharpoons \text{billing}[Y_b]$$

$$\text{rck}_2\text{: card[email]} = \text{billing[email]} \wedge \text{card[addr]} = \text{billing[post]} \rightarrow \text{card}[Y_c] \rightleftharpoons \text{billing}[Y_b]$$

$$\text{rck}_3\text{: card[LN]} = \text{billing[SN]} \wedge \text{card[tel]} = \text{billing[phn]} \wedge \text{card[FN]} \approx \text{billing[FN]} \rightarrow \text{card}[Y_c] \rightleftharpoons \text{billing}[Y_b]$$

Here rck_1 asserts that if $t[\text{LN, addr}]$ and $t'[\text{SN, post}]$ are identical and if $t[\text{FN}]$ and $t'[\text{FN}]$ are similar *w.r.t.* a similarity operator $\approx$, then $t[Y_c]$ and $t'[Y_b]$ match, *i.e.*, they refer to the same person; similarly for rck_2 and rck_3. Hence instead of comparing the entire Y_c and Y_b lists of t and t', one can inspect the attributes in rck_1–rck_3. If t and t' match on any of rck_1–rck_3, then $t[Y_c]$ and $t'[Y_b]$ *match.*

Better still, one can automatically derive RCKs from given matching rules. For example, suppose that rck_1 and the following matching rules are known, developed either by human experts or via learning from data samples: (a) if $t[\text{tel}]$ and $t'[\text{phn}]$ match, then $t[\text{addr}]$ and $t'[\text{post}]$ should refer to the same address (even if $t[\text{addr}]$ and $t'[\text{post}]$ might be radically different); and (b) if $t[\text{email}]$ and $t'[\text{email}]$ match, then $t[\text{FN, LN}]$ and $t'[\text{FN, SN}]$ match. Then rck_2 and rck_3 can be derived from these rules via automated reasoning.

The derived RCKs, when used as matching keys, can improve match quality: when t and t' differ in some pairs of attributes, *e.g.*, ([addr], [post]), they can still be matched via other, more reliable attributes, *e.g.*, ([LN,tel,FN], [SN,phn,FN]). In other words, true matches may be identified by derived RCKs, even when they cannot be found by the given matching rules from which the RCKs are derived.

In contrast to traditional constraints, RCKs are defined in terms of both equality and similarity; further, they are defined across multiple relations, rather than on a single relation. Moreover, to cope with unreliable data, RCKs adopt a dynamic semantics very different from its traditional counterpart.

Several results have been established for RCKs [14]. (1) A finite inference system has been proposed for deriving new RCKs from matching rules. (2) A quadratic-time algorithm has been developed for deriving RCKs. (3) There are effective algorithms for matching records based on RCKs.

RCKs have also proved effective in improving the performance of record matching processes. It is often prohibitively expensive to compare every pair of tuples even for moderately sized relations [11]. To handle large data sources one often needs to adopt (a) blocking: partitioning the relations into blocks based on certain keys such that only tuples in the same block are compared, or (b) windowing: first sorting tuples using a key, and then comparing the tuples using a sliding window of a fixed size, such that only tuples within the same window are compared (see, *e.g.*, [11]). The match quality is highly dependent on *the choice of keys*. It has been experimentally verified that blocking and windowing can be effectively conducted by grouping similar tuples by (part of) RCKs.

4 Improving Data Quality with Dependencies

While constraints should logically become an essential part of data quality tools, we are not aware of any commercial tools with this facility. Nevertheless, we have developed a prototype system, referred to as SEMANDAQ, for improving the quality of relational data [17]. Based on conditional dependencies and relative candidate keys, the system has proved effective in repairing inconsistent data and matching non-error-free records when processing real-life data from two large US companies. Below we present some functionalities supported by SEMANDAQ.

Discovering data quality rules. To use dependencies as data quality rules, it is necessary to have techniques in place that can *automatically discover* dependencies from sample data. Indeed, it is often unrealistic to rely solely on human experts to design data quality rules via an expensive and long manual process.

This suggests that we settle *the profiling problem*. Given a database instance D, it is to find a *minimal cover* of all dependencies (*e.g.*, CFDs, CINDs) that hold on D, *i.e.*, a non-redundant set of dependencies that is logically equivalent to the set of all dependencies that hold on D. That is, we want to learn informative and interesting data quality rules from data, and prune away trivial and insignificant rules based on a threshold specified by users.

Several algorithms have been developed for discovering CFDs [6,18,22]. SEMANDAQ has implemented the discovery algorithms of [18].

Reasoning about data quality rules. A given set S of dependencies, either automatically discovered or manually designed by domain experts, may be dirty itself. In light of this we have to identify consistent dependencies from S, to be used as data quality rules. This problem is, however, nontrivial. As remarked in Section 2.4, it is already intractable to determine whether a given set S is consistent when S consists of CFDs only. It is also intractable to find a maximum subset of consistent rules from S.

Nevertheless, we have developed an approximation algorithm for finding a set S' of consistent rules from a set S of possibly inconsistent CFDs [16], while guaranteeing that S' is within a constant bound of the maximum consistent subset of S. SEMANDAQ supports this reasoning functionality.

Detecting errors. After a consistent set of data quality rules is identified, the next question concerns how to effectively catch errors in a database by using the rules. Given a set Σ of data quality rules and a database D, we want to *detect inconsistencies* in D, *i.e.*, to find all tuples in D that violate some rule in Σ.

We have shown that given a set Σ of CFDs and CINDs, a fixed number of SQL queries can be *automatically* generated such that, when being evaluated against a database D, the queries return all and only those tuples in D that violate Σ [4,16]. That is, we can effectively detect inconsistencies by leveraging existing facility of commercial relational database systems. This is another feature of SEMANDAQ.

Repairing errors. After the errors are detected, we want to automatically edit the data, fix the errors and make the data consistent. This is known as *data repairing* as formalized in [1]. Given a set Σ of dependencies and an instance D of a database schema $\mathcal{R}$, it is to find a candidate *repair* of D, *i.e.*, an instance D' of $\mathcal{R}$ such that D' satisfies Σ and D' *minimally differs* from the original database D [1]. This is the method that US national statistical agencies, among others, have been practicing for decades for cleaning census data [20,23].

Several repair models have been studied to assess the accuracy of repairs [1,3,8,30] (see [7] for a survey). It is shown, however, that the repairing problem is already intractable when traditional FDs or INDs are considered [3]. Nevertheless, repairing algorithms have been developed for FDs and INDs [3] and for CFDs [9].

SEMANDAQ supports these methods. It automatically generates candidate repairs and presents them to users for inspection, who may suggest changes to the repairs and the data quality rules. Based on users input, SEMANDAQ further improves the repairs until the users are satisfied with the quality of the repaired data. This interactive nature guarantees the *accuracy* of the repairs found.

Record matching. SEMANDAQ supports RCK-based methods outlined in Section 3, including (a) specification of matching rules, (b) automatic derivation of top k quality RCKs from a set of matching rules, for any given k, (c) record matching based on RCKs, and (d) blocking and windowing via RCKs. Compared to record matching facilities found in commercial tools, SEMANDAQ leverages RCKs to explore the semantics of the data and is able to find more accurate matches, while significantly reducing manual effort from domain experts.

5 Open Problems and Emerging Applications

The study of constraint-based data cleaning has raised as many questions as it has answered. While it yields a promising approach to improving data quality and will lead to practical data-quality tools, a number of open questions need to be settled. Below we address some of the open research issues.

The interaction between data repairing and record matching. Most commercial tools either support only one of these, or separate the two processes. However, these processes interact with each other and often need to be combined. The need is particularly evident in master data management (MDM), one of the fastest growing software markets [24]. In MDM for an enterprise, there is typically a collection of data that has already been cleaned, referred to as *master data* or *reference data.* To repair databases by capitalizing on available master data it is necessary to conduct record matching and data repairing at the same time.

To this end conditional dependencies and relative candidate keys allow us to conduct data repairing and record matching in a uniform constraint-based framework. This calls for the study of reasoning about conditional dependencies and relative candidate keys taken together, among other things.

Incomplete information. Incomplete information introduces serious problems to enterprises: it routinely leads to misleading analytical results and biased decisions, and accounts for loss of revenues, credibility and customers. Previous work either assumes a database to be closed (the Closed World Assumption, CWA, *i.e.,* all the tuples representing real-world entities are assumed already in place, but some data elements of the tuples may be missing), or open (the Open World Assumption, OWA, *i.e.,* a database may only be a proper subset of the set of tuples that represent real-world entities; see, *e.g.,* [25,26]). In practice, however, a database is often neither entirely closed nor entirely open. In MDM environment, for instance, master data is a closed database. Meanwhile a number of other databases may be in use, which may have missing tuples or missing data elements, but certain parts of the databases are *constrained by* the master data and are closed. To capture this we have proposed a notion of relative information completeness [15]. Nevertheless practical algorithms are yet to be developed to quantitatively assess the completeness of information *w.r.t.* user queries.

Repairing distributed data. In practice a relation is often fragmented, vertically or horizontally, and is distributed across different sites. In this setting, even inconsistency detection becomes nontrivial: it necessarily requires certain data to be shipped from one site to another. In other words, SQL-based techniques for detecting CFD violations no longer work. It is necessary to develop error detection and repairing methods for distributed data, to minimize data shipment. Another important yet challenging issue concerns the quality of data that is integrated from distributed, unreliable sources. While there has been work on automatically propagating data quality rules (CFDs) from data sources to integrated data [19], much more needs to be done to effectively detect errors during data integration and to propagate corrections from the integrated data to sources.

The quality of Web data. Data quality issues are on an even larger scale for data on the Web, *e.g.,* XML and semistructured data. Already hard for relational data, error detection and repairing are far more challenging for data on the Web. In the context of XML, for example, the constraints involved and their interaction with XML Schema are far more intriguing than their relational counterparts, even for static analysis, let alone for data repairing. In this setting data quality remains, by and large, unexplored (see, *e.g.,* [21]). Another open issue concerns object identification, *i.e.,* to identify complex objects that refer to the same real-world entity, when the objects do not have a regular structure. This is critical not only to data quality, but also to Web page clustering, schema matching, pattern recognition, plagiarism detection and spam detection, among other things. Efficient techniques for identifying complex objects deserve a full exploration.

References

1. Arenas, M., Bertossi, L.E., Chomicki, J.: Consistent query answers in inconsistent databases. In: PODS (1999)
2. Batini, C., Scannapieco, M.: Data Quality: Concepts, Methodologies and Techniques. Springer, Heidelberg (2006)
3. Bohannon, P., Fan, W., Flaster, M., Rastogi, R.: A cost-based model and effective heuristic for repairing constraints by value modification. In: SIGMOD (2005)
4. Bravo, L., Fan, W., Geerts, F., Ma, S.: Increasing the expressivity of conditional functional dependencies without extra complexity. In: ICDE (2008)
5. Bravo, L., Fan, W., Ma, S.: Extending dependencies with conditions. In: VLDB (2007)
6. Chiang, F., Miller, R.: Discovering data quality rules. In: VLDB (2008)
7. Chomicki, J.: Consistent query answering: Five easy pieces. In: Schwentick, T., Suciu, D. (eds.) ICDT 2007. LNCS, vol. 4353, pp. 1–17. Springer, Heidelberg (2006)
8. Chomicki, J., Marcinkowski, J.: Minimal-change integrity maintenance using tuple deletions. Inf. Comput. 197(1-2), 90–121 (2005)
9. Cong, G., Fan, W., Geerts, F., Jia, X., Ma, S.: Improving data quality: Consistency and accuracy. In: VLDB (2007)
10. Eckerson, W.: Data quality and the bottom line: Achieving business success through a commitment to high quality data. The Data Warehousing Institute (2002)
11. Elmagarmid, A.K., Ipeirotis, P.G., Verykios, V.S.: Duplicate record detection: A survey. TKDE 19(1), 1–16 (2007)
12. English, L.: Plain English on data quality: Information quality management: The next frontier. DM Review Magazine (April 2000)
13. Fagin, R., Vardi, M.Y.: The theory of data dependencies - An overview. In: ICALP (1984)
14. Fan, W.: Dependencies revisited for improving data quality. In: PODS (2008)
15. Fan, W., Geerts, F.: Relative information completeness. In: PODS (2009)
16. Fan, W., Geerts, F., Jia, X., Kementsietsidis, A.: Conditional functional dependencies for capturing data inconsistencies. TODS 33(2) (June 2008)
17. Fan, W., Geerts, F., Jia, X.: SEMANDAQ: A data quality system. based on conditional functional dependencies. In: VLDB, demo (2008)
18. Fan, W., Geerts, F., Lakshmanan, L., Xiong, M.: Discovering conditional functional dependencies. In: ICDE (2009)
19. Fan, W., Ma, S., Hu, Y., Liu, J., Wu, Y.: Propagating functional dependencies with conditions. In: VLDB (2008)
20. Fellegi, I., Holt, D.: A systematic approach to automatic edit and imputation. J. American Statistical Association 71(353), 17–35 (1976)

21. Flesca, S., Furfaro, F., Greco, S., Zumpano, E.: Querying and repairing inconsistent XML data. In: Ngu, A.H.H., Kitsuregawa, M., Neuhold, E.J., Chung, J.-Y., Sheng, Q.Z. (eds.) WISE 2005. LNCS, vol. 3806, pp. 175–188. Springer, Heidelberg (2005)
22. Golab, L., Karloff, H., Korn, F., Srivastava, D., Yu, B.: On generating near-optimal tableaux for conditional functional dependencies. In: VLDB (2008)
23. Herzog, T.N., Scheuren, F.J., Winkler, W.E.: Data Quality and Record Linkage Techniques. Springer, Heidelberg (2007)
24. Loshin, D.: Master Data Management, Knowledge Integrity Inc. (2009)
25. Imieliński, T., Lipski Jr., W.: Incomplete information in relational databases. J. ACM 31(4), 761–791 (1984)
26. van der Meyden, R.: Logical approaches to incomplete information: A survey. In: Chomicki, J., Saake, G. (eds.) Logics for Databases and Information Systems, pp. 307–356 (1998)
27. Rahm, E., Do, H.H.: Data cleaning: Problems and current approaches. IEEE Data Eng. Bull. 23(4), 3–13 (2000)
28. Redman, T.: The impact of poor data quality on the typical enterprise. Commun. ACM 41(2), 79–82 (1998)
29. Shilakes, C., Tylman, J.: Enterprise information portals. Merrill Lynch (1998)
30. Wijsen, J.: Database repairing using updates. TODS 30(3), 722–768 (2005)

A Prioritized Collective Selection Strategy for Schema Matching across Query Interfaces

Zhongtian He, Jun Hong, and David A. Bell

School of Electronics, Electrical Engineering and Computer Science,
Queen's University Belfast, Belfast, BT7 1NN, UK
{zhe01,j.hong,da.bell}@qub.ac.uk

Abstract. Schema matching is a crucial step in data integration. Many approaches to schema matching have been proposed. These approaches make use of different types of information about schemas, including structures, linguistic features and data types etc, to measure different types of similarity between the attributes of two schemas. They then combine different types of similarity and use combined similarity to select a collection of attribute correspondences for every source attribute. Thresholds are usually used for filtering out likely incorrect attribute correspondences, which have to be set manually and are matcher and domain dependent. A selection strategy is also used to resolve any conflicts between attribute correspondences of different source attributes. In this paper, we propose a new prioritized collective selection strategy that has two distinct characteristics. First, this strategy clusters a set of attribute correspondences into a number of clusters and collectively selects attribute correspondences from each of these clusters in a prioritized order. Second, it introduces use of a null correspondence for each source attribute, which represents the option that the source attribute has no attribute correspondence. By considering this option, our strategy does not need a threshold to filter out likely incorrect attribute correspondences. Our experimental results show that our approach is highly effective.

1 Introduction

There are tens of thousands of searchable databases on the Web. These databases are accessed through queries formulated on their query interfaces only, which are usually query forms. Query results from these databases are dynamically generated Web pages in response to form-based queries. In many application domains, users are often interested in obtaining information from multiple data sources. Thus, they have to access different Web databases individually via their query interfaces. Web data integration aims to develop systems that allow users to make queries on a uniform query interface and be responded with a set of combined results from multiple data sources automatically.

Schema matching across query interfaces is a crucial step in Web data integration, which finds attribute correspondences between the uniform query interface and a local query interface. In general, schema matching takes two schemas

A.P. Sexton (Ed.): BNCOD 2009, LNCS 5588, pp. 21–32, 2009.

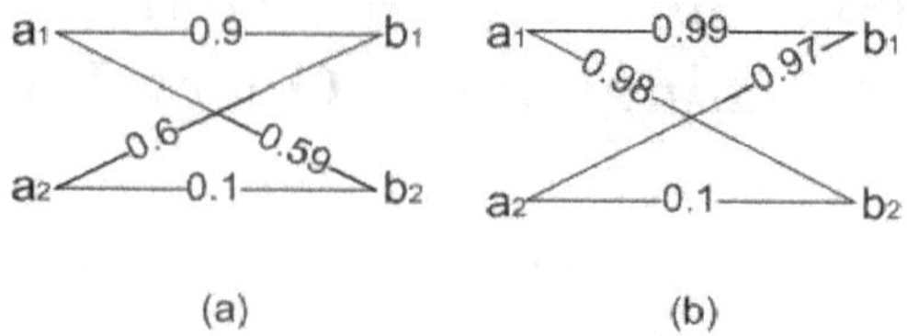

Fig. 1. Resolving Conflicts between Attribute Correspondences

as input and produces a set of attribute correspondences between them [1,2]. The problem of schema matching has been extensively studied and many approaches have been proposed [2,3,5,6,7,8,9,10,11]. These approaches make use of whatever types of information they can find about attributes of schemas to match attributes between them. Typically different types of similarity between the attributes of two schemas are measured and combined. Based on combined similarity between attributes, a collection of attribute correspondences are selected for every source attribute. Sometimes a selection strategy is also used to resolve any conflicts between the candidate attribute correspondences of different source attributes, that is, when the best candidate attribute correspondences of two or more source attributes are the same.

A number of selection strategies have been proposed [4,5,7,17]. Three common selection strategies are *maximal sum* [17], *greedy* [4,7] and *stable marriage* [5]. The *maximal sum* selection strategy collectively selects a collection of attribute correspondences for every source attribute, which has the maximal sum of similarity values. As shown in Figure 1(a), there are a set of source attributes $\{a_1, a_2\}$ and a set of target attributes $\{b_1, b_2\}$. Links between source and target attributes represent attribute correspondences and numbers on links represent similarity values between source and target attributes. Using the maximal sum selection strategy, a set of attribute correspondences $\{\langle a_1, b_2\rangle, \langle a_2, b_1\rangle\}$ is selected since it has the maximal sum of similarity values. However, the strategy fails to select attribute correspondence $\langle a_1, b_1\rangle$ which has the highest similarity value and is most likely to be a correct attribute correspondence. This selection strategy maximizes collective benefits of a set of attribute correspondences at the expense of individual benefits of attribute correspondences.

The *greedy* selection strategy selects, at a time, an attribute correspondence with the highest similarity value from a set of attribute correspondences and removes it and those attribute correspondences which are in conflict with it from the set. This is then repeated with the remaining attribute correspondences in the set until the set becomes empty. As shown in Figure 1(b), using the greedy selection strategy, first, $\langle a_1, b_1\rangle$ is selected, and $\langle a_1, b_1\rangle$, $\langle a_1, b_2\rangle$ and $\langle a_2, b_1\rangle$ are removed from the set; Second, $\langle a_2, b_2\rangle$ is selected and removed from the set. The set now becomes empty. However, $\{\langle a_1, b_2\rangle, \langle a_2, b_1\rangle\}$ instead of $\{\langle a_1, b_1\rangle, \langle a_2, b_2\rangle\}$ seems to be a reasonable set of attribute correspondences. In this case, collective benefits of a set of attribute correspondences should have been considered.

The *stable marriage* selection strategy is a stronger case of the greedy selection strategy, which selects those attribute correspondences that satisfy the following condition, from a set of attribute correspondences. An attribute correspondence is selected if the two attributes in it are a better choice for each other than any other attribute correspondences which are in conflict with it in terms of similarity values. As shown in Figure 1(b), using the stable marriage selection strategy, only one attribute correspondence $\{\langle a_1, b_1\rangle\}$ is selected. However, $\{\langle a_1, b_2\rangle, \langle a_2, b_1\rangle\}$ seems to be a reasonable set of attribute correspondences. In this case, collective benefits of a set of attribute correspondences should also have been considered.

In order to filter out those attribute correspondences with low similarity values, which are likely to be incorrect but sometimes get in the way, thresholds are used [16]. Using a threshold, attribute correspondences $\langle a_2, b_2\rangle$ in Figure 1(a) and $\langle a_2, b_2\rangle$ in Figure 1(b) can both be filtered out. However, even with a threshold, using the maximal sum strategy in Figure 1(a), we still get $\{\langle a_1, b_2\rangle, \langle a_2, b_1\rangle\}$ as the result though $\{\langle a_1, b_1\rangle\}$ seems more reasonable. Similarly, using the greedy selection strategy in Figure 1(b), we still get $\{\langle a_1, b_1\rangle\}$ as the result, though $\{\langle a_2, b_1\rangle, \langle a_1, b_2\rangle\}$ seems more reasonable. Using the greedy selection strategy in Figure 1(a) and the maximal sum selection the in Figure 1(b) instead will solve the problem. However, no single selection strategy suits both.

In addition, thresholds have to be set manually, and are matcher and domain dependent. Sometimes it may be very difficult to set a threshold to distinguish correct attribute correspondences from incorrect ones. In some cases, though the similarity value of an attribute correspondence is very high, it is not a correct attribute correspondence. For example, when we use a semantic similarity-based matcher to match two query interfaces in the flight booking domain, attribute correspondence ⟨*Departure from, Departure date*⟩, has a very high similarity value, but it is not a correct attribute correspondence. In some other cases, though the similarity value of an attribute correspondence is low, it could still be a correct one. For instance, though attribute correspondence ⟨*Number of tickets, adults*⟩ has a low semantic similarity value, it is still a correct attribute correspondence.

To address these issues, we have proposed a new selection strategy, called *prioritized collective selection strategy*. We cluster a set of attribute correspondences by their similarity values. We use each of these clusters in a prioritized order, in the spirit of the greedy selection strategy. In each cluster, we first add a null correspondence for each source attribute in the cluster, and then use the maximal sum selection strategy to select attribute correspondences in a collective manner. A null correspondence for a source attribute in a cluster represents the option that the source attribute has no attribute correspondence in the cluster. Therefore, our strategy is intended to take the advantages of both the greedy selection strategy and the maximal sum selection strategy, overcoming their disadvantages at the same time. By using null correspondences, we no longer need a threshold to filter out likely incorrect attribute correspondences and reasoning about a source attribute having no attribute correspondence becomes an explicit process.

For example, we apply our prioritized collective selection strategy to Figure 1(a). We cluster attribute correspondences into three clusters: $C_1 = \{\langle a_1, b_1\rangle\}$, $C_2 = \{\langle a_1, b_2\rangle, \langle a_2, b_1\rangle\}$, $C_3 = \{\langle a_2, b_2\rangle\}$. We rank these three clusters by their average similarity values. We now use these three clusters in their ranked order. First, we add a null correspondence for source attribute a_1, $\langle a_1, null\rangle$, to C_1 which becomes $C_1 = \{\langle a_1, b_1\rangle, \langle a_1, null\rangle\}$. The similarity value of $\langle a_1, null\rangle$ is $Sim(\langle a_1, null\rangle) = 0.1$ (How to calculate the similarity value of a null correspondence is described in Section 3.2). Applying the maximal sum selection strategy to C_1, $\langle a_1, b_1\rangle$ is selected. $\langle a_1, b_2\rangle$ and $\langle a_2, b_1\rangle$ that are in conflict with $\langle a_1, b_1\rangle$, are removed from C_2 and C_3. C_2 becomes empty. We are left with C_3 only. Adding a null correspondence for source attribute a_2, $\langle a_2, null\rangle$, to C_3 which becomes $C_3 = \{\langle a_2, b_2\rangle, \langle a_2, null\rangle\}$. Since we have the following similarity values, $Sim(\langle a_2, b_2\rangle) = 0.1$ and $Sim(\langle a_2, null\rangle) = 0.9$, $\langle a_2, null\rangle$ is selected. The final result is $\{\langle a_1, b_1\rangle, \langle a_2, null\rangle\}$ which is very reasonable.

We now apply our strategy to Figure 1(b). Attribute correspondences are clustered into two clusters $C_1 = \{\langle a_1, b_1\rangle, \langle a_1, b_2\rangle, \langle a_2, b_1\rangle\}$ and $C_2 = \{\langle a_2, b_2\rangle\}$, which are ranked by their average similarity values. C_1 becomes $C_1 = \{\langle a_1, b_1\rangle, \langle a_2, b_1\rangle, \langle a_1, b_2\rangle, \langle a_1, null\rangle, \langle a_2, null\rangle\}$ after null correspondences for source attributes a_1 and a_2 are added, with $Sim(\langle a_1, null\rangle) = 0.0002$ and $Sim(\langle a_2, null\rangle) = 0.0003$. The final result is $\{\langle a_1, b_2\rangle, \langle a_2, b_1\rangle\}$,which is also very reasonable.

The rest of this paper is organized as follows. Section 2 provides an overview of the schema matching process. Section 3 describes our prioritized collective selection strategy in detail. In Section 4, we report our experiments on evaluation of our algorithms that implement the prioritized collection selection strategy. Section 5 compares our work with related work. Section 6 concludes the paper.

2 Overview of Schema Matching Process

2.1 Combining Different Types of Similarity between Attributes

Individual matchers [7,8,10] have been developed to make use of various types of information about schemas, including structures, linguistic features, data types, value ranges, etc. In this paper, we use three of the most frequently used matchers to measure three types of similarity between attributes: semantic, string distance and data type.

Semantic similarity-based matcher: Suppose that two attribute names have two sets of words $A_1 = \{w_1, w_2, ..., w_m\}$ and $A_2 = \{w'_1, w'_2, ..., w'_n\}$. We use WordNet[1], an ontology database to compute semantic similarity between each word w_i in A_1 with every word in A_2 and find the highest semantic similarity value, v_i, for w_i. We then get a similarity value set $Sim_1 = \{v_1, v_2, ..., v_m\}$ for A_1. Similarly, we get a similarity value set $Sim_2 = \{v'_1, v'_2, ..., v'_n\}$ for A_2. From these two similarity value sets we calculate the similarity value between two attribute names A_1 and A_2 as:

$$Sim(A_1, A_2) = \frac{\sum_{i=1}^{m} v_i + \sum_{i=1}^{n} v'_i}{m+n} \quad (1)$$

where m is the number of the words in A_1, n is the number of the words in A_2.

[1] http://wordnet.princeton.edu/

Edit distance-based matcher: Edit distance is the number of edit operations necessary to transform one string to another [12]. We use edit distance-based string similarity to measure similarity between words in two attribute names. Similar to the semantic similarity matcher, we get two similarity value sets for two attribute names and then calculate the similarity value between two attribute names based on the two similarity value sets using Equation 1.

Data types-based matcher: We define that two data types are compatible if they are the same or one subsumes another. If two attributes are compatible then the similarity value between them is 1; Otherwise it is 0.

Combining similarity values by individual matchers: After similarity values are calculated by individual matchers, a combination strategy is used to combine these similarity values. A number of combination strategies have been proposed, such as max, min, average, weighing [2,10] and evidential combination [15]. In this paper, we use the weighing strategy. We assign a weight to each individual matcher and combine their similarity values using Equation 2:

$$Sim = w_1 * Sim_1 + ... + w_n * Sim_n \quad (2)$$

where w_i is a weight assigned to the ith matcher, Sim_i is the similarity value by the ith matcher and Sim is the combined similarity value.

2.2 Outline of Prioritized Collective Selection Strategy

In this paper, we propose a new prioritized collective selection strategy to resolve conflicts among the attribute correspondences of different source attributes. First, we cluster a set of attribute correspondences by their similarity values. Second, we rank these clusters by their average similarity values. Third, each of these ranked clusters is used in their ranked order. For each cluster, we add a null correspondence for each source attribute that is present in the cluster, which represents the option that the source attribute has no attribute correspondence. We then apply the maximal sum selection strategy to the cluster. Those attribute correspondences that are in conflict with selected attribute correspondences are removed from the subsequent clusters. A selected null correspondence for a source attribute is treated as an intermediate result. If an attribute correspondence for the same source attribute is selected in a subsequent cluster, the null correspondence will then be replaced by the newly selected attribute correspondence.

3 Prioritized Collective Selection Strategy

3.1 Clustering Attribute Correspondences

We use a density-based clustering algorithm, DBSCAN (Density-Based Spatial Clustering of Applications with Noise) [18], which grows regions with sufficiently high density into clusters. In DBSCAN, two parameters are used: a radius, ϵ,

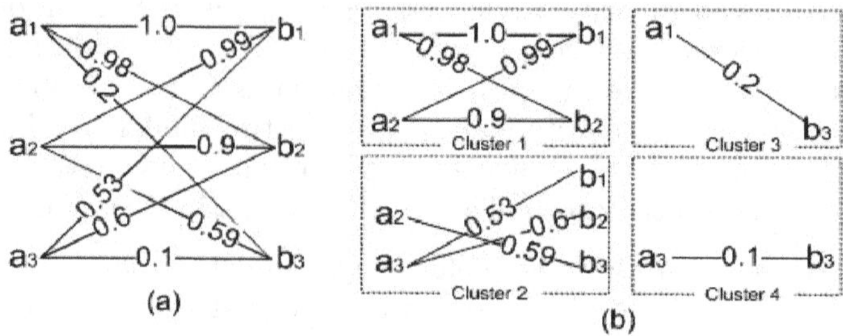

Fig. 2. Resolving Conflicts by Collectively Selecting Attribute Correspondences from a Prioritized Sequence of Clusters

to define the neighborhood of an object and a minimum number, $MinPts$, to determine the density of regions to be grown into clusters. The ϵ-neighborhood of an object is defined as the neighborhood within a radius ϵ of the object. An object is called a core object if the number of objects in its ϵ-neighborhood is the same as or greater than $MinPts$.

Given a set of objects, Θ, DBSCAN grows regions in Θ with sufficiently high density into clusters by checking each object in it. If an object is a core object, a new cluster is created with its ϵ-neighborhood. DBSCAN then iteratively adds the objects in the ϵ-neighborhood of each core object in the cluster to the cluster. Whenever an object is added to a cluster, it is removed from Θ. The process terminates when no new objects can be added to any cluster.

After DBSCAN is applied, Θ may become empty, that is, every object in Θ is now in a cluster. On the other hand, there may be still objects in Θ, which cannot be added to any cluster. We create a singleton cluster with each of these objects. Therefore, Θ is eventually divided into a number of clusters, though some of the clusters may be singleton ones.

We now apply DBSCAN to Fig. 2(a), letting ϵ be the average difference between similarity values of two neighboring attribute correspondences and $MinPts = 3$. As shown in Fig. 2(b), DBSCAN grows two regions with sufficiently high density into two clusters, Clusters 1 and 2. In addition, there are also two singleton clusters, Clusters 3 and 4.

There are two special cases. The first is that Θ itself becomes one cluster only, that is, the similarity values of all attribute correspondences are in one sufficiently high density region. The second is that every attribute correspondence in Θ becomes a singleton cluster, that is, there exists no sufficiently high density region in Θ.

3.2 Null Correspondences

Sometimes it is possible that a source attribute has no attribute correspondence. A threshold can be used to filter out likely incorrect attribute correspondences. Therefore, if no attribute correspondence above a threshold has been selected for a source attribute, we can conclude that the source attribute has no correspondence. Our prioritized collective selection strategy does not need a threshold. We instead use a special attribute correspondence, called null correspondence, to represent the option that a source attribute has no attribute correspondence.

Given a source attribute in a cluster, the similarity value of its null correspondence is calculated on the basis of the similarity values of its other attribute correspondences in the cluster as follows:

$$Sim(\langle s_i, null\rangle) = \Pi_{j=1}^{m}(1 - Sim(\langle s_i, t_j\rangle)) \tag{3}$$

where $Sim(\langle s_i, null\rangle)$ is the similarity value of the null correspondence for source attribute s_i in a cluster, $\langle s_i, t_j\rangle$ is an attribute correspondence for s_i in the cluster.

For example, as shown in Fig. 2(b), we have a cluster, $C_1 = \{\langle a_1, b_1\rangle, \langle a_1, b_2\rangle, \langle a_2, b_1\rangle, \langle a_2, b_2\rangle\}$. We add two null correspondences, $\langle a_1, null\rangle$ and $\langle a_2, null\rangle$, into this cluster and C_1 becomes $C_1 = \{\langle a_1, b_1\rangle, \langle a_1, b_2\rangle, \langle a_2, b_1\rangle, \langle a_2, b_2\rangle, \langle a_1, null\rangle, \langle a_2, null\rangle\}$. The similarity values of the null correspondences are $Sim(\langle a_1, null\rangle) = 0$ and $Sim(\langle a_2, null\rangle) = 0.001$.

3.3 Collectively Selecting Attribute Correspondences from a Cluster

After null correspondences have been added into a cluster, we use the maximal sum selection strategy to collectively select a collection of attribute correspondence every source attribute. A collection of attribute correspondences with the maximal sum of similarity values is selected. The algorithm is shown in Algorithm 1.

Algorithm 1. Maximal Sum Selection

Input: A set of all possible collections of attribute correspondences for every source attribute $\Sigma = \{C | C = \{\langle a_1, b'_1\rangle, \langle a_2, b'_2\rangle, ..., \langle a_m, b'_m\rangle\}\}$,where $\langle a_i, b'_i\rangle \in \{\langle a_i, b_{i1}\rangle, \langle a_i, b_{i2}\rangle, ..., \langle a_i, b_{ik}\rangle\}$ for $i = 1, 2, ..., m$

Output: A collection of attribute correspondences for every source attribute, with the maximal sum of similarity values

Set Max to 0, $Best$ to $\emptyset$
for each $C \in \Sigma$ **do**
  Set Sum to $\Sigma_{i=1}^{m} Sim(\langle a_i, b'_i\rangle)$, where $Sim(\langle a_i, b'_i\rangle)$ is the similarity value of $\langle a_i, b'_i\rangle$
  if $Sum > Max$ **then**
    Set Max to Sum, $Best$ to C
return Best

For example, applying Algorithm 1 to $C_1 = \{\langle a_1, b_1\rangle, \langle a_1, b_2\rangle, \langle a_2, b_1\rangle, \langle a_2, b_2\rangle, \langle a_1, null\rangle, \langle a_2, null\rangle\}$, as shown in Section 3.2, the following collection of attribute correspondences is collectively selected $\{\langle a_1, b_2\rangle, \langle a_2, b_1\rangle\}$.

3.4 Prioritized Collective Selection Strategy

The prioritized collective selection strategy is implemented in Algorithm 2. If a selected attribute correspondence is not a null correspondence, those attribute correspondences in the subsequent clusters, which are in conflict with the selected attribute correspondence, are removed from these clusters. If a selected attribute is a null correspondence, we know that the source attribute has no attribute correspondence in the cluster. In a subsequent cluster, we may find an attribute correspondence for the source attribute. If so, the null correspondence will be replaced by the newly found attribute correspondence. If after all the clusters have been used in the selection, a null correspondence still remains, we know that the source attribute has no attribute correspondence after all.

Algorithm 2. Prioritized Collective Selection

Input: A set of attribute correspondences $\Omega = \{\langle a_i, b_j \rangle | a_i \text{ is a source attribute and } b_j \text{ is a target attribute}\}$
Output: A collection of attribute correspondences for every source attribute, Ψ

Cluster Ω into a ranked list of clusters $C_1, C_2, ..., C_k$
for each C_i **do**
  for each source attribute a_x **do**
    if there exists $\langle a_x, b_y \rangle \in C_i$ **then**
      Add $\langle a_x, null \rangle$ to C_i
  Collectively select a set of attribute correspondences, Θ_i, from C_i, by Algorithm 1
  for each $A \in \Theta_i$ **do**
    if A is not a null correspondence **then**
      Remove those attribute correspondences from clusters, $C_{i+1}, C_{i+2}, ..., C_k$, which are in conflict with A
      if there exists a null correspondence, ϕ, in Ψ, which has the same source attribute as A **then**
        Remove ϕ from Ψ
    Add A to Ψ
return Ψ

For example, after a collection of attribute correspondences, $\{\langle a_1, b_2 \rangle, \langle a_2, b_1 \rangle\}$, has been selected from C_1 as shown in Section 3.3, those attribute correspondences in the subsequent clusters, $C_2 = \{\langle a_2, b_3 \rangle, \langle a_3, b_1 \rangle, \langle a_3, b_2 \rangle\}$, $C_3 = \{\langle a_1, b_3 \rangle\}$ and $C_4 = \{\langle a_3, b_3 \rangle\}$, which are in conflict with the selected attribute correspondences, are removed from these clusters. Both C_2 and C_3 become empty. We are left with C_4 only, adding a null correspondence for source attribute a_3 to C_4, we have $C_4 = \{\langle a_3, b_3 \rangle, \langle a_3, null \rangle\}$. Applying Algorithm 1 to C_4, the null correspondence, $\langle a_3, null \rangle$, is selected as the result. Adding this to the collection of attribute correspondences collectively selected from C_1, we have the final collection of attribute correspondences collectedly selected from C_1, C_2, C_3 and C_4 in a prioritized order, $\{\langle a_1, b_2 \rangle, \langle a_2, b_1 \rangle, \langle a_3, null \rangle\}$.

4 Experimental Results

To evaluate our new prioritized collective selection strategy, we tested it in schema matching using a set of schemas extracted from real-world Web sites. We compared our strategy with three other selection strategies. Our goal was to evaluate the accuracy of schema matching using our strategy in comparison with using the other strategies.

4.1 Data Set

We selected a set of query interfaces on the real-world Web sites from the ICQ Query Interfaces data set at UIUC, which contains manually extracted schemas of 100 interfaces in 5 different domains: Airfares, Automobiles, Books, Jobs, and Real Estates. In this paper we have focused on 1:1 matching only, so only 88 interfaces have been chosen from the data set. In each domain we chose an interface as the source schema and others as the target schemas.

4.2 Performance Metrics

We used three metrics: precision, recall, and F-measure [8,13]. Precision is the percentage of correct matches over all matches identified by a matcher. Recall is the percentage of correct matches by a matcher over all the matches by domain experts. F-measure is the incorporation of precision and recall [7], defined as $F = \frac{2PR}{P+R}$, where P is the precision and R is the recall. When using our selection strategy, for each source attribute, we have a null correspondence that represents the option that the source attribute has no attribute correspondence. Therefore, it is always possible to find an attribute correspondence for each source attribute though the attribute correspondence may be a null correspondence. Therefore, the number of matches using our strategy equals to the number of matches identified by domain experts. So when using our strategy, precision, recall and F-measure are all the same.

4.3 Discussion on Experimental Results

First, in each domain we performed four experiments. We used three other selection strategies (maximal sum, stable marriage and greedy) and our own prioritized collective selection strategy to select a collection of attribute correspondences for every source attribute. Table 1 shows precision, recall and F-measure of different selection strategies in each domain and their average precision, recall and F-measure across different domains. Our strategy consistently gets higher average precision, recall and F-measure than the greedy, stable marriage[2], and maximal sum strategies.

[2] The stable marriage is first used as much as possible. When it becomes not possible to use the stable marriage, the maximal sum selection strategy is used instead.

Table 1. Precision (P), recall (R) and F-measure (F) of different selection strategies

	Greedy			Stable Marriage			Maximal Sum			Ours		
	P	R	F	P	R	F	P	R	F	P	R	F
Airfares	93.8%	91.7%	92.7%	88.9%	84.2%	86.3%	92.6%	90.5%	91.5%	95.1%	95.1%	95.1%
Automobiles	96.3%	94.4%	95.2%	97.5%	95.7%	96.5%	96.3%	93.2%	94.6%	96.3%	96.3%	96.3%
Books	90.6%	83.8%	86.7%	91.8%	85.6%	88.2%	86.6%	76.2%	80.4%	90.6%	90.6%	90.6%
Jobs	90.1%	86.4%	87.5%	91.2%	88.8%	89.3%	84.0%	79.8%	81.3%	87.0%	87.0%	87.0%
Real Estates	91.5%	91.3%	91.1%	93.5%	92.9%	92.9%	91.5%	90.4%	90.6%	93.5%	93.5%	93.5%
Average	91.9%	89.5%	90.7%	92.0%	89.4%	90.6%	90.3%	86.0%	87.7%	92.5%	92.5%	92.5%

Second, if we consider other factors, such as time complexity, space complexity and manual efforts, our strategy is better in real-world applications. As shown in Table 2, both greedy and stable marriage selection strategies have low time complexity and space complexity though thresholds for filtering out incorrect attribute correspondences have to be set manually. The maximal sum selection strategy has high time complexity and space complexity, which restrict its use in real-world applications. Not only does our new selection strategy have low time complexity and space complexity, but also there is no need for a threshold, making it a good choice for real-world applications.

Table 2. Comparison of different selection strategies in terms of time, space and manual efforts

	Time Complexity	Space Complexity	Manual Efforts
Greedy	low	low	thresholds
Stable Marriage	low	low	thresholds
Maximal Sum	high	high	thresholds
Ours	low	low	no threshold

5 Related Work

A number of selection strategies, including maximum cardinality, maximum total weights and stable marriage, have been described in [14]. Maximum cardinality aims at achieving the largest number of matches. Maximum total weights selects a set of matches that has the maximum total weights. The stable marriage selects those matches $\langle x, y \rangle$, where x and y are the best choices for each other.

In [5], a mixture of stable marriage and maximal sum selection strategies has been used. First attribute correspondences are selected using the stable marriage strategy. Attribute correspondences are then further selected from the remaining attribute correspondences, using the maximal sum selection strategy. As we shown in Table 2, this strategy may have higher time complexity and space complexity if a large number of attribute correspondences remain after the stable marriage has been used.

In [4,7], the greedy strategy is used. This strategy keeps selecting the best correspondence, and removes it and those that are in conflict with it from the set of attribute correspondences until the set becomes empty. Though this strategy is very efficient, it sometimes get wrong results.

In [17], the maximal sum strategy is used. A collection of attribute correspondences for every source attribute is collectively selected, which has the maximal sum of similarity values. As we discussed in Section 4.3, the accuracy of this strategy is not ideal, and it has high time complexity and space complexity.

In addition, the other selection strategies need a threshold to filter out incorrect attribute correspondences. However, these thresholds have to be set manually, and they are matcher and domain dependent, restricting their use in fully automated approaches to schema matching.

6 Conclusions

In this paper we proposed a new prioritized collective selection strategy, in which attribute correspondences are clustered and prioritized. Our strategy selects attribute correspondences from each of these clusters in a prioritized order, in the spirit of the greedy selection strategy. In each of the clusters, attribute correspondences are selected using the maximal sum selection strategy, making sure that attribute correspondences are selected in a collective manner. Therefore, our strategy takes the advantages of both the greedy selection strategy and the maximal sum selection strategy, overcoming their disadvantages at the same time. In our new selection strategy, we have also introduced use of a null correspondence in the process of selecting attribute correspondences, that represents the option that a source attribute has no attribute correspondence. In doing so, we no longer need a threshold to filter out likely incorrect attribute correspondences and reasoning about a source attribute having no attribute correspondence becomes an explicit process.

We have tested our strategy using a large data set that contains the schemas extracted from real-world Web interfaces in five different domains. We have also compared our new selection strategy with three other selection strategies. Our experimental results show that our selection strategy works effectively and efficiently in automated approaches to schema matching.

In this paper we have mainly focused on selecting attribute correspondences in the context of schema matching. It is evident that our new selection strategy can also be used in other selection problems, where conflict resolution is needed in the selection process.

References

1. Shvaiko, P., Euzenat, J.: A survey of schema-based matching approaches. Journal of Data Semantics, 146–171 (2005)
2. Rahm, E., Bernstein, P.A.: A survey of approaches to automatic schema matching. VLDB Journal 10(4), 334–350 (2001)
3. He, B., Chang, K.C.C.: Statistical schema matching across web query interfaces. In: Proceedings of the 22th ACM International Conference on Management of Data (SIGMOD 2003), pp. 217–228 (2003)

4. He, B., Chang, K.C.C.: Automatic complex schema matching across Web query interfaces: A correlation mining approach. ACM Transactions on Database Systems (TODS) 31(1), 346–395 (2006)
5. Melnik, S., Garcia-Molina, H., Rahm, E.: Similarity flooding: A versatile graph matching algorithm and its application to schema matching. In: Proceedings of the 18th International Conference on Data Engineering (ICDE 2002), pp. 117–128 (2002)
6. He, B., Chang, K.C.C., Han, J.: Discovering complex matchings across web query interfaces: a correlation mining approach. In: Proceedings of the 10th ACM SIGKDD International Conference on Knowledge Discovery and Data Mining (KDD 2004), pp. 148–157 (2004)
7. Wu, W., Yu, C.T., Doan, A., Meng, W.: An interactive clustering-based approach to integrating source query interfaces on the deep web. In: Proceedings of the 23rd ACM International Conference on Management of Data (SIGMOD 2004), pp. 95–106 (2004)
8. Madhavan, J., Bernstein, P.A., Rahm, E.: Generic schema matching with cupid. In: Proceedings of the 27th International Conference on Very Large Data Bases (VLDB 2001), pp. 49–58 (2001)
9. Wang, J., Wen, J.R., Lochovsky, F.H., Ma, W.Y.: Instance-based schema matching for web databases by domain-specific query probing. In: Proceedings of the 30th International Conference on Very Large Data Bases (VLDB 2004), pp. 408–419 (2004)
10. Do, H.H., Rahm, E.: Coma - a system for flexible combination of schema matching approaches. In: Proceedings of the 28th International Conference on Very Large Data Bases (VLDB 2002), pp. 610–621 (2002)
11. Doan, A., Domingos, P., Halevy, A.Y.: Reconciling schemas of disparate data sources: A machine-learning approach. In: Proceedings of the 20th ACM International Conference on Management of Data (SIGMOD 2001), pp. 509–520 (2001)
12. Hall, P., Dowling, G.: Approximate string matching. Computing Surveys, 381–402 (1980)
13. Halevy, A.Y., Madhavan, J.: Corpus-Based Knowledge Representation. In: Proceedings of the 18th International Joint Conference on Artificial Intelligence (IJCAI 2003), pp. 1567–1572 (2003)
14. Lovasz, L., Plummer, M.: Matching Theory. North-Holland, Amsterdam (1986)
15. He, Z., Hong, J., Bell, D.: Schema Matching across Query Interfaces on the Deep Web. In: Gray, A., Jeffery, K., Shao, J. (eds.) BNCOD 2008. LNCS, vol. 5071, pp. 51–62. Springer, Heidelberg (2008)
16. Gal, A.: Managing Uncertainty in Schema Matching with Top-K Schema Mappings. In: Spaccapietra, S., Aberer, K., Cudré-Mauroux, P. (eds.) Journal on Data Semantics VI. LNCS, vol. 4090, pp. 90–114. Springer, Heidelberg (2006)
17. Bilke, A., Naumann, F.: Schema Matching using Duplicates. In: Proceedings of the 21st International Conference on Data Engineering (ICDE 2005), pp. 69–80 (2005)
18. Han, J., Kamber, M.: Data Mining: Concepts and Techniques. Morgan Kaufmann, San Francisco (2006)

An Alternative Data Warehouse Reference Architectural Configuration

Victor Gonzalez-Castro[1], Lachlan M. MacKinnon[2], and María del Pilar Angeles[3]

[1] School of Mathematical and Computer Sciences, Heriot-Watt University, Edinburgh, EH14 4AS, Scotland
victor@macs.hw.ac.uk

[2] School of Computing and Creative Technologies, University of Abertay Dundee, Dundee, DD1 1HG, Scotland
l.mackinnon@abertay.ac.uk

[3] Facultad de Ingeniería, Universidad Nacional Autónoma de México, Ciudad Universitaria, CP04510, México
pilarang@unam.mx

Abstract. In the last few years the amount of data stored on computer systems is growing at an accelerated rate. These data are frequently managed within data warehouses. However, the current data warehouse architectures based on n-ary-Relational DBMSs are overcoming their limits in order to efficiently manage such large amounts of data. Some DBMS are able to load huge amounts of data nevertheless; the response times become unacceptable for business users during information retrieval. In this paper we describe an alternative data warehouse reference architectural configuration (ADW) which addresses many issues that organisations are facing. The ADW approach considers a Binary-Relational DBMS as an underlying data repository. Therefore, a number of improvements have been achieved, such as data density increment, reduction of data sparsity, query response times dramatically decreased, and significant workload reduction with data loading, backup and restore tasks.

Keywords: Alternative Data Model, Data Warehouse, Binary-Relational, Architectural Configuration, Database Explosion, Column DBMS.

1 Introduction

It has been demonstrated that a binary-relational data model can be more appropriate model for data warehouses than the n-ary relational model. The binary relational data model addresses a number of issues shown in [1] and presented in Section 2.2. The main intension of this paper is to present the Alternative Data Warehouse (ADW) Reference Architectural configuration which is based on an alternative data model, the Binary-Relational model. Some of the problems ADW solves are the management of large amount of data, long response times for business queries. An efficient Decision Support System (DSS) should provide fast response times in order to support fast making business decisions. An important aspect that has been considered when designing the reference configuration architecture is to keep compatibility with the

A.P. Sexton (Ed.): BNCOD 2009, LNCS 5588, pp. 33–41, 2009.

installed base as vast sums of money had been invested by organisations on ETL, analysis and reporting tools. As larger the data warehouse is as the slower the updates are. Therefore, the ETL processes required for the data warehouse maintenance tasks are unable to be completed timely. Nowadays, the budget Organisations usually estimates expenses as a result of increasing the storage capacity. The proposed ADW is aimed to reduce disk storage, representing saving in the organisation's budget.

2 The ADW Reference Architectural Configuration

The Relational model is the predominant model used in commercial RDBMS which have been demonstrated to be very successful in transactional environments. However, RDBMS have also been used to create Data Warehouses without questioning their suitability. Therefore, there are a number of issues regarding the use of a relational data model within Data Warehouses. In order to benchmark some alternative models, the extended TPC-H benchmark [1] was chosen. The tests were executed with two database sizes, called scale factors (SF=100 MB and SF=1GB) where the Binary-Relational model was found as the best suited for its use in Data warehouse environments. The tests and results demonstrating the suitability of the Binary-Relational model were further detailed in [1]. Table 1 presents a score card summarising such results of the evaluated metrics for the SF=1GB.

Based on these findings, an Alternative Configuration is proposed for Data Warehouses which includes a Binary-Relational repository. The purpose and issues solved by this configuration are presented in the following section.

Table 1. Results of the evaluated metrics for SF=1GB

Metric #	Extended TPC-H metrics (SF=1GB)	Unit of measure	Relational	Score	Binary-Relational	Score
1 and 2	Extraction from transactional systems and Transformation times	Minutes	27.6	√	29.5	
3	Input file sizes measurement	MegaBytes	1,049.58		1,004.95	√
4	Load data times in to the Data Warehouse	Minutes	19.9		5.1	√
5	Size of the Database tables after load (Data Base size)	MegaBytes	1,227.10		838.1	√
6	Data Density Measurement	Rows/Megabyte	7,058		10,334	√
7	Query execution times (Pristine mode)	Minutes	63,152.69		3.81	√
8	Data Warehouse Backup time	Minutes	6.68		1.86	√
9	Backup size	MegaBytes	1,157		838.1	√
10	Data Warehouse restore time	Minutes	8.41		1.72	√
11	Index creation times	Minutes	8.52		0	√
12	Index sizes	MegaBytes	484.6		0	√
13	Query times (with indexes)	Minutes	1,682.31		3.81	√
14	Statistics computation times	Minutes	15.63		0	√
15	Query times (with indexes & statistics)	Minutes	32.35		3.81	√
16	Query times (with statistics without indexes)	Minutes	29.57		3.81	√

2.1 The Alternative Data Warehouse Purpose and Problems Solved

This section presents how the Alternative Data Warehouse Reference architectural configuration (ADW) addresses the ten points identified in [1] as the main problems of current Relational Data Warehouses. These points are listed below together with the explanation of how ADW addresses each point.

1. *Data Warehouses are growing at an exponential rate* [2]. This problem is remediated in ADW by using a Binary-Relational repository, as it has been demonstrated in [1], where the Data warehouse size was measured. Savings of 32% are achieved when compared with an n-ary-Relational Data warehouse. These measurements considered base tables only, but usually in n-ary-Relational Data warehouses, consolidated tables or materialised views are created as well. These extra tables are reduced or even completely eliminated in an ADW, because of its fast query response time, it is not necessary to build these summary tables.
2. *The Database explosion phenomenon is difficult to control or eliminate* [3]. In the case of n-ary-Relational Data Warehouse architectures, when computing the cross product of all levels of the dimensions against each other -the cube computation- is one of the main contributors to the database explosion. In the case of an ADW and its Binary-Relational DBMS, such cube computation is not necessary because of the speed to answer queries.
3. *Poor management of data sparsity* [4]. n-ary-Relational Data warehouses have poor management of data sparsity because they are based on the n-ary Relational model which has always suffered in the handling of missing information [5]. However, and even worse, the relational implementations based on SQL have poor Nulls management that is directly related to data sparsity. In contrast, the Binary-Relational has a better Data sparsity management by the model definition of the Binary-Relational model.
4. *Low Data Density* [6]. As demonstrated by the measurements presented in [1] that the Binary-Relational model has better data density (10,000 rows/MB) than the n-ary-Relational model (7,000 rows/MB). Hence the use of a Binary-Relational DBMS will improve the data density of the Data warehouses, which is precisely the kind of DBMS to be used in the ADW Reference Architectural configuration.
5. *Huge amounts of disk storage are required* [7]. When using the Binary-Relational model, savings of 65% of total disk space are achieved when compared against n-ary-Relational. This has been measured and shown in [1]).
6. *The cost of storage and its maintenance is not negligible* [2] *and the storage itself could be up to 80% of the entire system cost* [7]. As stated in [1] savings of 65% in disk space can be achieved by having ADW implemented. These savings are directly reflected in the total cost of the disk and its maintenance. Some other savings are listed below.
    a. 65% less disk space to be acquired.
    b. Electricity savings achieved by having fewer disks to be powered.

   c. Reduction in energy costs by reducing the cooling needed by the computer centre having fewer disks generating heat, and even savings when buying a smaller cooler machine.
   d. Savings in computer room floor space by having to accommodate fewer disk arrays.
   e. Savings in secondary storage (tapes) in the same proportion by having fewer disks to be backed up.
   f. By having fewer tapes to store, reductions in the size of the room to keep the historical tapes can also be considered.
   g. By having 65% less disks, the payment of the corresponding maintenance fee to the hardware vendors is also reduced.
7. *One of the main complaints from users is the high query processing times* [8]. This aspect can be dramatically improved when implementing ADW, because the Binary-Relational model beneath ADW has achieved query performance improvements in many orders of magnitude - 3.8 minutes versus the 43.8 days of the n-ary-Relational model (refer to [1] where the actual measurements were presented).
8. *Long periods of time to Extract, Transform and Load (ETL) data* [1]. Based on the measurements and the analysis presented in [1], ADW reduces the ETL time required to Load the Data warehouse.
9. *Big batch processing windows to backup and restore the environment* [1]. This aspect can also benefit from implementing ADW based on the experimental results presented in [1], where savings of between 70% and 85% were achieved in these aspects of the Data warehouse maintenance.
10. *High complexity of the Database Administration tasks, including Index creation and its maintenance and long times required to compute statistics* [1]. ADW is able to reduce the Database administration tasks because it does not need to create extra indexes, or to compute statistics. The reasons for these tasks not being required by the Binary-Relational model were analysed in [1]. When implementing ADW, organisations could forget about creating and maintaining extra indexes and computing statistics, hence reducing the database administration tasks and even reducing the maintenance time required by the whole environment.

2.2 Keeping Compatibility with the Installed Base

Keeping compatibility with the installed base is important because organisations have invested vast sums of money in their current technology. Such technology is compatible mainly with the n-ary-Relational model approach, or more precisely with the SQL language.

The majority of the existing tools use SQL, no matter which part of the Full Data Warehouse Cycle they are oriented to. ADW needs to consider such architectural pieces and the way to integrate them, and not how to replace them.

Therefore it is important for any Binary-Relational DBMS to have a SQL interface to enable existing commercial tools to be integrated. This could be named as backward compatibility, but future Binary-Relational based DBMS could also offer their own Binary-Relational language to enable the evolution of the exploitation or loading phases.

2.3 The Reference Architectural Configuration Description

An alternative Data Warehouse Reference Architectural configuration (ADW) has been defined. It is depicted in Figure 1, and a description of the architecture is given as follows:

- The ADW Reference Architectural configuration considers an ELT stage, a storage stage (or the Data Warehouse) plus an exploitation stage.
- Organisations' information has its origin on the working transactional systems, sometimes called 'operational systems' or 'OLTP environments'. From here data must be Extracted, Transformed and Loaded according to the specifications of the logical Data Warehouse model (star or snowflake) and with the requirements of the data model of the repository, which in the case of ADW is a Binary-Relational DBMS.
- It is important to stress that the ADW Reference Architectural configuration supports both the ETL or ELT approaches to acquiring data and feeding them into the Data Warehouse, but in the case of Data Warehouses it is better to do bulk loads and perform bulk transformations; hence the ELT approach is preferred.
- Once data has been loaded into the Binary-Relational repository, it is necessary to enable such DBMS to be accessed by existing reporting and analysis tools based mainly on the SQL language. Therefore ADW considers a Query interface Module which will manage the translation of SQL statements to the Binary-Relational language supported by the DBMS; the MonetDB DBMS [9] which was used during the research, uses a language called MIL, but other DBMSs can have their own language.
- ADW is not limited to SQL language tools; if in the future vendors develop exploitation tools that support the Binary-Relational language, and therefore the translation between SQL and the binary-relational language will no longer be required.
- In the exploitation phase, different kinds of tools are supported, e.g. Report Writers, CRM, Molap and Rolap tools or Data mining tools, depending on the kind and sophistication of the analyses required by the end user.

2.3.1 ADW Logical Modelling Support

Logical modelling design is an important part of a Data Warehouse. ADW is an open architectural configuration which supports different types of logical data warehouse models, including Star and Snowflake modelling techniques, or even traditional Entity-Relationship models, in which it still offers the advantages of using Binary-Relational-like disk space reduction and Query performance improvements.

2.3.2 ADW Benefits to Query Performance

As ADW is based on a Binary-Relational repository, it will provide the query performance improvements achieved by this alternative data model.

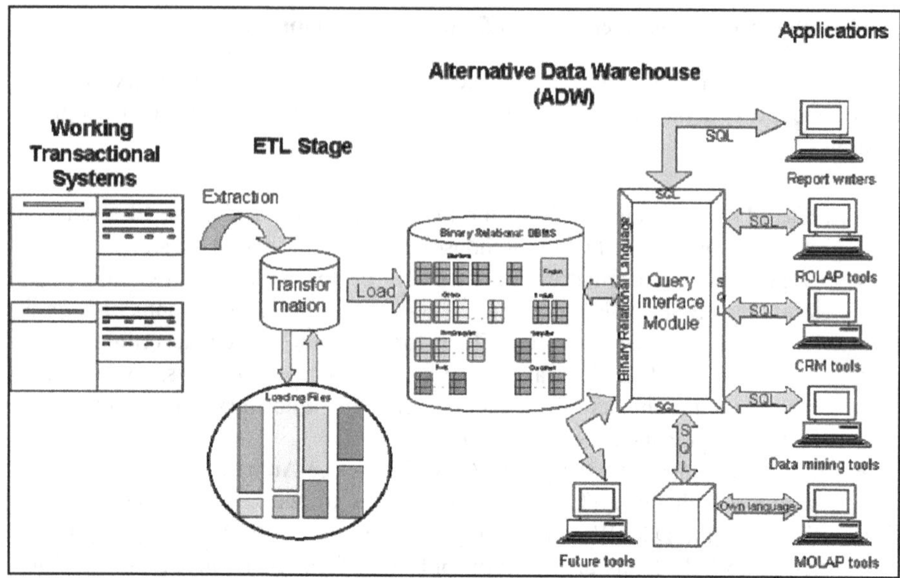

Fig. 1. The ADW Reference Architectural Configuration

ADW will answer queries faster if the Binary-Relational repository is accessed by tools which generate direct queries to the DBMS, but special care must be taken when using cube-based tools. The cube building must avoid the kind of queries that select all rows and columns in order to make the cross product operation to build the cube. The operation of retrieving all columns of a row and then retrieving all rows of the table will force the Binary-Relational DBMS to rebuild all records and then compute the Cartesian product, which will be an extremely time-consuming task; consequently the benefit of fast query performance achieved by the Binary-Relational based DBMS would be diminished by the cube construction.

2.3.3 ADW Support for the Existing Query Performance Approaches

ADW is a reference architectural configuration that can support any of the existing approaches to improving query performance, such approaches are listed below and include how ADW can support or improve such approaches.

- ADW can support the use of summarized tables, as these tables are other tables which do not have any difference from the point of view of the DBMS; however, even the Binary-Relational model will help to build these summarized tables as it will compute the summary information in a faster fashion. Owing to the query performance of the Binary-Relational DBMS, the use of summarized tables will be reduced or even eliminated in ADW. This aspect will contribute to the disk and time savings of the whole ADW architectural configuration.
- ADW can also support the materialized views approach [10] to improve query performance this approach and, as stated, the materialized views are similar to the summarized tables. Hence Binary-Relational DBMSs can inte-

grate this technology, but again, owing to the query performance of the Binary-Relational model, materialized views should be not necessary.
- Approximate Queries [11], [12] can be used over ADW, but even though the Binary-Relational performance is superior, queries can be answered completely instead of computing just an approximation.
- Finally, the use of cubes to improve query performance is not excluded from ADW, as can be appreciated from Figure 1. If the decision is made to continue using cubes on ADW, then the benefit that ADW will provide is that the cube will be built faster than traditional configurations.

ADW will be more beneficial for ROLAP tools as they send queries directly to the DBMS. In some situations when ROLAP tools are used, users complain about the slow performance of the system, but usually the slow performance is caused mainly by the RDBMS used and not to the exploitation tool. Therefore in ADW, the data warehouse repository has been improved by using a Binary-Relational DBMS, which is faster.

2.3.4 ADW Approach to Increase the Data Density

According to the results achieved by the Binary-Relational model regarding the increment of Data Density, and, as ADW incorporates a Binary-Relational DBMS, then ADW will help to increase Data Density.

Data density has been measured considering only data from base tables, but if a generalisation is made and not only data but information is considered, then the density will be increased in a larger proportion, as ADW will not use summary tables or materialized views or even cube structures, as they are not necessary; then the amount of useful information stored per disk Megabyte is even higher.

The highest data density of ADW is achieved because it stores only different values at column level and then it relates each value to as many records as required.

The work conducted in this research differs from the current approaches to increase data density (which basically are data compression techniques), in the sense that in this research a change in the underlying model is made and models which abandon the classical record storage structure have been considered.

Recent work in commercial DBMS [13] considers the use of compression techniques on the DBMS. In [13], it compresses data which are clustered in near disk areas.

3 Conclusions

We have been able to design an Alternative Data Warehouse (ADW) reference architectural configuration, which uses the data model best suited for Data Warehouse Environments.

The ADW maintains compatibility with the widely deployed relational technology. By using the alternative data model and the ADW reference architectural configuration, improvements are achieved in data storage. Also querying times are improved, and at the same time maintenance tasks are benefit. The ADW reference architectural configuration challenges the actual architecture based on n-ary-Relational DBMSs and the ETL approach currently used.

The ADW reference architectural configuration serves as the reference to achieve:

- Reduction in the time required to Load data into the Data Warehouse
- Increase in the Data Density
- Better management of Data Sparsity
- Reduction of disk space, and consequently:
- Fewer disks to be acquired.
- Electricity savings by having fewer disks to be powered and less heat to be dissipated by the cooling equipment in the computer room.
- Savings in the facilities space required by computer room and the secondary storage space (fewer tapes are required to backup the disks).
- By having fewer disks, the payment of the corresponding maintenance fee to the hardware vendors is also reduced.
- The query execution time is dramatically improved
- No need or at least fewer summary tables to be maintained, owing to the query response time. This has positive impact on:
- Less disk space
- Less processing time to compute the summary tables or materialised views.
- Fewer indexes on such summary tables
- Reduction in the number of tables to be maintained, simplifying the DBA maintenance task.

The measurable benefits of using the Binary-Relational model within the context of Data Warehouse have been mentioned, but when it is considered within a broader scope as the one considered in the Alternative Data Warehouse reference configuration architecture, other benefits are achieved, as the whole Data Warehouse cycle has positive impacts:

- By changing the ETL approach for an ELT approach, the total time of feeding the Data Warehouse is reduced, as the Transformation is made in bulk instead of doing it row by row.
- The Data Warehouse based on a Binary-Relational DBMS has all the benefits listed before.
- The exploitation phase benefits by having faster queries for different analysis tools, including direct SQL statements, Report Writers, OLAP, CRM and Data Mining tools. In the case of tools which use cubes, the cube construction also benefits as the Binary-Relational Repository provides data faster to the cube builder engine.
- Hence this data model is ideal for ad hoc queries which are a common requirement in Data warehousing environments.
- Some other benefits are provided by ADW, but these are less tangible or hard to measure:

The IT personnel involved in the operation of the Data Warehouse can improve their quality of life by having fewer problems to operate the whole architecture and giving them more time for their own lives. This positive feedback has been obtained from the personnel of two of the organisations which had implemented ADW.

By having less power consumption (details have been listed before), ADW is contributing positively to environmental issues, as it has less CO_2 emissions to the atmosphere. This is an aspect that is not in the main focus of computer science research, but it is necessary that modern architectures consider their impact over the environment; therefore it can be said that ADW is an environmentally friendlier architecture.

References

1. Gonzalez-Castro, V., MacKinnon, L., Marwick, D.: A Performance Evaluation of Vertical and Horizontal Data Models in Data Warehousing. In: Olivares Ceja, J., Guzman Arenas, A. (eds.) Research in Computing Science, vol. 22, pp. 67–78 (2006), Special issue: Data Mining and Information Systems. Instituto Politécnico Nacional, Centro de Investigación en Computación, México, ISSN: 1870-4069
2. Datta, A., et al.: Curio: A Novel Solution for efficient Storage and Indexing in Data Warehouses. In: Proceedings 25th VLDB Conference, Edinburgh, Scotland, pp. 730–733 (2004)
3. Pendse, N.: Database Explosion. The OLAP Report, `http://www.olapreport.com/DatabaseExplosion.htm` (updated 10-March-2009)
4. Gonzalez-Castro, V., MacKinnon, L.: A Survey "Off the Record" –Using Alternative Data Models to Increase Data Density in Data Warehouse Environments. In: Lachlan, M., Burger Albert, G., Trinder Philip, W. (eds.) Proceedings of the 21st British National Conference on Databases BNCOD, Edinburgh, Scotland, vol. 2 (2004)
5. Codd, E.F.: The Relational Model for database Management Version 2. Addison Wesley Publishing Company, Reading (1990)
6. Gonzalez-Castro, V., MacKinnon, L.: Using Alternative Data Models in the context of Data Warehousing. In: Petratos, P., Michalopoulos, D. (eds.) 1st International Conference on Computer Science and Information Systems by the Athens Institute of Education and Research ATINER, Athens, Greece, pp. 83–100 (2005), ISBN-960-88672-3-1
7. Zukowski, M.: Improving I/O Bandwidth for Data-Intensive Applications. In: Proceedings BNCOD 2005, Sunderland, England U.K., vol. 2, pp. 33–39 (2005)
8. Pendse, N.: The OLAP survey 4, `http://www.olapreport.com`
9. MonetDB web site, `http://monetdb.cwi.nl`
10. Akadia website, `http://www.akadia.com/services/ora_materialized_views.html`
11. Acharya, S., et al.: Aqua: A Fast Decision Support System Using Approximate Query Answers. In: Proceedings 25 VLDB, Edinburgh, Scotland, pp. 754–757 (1999)
12. Shanmugasundaram, J., et al.: Compressed Data Cubes for OLAP aggregate Query Approximation on Continuous Dimensions. In: ACM KDD-1999, pp. 223–232 (1999)
13. Poss, M., Potapov, D.: Data Compression in Oracle. In: Proceedings of the 29th VLDB conference, Berlin Germany (2003)

A Data Privacy Taxonomy

Ken Barker, Mina Askari, Mishtu Banerjee, Kambiz Ghazinour, Brenan Mackas, Maryam Majedi, Sampson Pun, and Adepele Williams

Advanced Database Systems and Applications Laboratory
University of Calgary, Canada
kbarker@ucalgary.ca

Abstract. Privacy has become increasingly important to the database community which is reflected by a noteworthy increase in research papers appearing in the literature. While researchers often assume that their definition of "privacy" is universally held by all readers, this is rarely the case; so many papers addressing key challenges in this domain have actually produced results that do not consider the same problem, even when using similar vocabularies. This paper provides an explicit definition of data privacy suitable for ongoing work in data repositories such as a DBMS or for data mining. The work contributes by briefly providing the larger context for the way privacy is defined legally and legislatively but primarily provides a taxonomy capable of thinking of data privacy technologically. We then demonstrate the taxonomy's utility by illustrating how this perspective makes it possible to understand the important contribution made by researchers to the issue of privacy. The conclusion of this paper is that privacy is indeed multi-faceted so no single current research effort adequately addresses the true breadth of the issues necessary to fully understand the scope of this important issue.

1 Introduction

Owners of data repositories collect information from various data suppliers and store it for various purposes. Unfortunately, once the supplier releases their data to the collector they must rely upon the collector to use it for the purposes for which it was provided. Most modern database management systems (DBMS) do not consider privacy as a first-order feature of their systems nor is privacy an explicit characteristic of the underlying data model upon which these systems are built. Since privacy is poorly defined technically in the literature, even when a DBMS vendor builds privacy features into their systems, they must do so using their own understanding of what constitutes privacy protection. Research efforts that address an aspect of privacy often fail to formally define what they mean by privacy or only consider a subset of the many dimensions of privacy so they often miss critical issues that should be addressed to fully understand the privacy implications.

This paper specifically addresses the key "definitional" problem. We identify four key technical dimensions to privacy that provide the most complete definition to appear to date in the literature. The key players when considering privacy issues in any data repository environment include the data provider, the data collector, the data users, and the data repository itself.

A.P. Sexton (Ed.): BNCOD 2009, LNCS 5588, pp. 42–54, 2009.

Our key contributions are: (1) A clear description of the key components that make up the privacy research space with respect to data repositories. (2) An easily understandable taxonomy of these features and how they relate to each other. (3) A demonstration the taxonomy's expressiveness in both the real-world and academic literature. (4) A tool for researchers who need to clearly define where in the total space their work contributes and what other work is more comparable to their own. (5) A tool for graduate students to understand the scope of this domain and to be able classify work appearing in the literature.

Before describing our privacy taxonomy it is important to provide some terminology that is either new or used in different ways elsewhere in the literature. The *provider* is the individual or organization providing data that is to be stored or used. The provider may or may not be the owner of the data (see Section 2.2). The *collector* is the individual or organization that initially collects, uses, or stores data received from the provider. The collector is the individual or organization that ultimately solicits the data from the provider even if this is through another party who actually acquires the data (i.e. via outsourcing). A *third-party* is any individual or organization that acquires the provided data from the collector. (Note that there is no distinction made between authorized or unauthorized data release to a third-party.)

2 Defining Privacy

Article 8 of the Hippocratic Oath states: "And about whatever I may see or hear in treatment, in the life of human beings – things that should not ever be blurted out outside – I will remain silent, holding such things to be unutterable." This well-known standard is the corner stone of privacy standards upon which doctor-patient confidentiality is established. Unfortunately, modern database technology is far from being a subscriber to such a high standard and, in fact, much of it has actively engaged in violating Article 8 in an attempt to expose or discover information that is deliberately being withheld by data providers.

The U.S. Privacy Act (1974) articulates six principles including: ensuring an individual can determine what data is collected; guarantee that data collected is only used for the purpose for which it is collected; provides access to your own data; and the information is current, correct and only used for legal purposes [1]. The Act provides explicit clauses to collect damages if privacy is violated but it also permits explicit statutory exemption if there is an important public policy need. The Organisation for Economic Cooperation and Development (OECD) have developed a list of eight principles including: limited collection, data quality assurance, purpose specification, limited use, security safeguards, openness, individual participation and accounatiblity to guide providers and collectors to undertake best practices. Several countries have adopted these principles either explicitly in law or through various ethical codes adopted by practitioners that utilize private data [3].

These have been collected into a set of ten principles by Agrawal *et al.* [2] in their development of the seminal work on Hippocratic databases. The ten principles mirror those indicated by the governmental definitions but are described so they can be operationalized as design principles for a privacy-aware database. The principles are: (1) a requirement to specify the purpose; (2) acquiring explicit consent; (3) limiting collection to only required data; (4) limiting use to only that specified; (5) limiting disclosure of data as much as possible; (6) retention of the data is limited and defined; (7) the data is accurate; (8) security safeguards are assured; (9) the data is open to the provider; and (10) the provider can verify compliance to these principles.

This paper's first contribution is to identify four dimensions that collectively operationalize the abstract privacy definitions above into a tool that can be used to categorize and classify privacy research appearing in the literature. Some aspects, such as the requirement for legal recourse, are beyond the scope of a technical paper because they belong rightly in the venue of legal and/or legislative domains. However, any privacy-aware system must be cognizant of the legal and legislative requirements to ensure systems can conform to and facilitate these transparently. Thus, although legal recourse rightfully belongs in a different domain, it is reflected technologically as a requirement to protect logs and provide auditability.

The four dimensions identified in this work are *purpose*, *visibility*, *graunlarity*, and *retention*; to which we now turn.

2.1 Purpose

Providers have various motivations for providing data to a collector. A patient may provide highly personal information to a health care provider to ensure that they receive appropriate medical care. If that patient is a student, then some aspect of the same information may be provided to a teacher to ensure that a missed exam is not unfairly penalized. The provision of such data is for very different purposes so the provider is likely to want to provide different levels of detail (we will return to this aspect in Section 2.3). However, in both scenarios the student/patient is releasing the information for a very specific purpose so it is critical that a privacy-aware system explicitly record and track the purpose for which a data item is collected.

Figure 1 depicts the *purpose* dimension of our privacy definition along the x-axis. There are a number of discrete points along this dimension that represent increasingly general purposes as you travel along the axis. The origin of the purpose axis could be

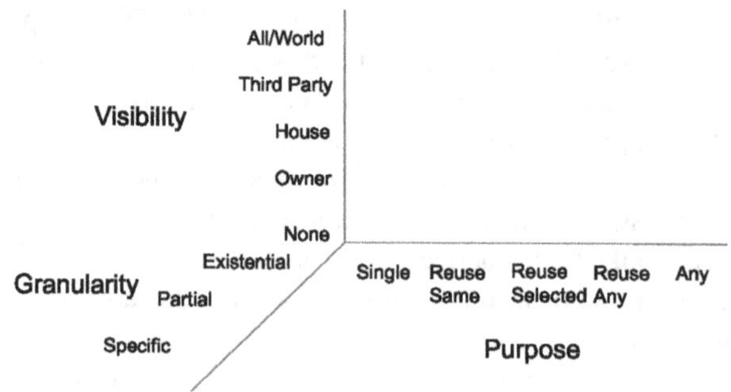

Fig. 1. Key Contributors to Data Privacy in a Data Repository

thought of as a provider providing data for no purpose whatsoever. This point is vacuous on this axis because it would mean that the data could not be used by the collector under any circumstances, for any reason, so there would be no need to collect it. In fact, we would argue that if the collector conforms to the spirit of the various acts and principles summarized in Section 2, they should refuse to collect any data that does not have an explicit purpose identified. Furthermore, a privacy-preserving database system should not permit the inclusion of any data that does not explicitly define the purpose of any data item to be collected or stored.

The first point beyond the origin is the most restrictive and explicit statement of data collection. This point identifies data that a provider supplies for an explicit *single* purpose and for a single use. The data can only be used for this one time single purpose and any other use would be considered a violation of privacy. The next point (labeled *Reuse Same*) allows the collector to reuse the data for the same purpose more than once so it represents a relaxation of privacy. Reuse of data can also permit a data item to be used for a selected set of explicitly defined other purposes (*Reuse Selected*). This scenario might occur when a patient provides medical information to a health-care provider for treatment and for the purpose of collecting payment from an insurance company. The next point on the x-axis (Reuse Selected) is intended to capture the case where the data collected can be used for purposes related to the primary purpose for which the data was provided. Thus, the *Reuse Any* point could be used when a patient is willing to let their data be used for un-foreseeable related purposes such as a disease research study or to monitor efficiency in the hospital itself to improve the hospital's operations. The final point on the x-axis captures the case where a provider does not care how the data might be used. The *Any* point is the least privacy protective point and probably should be discouraged as a matter of normal data collection. However, there are scenarios that could be envisioned where the data is either of such low value or the information already exists in the public domain, where this might be desirable.

2.2 Visibility

Visibility refers to who is permitted to access or use data provided by a provider. This dimension is depicted in Figure 1 on the y-axis. Once again the origin represents the case where the data should not be visible to anyone. As with the purpose dimension, this point should be a clear indication that the data should not be collected or retained because it should not be made available so there is no reason to acquire the data.

The first point on the y-axis introduces important new terminology as we use the term *Owner* to indicate that only the "owner" has access to the data provided. We have deliberately used the term *provider* to this point because the literature is extremely conflicted over who "owns" data in a data repository. This is a critical semantic because two possible definitions could be imposed:

1. Provider is owner: if the provider retains ownership of their private information then this point would indicate that only the provider would have access to their data at this point.
2. Data repository is owner: if the data repository becomes the owner of the data stored on the system then subsequent use of it (i.e. permission to access by others) will be that of the repository.

OECD, US Privacy Act (1974), and the Hippocratic Oath are silent on this critical issue largely because they were developed in a time when it was more difficult to quickly transmit private information in such a way that its original ownership could become obscured. Most data collectors consider the data stored in their systems to be their property and this has consequences on modern high speed networks. Thus, we argue that the ownership of data should be retained by the original provider no matter how many transitive hands it may have passed through. The implication of this on our visibility dimension is that data stored in a repository should only be accessible by the data provider at the *owner* point on the y-axis.

The next point on this dimension permits the repository to access the data provided. We have introduced the term *House* indicating that the the data repository housing the data can view the data provided. The current psyche of most data collectors is that they can view data provided to them by default. However, since storage is now considered a "service", and is often sold as such, it is critical that this distinction be made explicit.

Two examples will help illustrate the need for the distinction between the owner and the house. Google™ [1] is known to utilize data provided to it so they can provide a "value add" for their users. Conversely, iomega™ [2] offers a product called iStorage™that permits online storage of subscriber data on their servers[3]. Google™ would be placed at the house point on the y-axis since they will utilize and access the data as they see fit because they have not made any promises to the provider of how their data might be used. In effect, they are behaving like an "owner" in the traditional sense. However, iomega™offers a service to securely protect a providers data in a convenient way so they have made, at least an implicit promise, to only allow the provider (as owner) to access their data. In short, Google™visibility is the house while iomega istorage™should only be visible to the owner.

Third-parties are data users that are authorized to do so by the house. Typically, such access is permitted under certain conditions and the third-party is required to conform to some kind of an agreement before access is granted. This might occur when the house wants to "outsource" a portion of their data analysis or by virtue of an explicit agreement to share collected data when the provider initially agreed to provide their information. Thus, third-parties are those that have access to data provided to the house but are covered by an explicit agreement with the house.

The final point on the y-axis is the *All/World* extreme where the data is offered to anyone with access to the data repository. Web search tools are the best known examples of such systems in that their fundamental business model is to provide data to those posing queries on their data repositories. This represents the least amount of privacy protection so it is depicted at the extreme point on the y-axis.

2.3 Granularity

The *granularity* dimension is intended to capture characteristics of data that could be used to facilitate appropriate use of the data when there could exist multiple valid

[1] http://www.google.ca/

[2] http://www.iomega.com/

[3] iomega™ is only one of many such storage/backup providers widely available currently (see http://www.iomega.com/istorage/).

accesses for different purposes. For example, a medical care-giver could require very specific information about a particular condition while an insurance company may only need a summary statement of the costs of the patients care. Once again we begin at the origin where no information is to be made available of any kind. We will return to why the origin is needed for each of these dimensions when we consider how to use the taxonomy shortly but the same caveat should be applied to this point on the granularity dimension as on the others where this could suggest that the data should not be collected or stored at all.

The first non-origin point is necessary to capture scenarios where privacy could be compromised by undertaking a set of existence tests. Thus, a query could be written to probe a data repository that may not return any actual data value but a response could indicate the existence of data and this may be sufficient to reveal private data. This actually occurs in a number of different ways so we only mention two here to justify the data point. A query which asks for the "count" of the number of occurrences of a particular data item may not provide the questioner with a value for an attribute but if the system returns a "0" response, information is clearly revealed. A sequence of such queries can narrow down the search space to a unique value without actually releasing private information. This may also occur in more complicated scenarios. For example, by posing queries that include generating the difference between two similar queries could result in information being leaked by virtue of returning results for known data values (i.e. individuals may permit their data to revealed) and subtracting these from the full set, thereby revealing information about the existence of data of a particular type. Techniques that perform output filtering as a privacy protection mechanism may be susceptible to such attacks. There are many other possible existence tests that may unwittingly reveal private information so there is a non-zero chance of privacy compromise when these kinds of queries are permitted. We have identified this point on the z-axis as *existential* because existence testing can be done in some way.

The extreme point on the granularity dimension occurs when *specific* data is released by the data repository. This is actually the most common scenario in that most queries will operate on the actual data stored in the data repository. A query for the data item returns its actual value so the specific accurate data is released in an unobscured way. This point clearly depicts the least amount of privacy protection.

The final point considered on this dimension also provides the best intuiton for the name *"granularity"*. Many systems attempt to protect data privacy by altering it in some non-destructive way so privacy is preserved but a value is still useful to the query poser. Several techniques including aggregation or summarization and categorizing have been used to protect data in some way. Classic examples of this have appeared in the literature such as changing "age" to a category such as "middle age" or an actual income value to "high income." Thus, the system is providing *partial* information but not specific data. We have grouped these techniques under a single heading because all of them, with respect to privacy, attempt to obscure the data in some way to make them resistant to various kinds of attacks. The taxonomy captures the nature of the protection provided by grouping related approaches into classes that ultimately exhibit similar privacy characteristics, although they may be implemented in very different ways. For example, the data ordinalization represented by the age example is in the same class (i.e. partial) as one that provides partial information using anonymization of quasi-keys [5] or its varieties. Many techniques have been developedto break down such methods including statistical attacks, inference, or query attacks all aimed at improving the confidence the attacker has about the actual value stored. In fact, the entire field of data mining is specifically intended to extract information from a data repository when the provider has specifically not included that information to the collector. Obviously, these techniques can (and will) be applied whenever partial results are applied so the point provides less protection than existential testing but more than simply giving the querier the specific value.

2.4 Retention

The privacy characteristics identified by the OECD and the 1974 US Privacy Act implicitly require that data be collected for a particular purpose and remain current. These documents imply that the data should be removed after it has been used for its intended purpose or it has become dated. Thus, a privacy definition must also include an explicit statement of its *retention* period so any privacy model developed must explicitly indicate for all data collected how long it can be used under any circumstances. Data that has passed its "best before" date must be deleted or the retention period must be refreshed by a renewed agreement with the owner.

2.5 Summary and Nomenclature

In the interest of brevity, a point on the taxonomy is described by a tuple $Pr =< p, v, g, r >$ where $p \in \{none, single, reusesame, reuseselected, reuseany, all\}$; $v \in \{none, owner, house, third-party, all-world\}$; $g \in \{none, existential, partial, specific\}$; and $r \in \{\infty, < date >\}$ where $< date >$ is an explicit date indicating when this data is to removed and ∞ indicates no epiration is specified. Thus, a privacy statement such as: $Pr_1 =< all, third-party, partial, \infty >$ would describe a privacy system that permitted third party(s) (visibility) to access modified non-specific data (i.e. partial) (granularity) for any purpose and there would be no requirement to remove this data from the repository.

3 Applying the Privacy Definition

This taxonomy provides the most complete definition of privacy to appear in the literature to date. The lack of a complete definition has meant researchers have generally provided a very brief, often intuitive, definition for privacy and then used the term liberally in their work as though it was a complete definition. Although many useful contributions have been made recently, the lack of a generally accepted definition that can be used in developing privacy research has undoubtedly slowed progress as researchers often "talk past" each other. The dimensional taxonomy provided here addresses this gap by providing a framework that can be used by researchers to place their work relative to an absolute definition (Section 3.1 provides some examples of how to use this framework). It can also help researchers understand how their work can legitimately be compared to other work thereby avoiding the miscommunication that often happens when a term carries different connotations (see Section 3.2).

3.1 Examples of Possible Taxonomizations

Figure 1 can be used in several ways to precisely define the aspect of privacy that is of interest to a particular research project or system. Before describing where on this taxonomy other work should be placed, we first illustrate how the taxonomy can be used in a more abstract way by providing a few examples.

Example 1 Online Storage as a Service: A relatively recent business model has developed where a company provides online storage to clients so they can access the data from anywhere. This is a valuable service as it provides access globally but there is a clear implication, at least on the provider's part, that the data is stored by the company for the provider's exclusive use[4]. The best "real life" analogy would be a safety deposit box. Thus, this scenario explicitly describes a point on all three dimensions where the visibility is only provided to the owner (y-axis), for the repeatable unique purpose (reuse same on the x-axis) that returns the specific data (z-axis) to the owner when requested. Furthermore, there is an implication that the data should only be retained for one year as this is the length of time that the provider agreed to pay for this service (we assume that the current year is 2008 for this example). Thus, this point on our taxonomy would be: $Pr_2 =< reusesame, owner, specific, 2009 >$ and represents the most explicit description of privacy in that it is a point (see Figure 2).

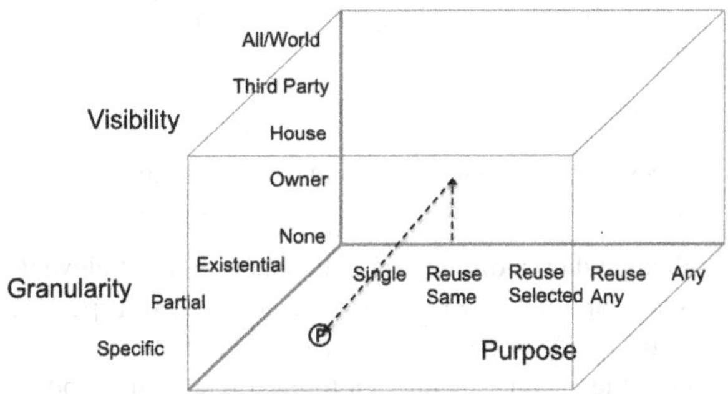

Fig. 2. Privacy point indicating expecation for online storage for clients

The degree of freedom is most restricted when all axes are specified explicitly by a privacy definition. If only one non-retention dimension is unspecified, a line is described in our 3-dimensional taxonomy. To illustrated this we consider two further examples. The first does not specify any aspect of one dimension while the second only partially specifies one of the dimensions and we use these to illustrate the flexibility/utility of our taxonomy.

Example 2 Facebook Social Network: Facebook[5] and other similar forms of social network provide users with a mechanism to present "individual" data that is largely

[4] This is similar to iomega iStorage™mention above.

[5] http://www.facebook.com/

accessible openly. The implied purpose for such social networks from the provider's perspective is to provide a public interface about themselves and as such the provider permits any such posted data to be revealed for "any" purpose so the *reuse any* point on the x-axis best describes its purpose. The data provided is not obscured in anyway so it is clearly *specific* in terms of its granularity. However, the provider is able to control to some varying extend how visible this data is made. For example, a group of "friends" can be defined that provides increased access restrictions to the specific data stored. Thus, the privacy description on the visibility dimension is not a specific point but rather something that must be defined by the provider so it represents a range of possible values on the y-axis but is restricted to the line defined by the x and z axes as depicted in Figure 3 (labeled (f)). Thus: $Pr_3 =< reuseany, \phi, specific, \infty >$.

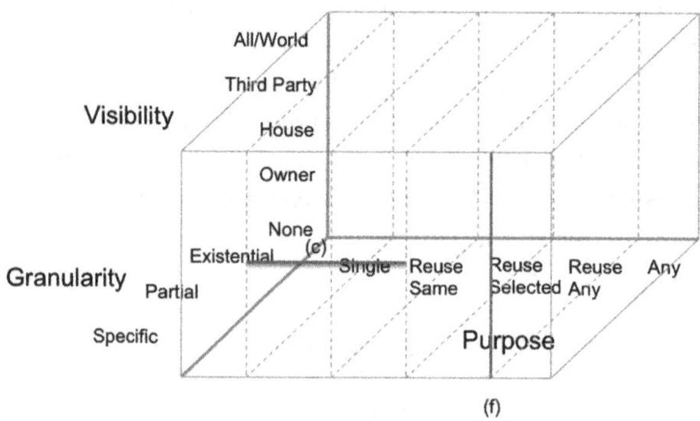

Fig. 3. Privacy point indicating expecation for online storage for clients

We will use ϕ to indicate when a dimension is unspecified or irrelevant for a particular dimension. Recall that our goal is to develop a method to properly capture privacy definitions and we will see in Section 3.2 that some authors do not explicit define a particular dimension or view as it is irrelevant to their research agenda.

Example 3 Online Credit Card Transactions: A primary payment mechanism for goods and services on the Internet is via credit cards. This is also probably the single greatest risk associated with online commerce from a user's perspective so it is clear that this information must be kept private. However, there is no way to meaningfully obscure or "roll-up" a credit card number so only specific data can be used meaningfully in a transaction. The online vendor is the house in this scenario so the visibility is clearly at this level. An argument could be made that a credit-card broker or some trusted-party could act as an intermediary between the vendor and the purchaser but this simply means that the broker is now acting as the house in that it is the holder of the private information. Such trusted third-parties clearly can increase security pragmatically; but from a definitional viewpoint, which is our purpose here, it really only changes who the "house" is but does not change the definitional relationship. In short, the house must have specific information to facilitate online transactions. The purpose may vary however. A credit card could be used for the purpose of a single purchase, it could be used for the same purpose repeatedly (eg. making monthly payments automatically); or for various selected purposes such as purchasing books a intermittent times into the future. The final point to consider is retention. It is easy to see that if the card is being used for a single transaction, then it should only be retained for this single purchase so the retention period should be "now". However, if it is to be retained for a recurring purpose, then it must be retained until the commitment period expires. Thus, the privacy description on the visibility dimension is not a specific point but rather something that must be defined by the provider so it represents a range of possible values on the x-axis but is restricted to the line defined by the y and z axes as depicted in Figure 3 (labeled (c)). Thus: $Pr_4 =< \{\text{single, reusesame, reuseselected}\}, house, specific, \{now, \infty\} >$.

We will use set notation to indicate when only a subset of a particular dimension is defined by the system. This will provide us with substantial flexibility to correctly describe systems and research efforts quite accurately either diagrammatically or with our tuple notation.

When only one of the dimensions is defined by a system, there are two degrees of freedom so a plane is described. To illustrate this we now turn to an all too common modern scenario.

Example 4 Online Data Collection for a Service: Many internet vendors often offer either software or a service in exchange for information about yourself. This may or may not involve also paying for that software or service in addition to the data collection. The privacy agreement typically indicates how this data can be used by the collector and usually limits its use to the company or its affiliates but this is not necessarily the case. Furthermore, for legal reasons the collector may need to make the data collected available to the provider as well to ensure its correctness. Thus, the visibility is essentially undefined or unconstrained. The collector will use this data in many different ways including using the specific data or by aggregating it in various ways so the granularity is often also undefined. They generally limit the purpose to only ones specified by them but they rarely limit how they can use it once it is collected so we somewhat generously identify the point on the x-axis as "reuse any". The retention period, if it is defined at all, is unlimited so this scenario can be defined as: $Pr_5 =< reuseany, \phi, \phi, \infty >$.

Figure 4 illustrates this scenario and demonstrate the lack of specificity and increased degree of freedom indicates that user has a much greater privacy exposure than a system such as the one depicted in Figure 2.

3.2 Placing Literature on the Taxonomy

The taxonomy helps us understand privacy pragmatically in various scenarios such as those described above. It is clear that the framework is able to capture several different scenarios and it highlights the orthogonality among the dimensions described. However, we also feel that the taxonomy is very valuable in classifying research that has appeared in the literature. It also forces us to be more precise about these contributions and leads to a deeper understanding of the strengths and limitations of a piece of work. We have tested the taxonomy's utility by classifying much of the literature on privacy using it.

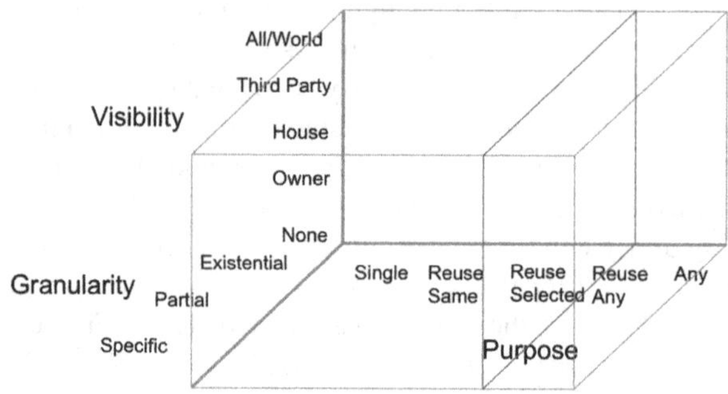

Fig. 4. Privacy point indicating expecation for online storage for clients

Table 1. Applying Privacy Taxonomy to Various Research Systems

Reference	Purpose	Visibility	Granularity
k_anonimity [5]	{RSpec, Rany, any, ϕ}	3rd Party	Partial
Domiongo-Ferrer [4]	ϕ	Provider $\rightarrow$ Owner, House $\rightarrow$ respondant, 3rd party $\rightarrow$ user	ϕ
Hippocratic [2]	Reuse Same <purpose/data item>	House $\rightarrow$ Access based on G and P	Roll up to facilitate privacy

However, we will limit our discussion here to three contributions that are particularly interesting or represent a seminal contribution to privacy in the data repository area and because each highlights a particular feature of the taxonomy that we believe is valuable. We understand that there are many excellent contributions that could have been included but space limitations prevent a full treatment. The reader is encouraged to test the taxonomy by placing their favourite contribution on it to convince themselves of the utility of the taxonomy. Furthermore, the purpose of this section is not to comment on the "value" of the contributions made by these researchers but rather to frame it. We feel that the greatest challenge in understanding and determining how to make future contributions in this area is the need to provide a clear statement of where these contributions are and this requires a clear understanding of the many dimensions identified in this paper. Table 1 provides a summary of some key contributions appearing in the literature and we briefly provide a rationale for this classification.[6]

The **Hippocratic database** [2] proposal is arguably the first that explicitly addresses the need to manage data privacy within a data repository. We do not consider the utility of the model described by the research but rather focus only on those aspects of privacy are explicitly identified by the work.Agrawal *et al.* [2] use the classical definition found in medical domains where data collected can be reused for the purpose for which it is collected. Thus, this point is identified on the x-axis as "reuse same" in our taxonomy.

[6] In all cases, the retention value is undefined (i.e. ϕ) so this column is omitted from the table.

The "House" can view, and may be able to grant access to others, based on a predefined granularity and purpose. Thus, it appears that this point can best be defined as the "House", though there is a strong implication that the "House" become the owner; so it can subsequently grant access to third-parties but this is not made explicit in their initial paper. Granularity is less explicitly defined in the work but it appears that the work suggests that a "roll-up" or "aggregation" mechanism could be used to further facilitate privacy while allowing for "provider" privacy that permits "collector" utility. Finally, retention is not explicitly mentioned in the data model though the paper indicates that this is a critical aspect of privacy protection. However, the lack of retention's explicit inclusion in the model means that we must assume that for all practical purposes, the data retention period is undefined: $Pr_H = <$reuse same, House, Partial,$\phi >$.

k-anonimity [5] is specifically developed to protect privacy by eliminating any form of identifying "keys" that may exist in the data. The obvious requirement to eliminate identifying information is assumed to have occurred so any primary or candidate key data is eliminated. However, the remaining data (or *quasi-keys*) may be sufficient to uniquely identify individuals with respect to some of their attributes so k-anonimity has been developed to ensure that the characteristics of the accessible data protects "provider" privacy. The assumption is that the data being protected with k-anonomity is either undefined or can be used for any purpose including those not necessarily anticipated by either provider or collector. This is in fact the primary motivation for the work on k-anonimity. The second motivation for this data "cleaning"[7] for the purpose of privacy is so it can be safely released to "third-parties" for subsequent analysis. Thus, the visibility assumed by this work is easily identified as 3rd-party. This approach is really intended to provide a form of pre-computed aggregation so the "partial" point best describes the granularity dimension. The retention dimension is not discussed at all but the work implicitly assumes that the data will be retained by both the house and third-partiies indefinitely. Thus: $Pr_k = <$any, 3rd-party, Partial,$\phi >$.

Domiongo-Ferrari [4] makes a valuable contribution by describing a "three-dimensional" framework that is an abstraction of our visibility dimension. The work specifically identifies privacy issues as they relate to our provider, house, and third-party. Thus, our work includes the definitional aspect of Domingo-Ferrari so it is included within our taxonomy as illustrated in Table 1. The contribution provides additional insight along this dimension but also demonstrates the unidimensional privacy definitions often assumed by researches when addressing a specific aspect of the issue. This results in the absence, in this particular case, of any issues associated with purpose, granularity, or retention.

4 Conclusions and Ongoing Work

The abstract claimed, and the paper shows, that the unavoidable conclusion of this paper is that privacy is indeed multifaceted so no single current research effort adequately addresses the true breadth of the issues to fully understand the scope of this important issue. The paper provides a thorough discussion of the privacy research space and

[7] "anonymity" might be a more accurate term but this can only be said to be as strong as "k-anonymous" so we use the weaker term "cleaning" imprecisely and with apologies.

demonstrates that many of the issues are orthogonal. The first contribution is an explicit characterization of the issues involved and their relationship. The paper demonstrates its applicability in a number of different "real-world" scenarios before using the taxonomy to understand the place a few key contributions. A critical value in the paper is a tool that can be used to frame future research in the area. Researchers can use it to identify precisely what aspect of privacy they are considering in their work and to more acurately compare their contributions to other work in the field. Some of the points in the taxonomy may not co-occur as they contradict one another. However, we argue that the importance is it produces a better understand of where contributions fit within the breadth of ongoing privacy research. Furthermore, we have demonstrated that the taxonomy is widely applicable to much of the current literature and have tested it against a wide-range of the literature currently available.

To create a complete system, capable of protecting user privacy, requires that all identified aspects be considered. This paper provides the most complete definition to appear in the literature to date and should be used in developing and using modern data repositories interested in protecting privacy. Fundamental to realizing this vision is the need to develop a data model that has privacy as first-order feature so we intend to build the aspects of this taxonomy into a new privacy-aware data model. In tandem with this formal underpinning we are also working to implement the model into the DBMS architecture at all levels from the query processor through to the cache management system. Thus, we are developing a novel data directory that incorporates this data model and extending SQL (and an XML derivative) to incorporate these privacy features into the query language.

References

1. The privacy act of 1974, September 26 (2003) (1974)
2. Agrawal, R., Kiernan, J., Srikant, R., Xu, Y.: Hippocratic databases. In: VLDB 2002: Proceedings of the 28th International Conference on Very Large Databases, VLDB Endowment, Hong Kong, China, vol. 28, pp. 143–154 (2002)
3. Bennett, C.J.: Regulating Privacy: Data Protection and Public Policy in Europe and the United States. Cornell University Press (April 1992)
4. Domiongo-Ferrer, J.: A three-dimensional conceptual framework for database privacy. In: Jonker, W., Petković, M. (eds.) SDM 2007. LNCS, vol. 4721, pp. 193–202. Springer, Heidelberg (2007)
5. Sweeney, L.: k-anonymity: a model for protecting privacy. International Journal on Uncertainty, Fuzziness and Knowledge-based Systems 10(5), 557–570 (2002)

Dimensions of Dataspaces

Cornelia Hedeler, Khalid Belhajjame, Alvaro A.A. Fernandes, Suzanne M. Embury, and Norman W. Paton

School of Computer Science,
The University of Manchester Oxford Road,
Manchester M13 9PL, UK
{chedeler,khalidb,alvaro,embury,norm}@cs.manchester.ac.uk

Abstract. The vision of dataspaces has been articulated as providing various of the benefits of classical data integration, but with reduced up-front costs, combined with opportunities for incremental refinement, enabling a "pay as you go" approach. However, results that seek to realise the vision exhibit considerable variety in their contexts, priorities and techniques, to the extent that the definitional characteristics of dataspaces are not necessarily becoming clearer over time. With a view to clarifying the key concepts in the area, encouraging the use of consistent terminology, and enabling systematic comparison of proposals, this paper defines a collection of dimensions that capture both the components that a dataspace management system may contain and the lifecycle it may support, and uses these dimensions to characterise representative proposals.

1 Introduction

Data integration, in various guises, has been the focus of ongoing research in the database community for over 20 years. The objective of this activity has generally been to provide the illusion that a single database is being accessed, when in fact data may be stored in a range of different locations and managed using a diverse collection of technologies. Providing this illusion typically involves the development of a single central schema to which the schemas of individual resources are related using some form of mapping. Given a query over the central schema, the mappings, and information about the capabilities of the resources, a distributed query processor optimizes and evaluates the query.

Data integration software is impressive when it works; declarative access is provided over heterogeneous resources, in a setting where the infrastructure takes responsibility for efficient evaluation of potentially complex requests. However, in a world in which there are ever more networked data resources, data integration technologies from the database community are far from ubiquitous. This stems in significant measure from the fact that the development and maintenance of mappings between schemas has proved to be labour intensive. Furthermore, it is often difficult to get the mappings right, due to the frequent occurrence of exceptions and special cases as well as autonomous changes in the sources that

A.P. Sexton (Ed.): BNCOD 2009, LNCS 5588, pp. 55–66, 2009.

require changes in the mappings. As a result, deployments are often most successful when integrating modest numbers of stable resources in carefully managed environments. That is, classical data integration technology occupies a position at the high-cost, high-quality end of the data access spectrum, and is less effective for numerous or rapidly changing resources, or for on-the-fly data integration.

The vision of *dataspaces* [1,2] is that various of the benefits provided by planned, resource-intensive data integration should be able to be realised at much lower cost, thereby supporting integration on demand but with lower quality of integration. As a result, dataspaces can be expected to make use of techniques that infer relationships between resources, that refine these relationships in the light of user or developer feedback, and that manage the fact that the relationships are intrinsically uncertain. However, to date, no dominant proposal or reference architecture has emerged. Indeed, the dataspace vision has given rise to a wide range of proposals either for specific dataspace components (e.g. [3,4]), or for complete dataspace management systems (e.g. [5,6]). These proposals often seem to have little in common, as technical contributions stem from very different underlying assumptions – for example, dataspace proposals may target collections of data resources as diverse as personal file collections, enterprise data resources or the web. It seems unlikely that similar design decisions will be reached by dataspace developers working in such diverse contexts. This means that understanding the relationships and potential synergies between different early results on dataspaces can be challenging; this paper provides a framework against which different proposals can be classified and compared, with a view to clarifying the key concepts in dataspace management systems (DSMS), enabling systematic comparison of results to date, and identifying significant gaps.

The remainder of the paper is structured as follows. Section 2 describes the classification framework. For the purpose of instantiating the framework Section 3 presents existing data integration proposals, and Section 4 presents existing dataspace proposals. Section 5 makes general observations, and Section 6 concludes the paper.

2 The Classification Framework

Low-cost, on-demand, automatic integration of data with the ability to search and query the integrated data can be of benefit in a variety of situations, be it the short-term integration of data from several rescue organisations to help manage a crisis, the medium-term integration of databases from two companies, one of which acquired the other until a new database containing all the data is in place, or the long-term integration of personal data that an individual collects over time, e.g., emails, papers, or music. Different application contexts result in different dataspace lifetimes, ranging from short-, medium- to long-term (*Lifetime* field in Table 1).

Figure 1 shows the conceptual life cycle of a dataspace consisting of phases that are introduced in the following. Dataspaces in different application contexts will only need a subset of the conceptual life cycle. The phases addressed are

Table 1. Properties of the initialisation and usage phase of existing data integration and dataspace proposals

Dimension	DB2 II[9]	Aladin [10]	SEMEX [11,12]	iMeMex[5], iTrails[13]	PayGo[6]	UDI[3]	Roomba [4]	Quarry [14]
Life time/Life cycle								
Lifetime	long	long	long	long	long	long	long	long
Life cycle	init/use/maint	init/use/maint	init/use	init/use/maint/impr	init/use/maint/impr	init	impr	init/use
Data sources								
Cont	gen	app_sp	app_sp	app_sp	gen	gen	gen	gen
Type	s_str/str	s_str/str	unstr/s_str/str	unstr/s_str/str	str	str		s_str
Location	distr	distr	loc/distr	distr	distr	loc		loc
Integration schema; design/derivation								
Type	union/merge	union	merge	union	union	merge	union	union
Scope	dom_sp	dom_sp	dom_sp	gen	dom_sp	dom_sp	gen	gen
Uncertainty						score		
Process	s_aut/man	s_aut	man	aut	aut	aut	aut	aut
Input						schema/match	schema/inst	schema/inst
Matchings; identification								
Endpoints	src-int	src-src	src-int	src-src	src-src	src-src, src-int	src-src	
Uncertainty				score	score	score	score	
Process	man	aut	aut	s_aut	aut	aut	aut	
Input		schema/inst	schema/inst	schema/inst	schema/inst/train	schema	schema/inst	
Mappings; identification								
Uncertainty				score		score		
Process	man		aut	man		aut		
Input			match	schema/inst		schema/match		
Resulting data resource; creation								
Materialis.	virt/p_mat	mat	mat	virt		virt	virt	f_mat
Uncertainty						score		
Reconcil.	NA	dupl	dupl			dupl	dupl	
Search/query; evaluation								
Specification	in_adv/run	run	run	run	run	run	in_adv	run
Type	SPJ/aggr	browse/key/SPJ	browse/key SP	browse/key/SPJ	key	SP(J)	key/S	browse/SP
Uncertainty		ranked			ranked	score		
Evaluation	compl	compl	compl	partial	compl	compl	partial	compl
Comb. res.					union	merge	merge	union

listed in *Life cycle* in Table 1 with the initialisation, test/evaluation, deployment, maintenance, use, and improvement phases denoted as *init*, *test*, *depl*, *maint*, *use*, and *impr*, respectively.

A dataspace, just like any traditional data integration software, is initialised, which may include the identification of the data resources to be accessed and the integration of those resources. Initialisation may be followed by an evaluation and testing phase, before deployment. The deployment phase, which may not be required, for example, in the case of a personal dataspace residing on a single desktop computer, could include enabling access to the dataspace for users or moving the dataspace infrastructure onto a server. As the initialisation of a DSMS should preferably require limited manual effort, the integration may be improved over time in a pay-as-you-go manner [6] while it is being used to

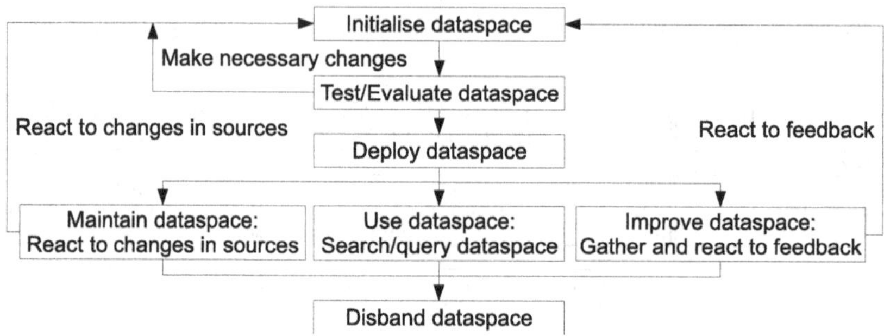

Fig. 1. Conceptual life cycle of a dataspace

search and query the integrated data resources. In ever-changing environments, a DSMS also needs to respond to changes, e.g., in the underlying data resources, which may require support for incremental integration. The phases *Use*, *Maintain* and *Improve* are depicted as coexisting, because carrying out maintenance and improvement off-line would not be desirable. For clarity, the figure does not show any information flow between the different phases, so the arrows denote transitions between phases.

In the remainder of this section, the initialisation, usage, maintenance and improvement phases are discussed in more detail with a view to eliciting the dimensions over which existing dataspace proposals have varied. The dimensions are partly based on the dataspace vision [1,2] and partly on the characteristics of dataspace proposals.

2.1 Initialisation Phase

Figure 2 presents a more detailed overview of the steps that may be part of the initialisation phase. In the following, each of these steps is discussed in more detail and the dimensions that are used to classify the existing proposals are introduced. For each step, the dimensions are either concerned with the process (e.g., identifying matchings) and its input, or with the output of the process (e.g., the matchings identified). As others have proposed (e.g., [7]) we distinguish between *matchings*, which we take to be correspondences between elements and attributes in different schemas, and *mappings*, which we take to be executable programs (e.g., view definitions) for translating data between schemas.

Identify data sources. A DSMS can either provide support for the integration of data sources with any kind of content (*Cont* field in Table 1) or it can provide support for a specific application (*app_sp*), making use of assumptions that apply for that particular application. General support is denoted by *gen* in Table 1. Examples of specific applications include the life sciences, personal information and enterprise data. Furthermore, the data sources to be integrated can be of different types (*Type* field in Table 1). Examples include unstructured (*unstr*), semi-structured (with no explicit schema) (*s_str*) or structured (with explicit

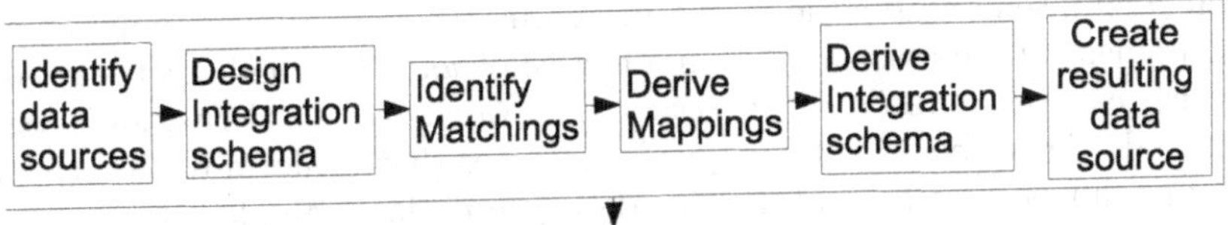

Fig. 2. Initialisation of a dataspace

schema) (*str*). The data sources can also differ in their *location*: they can be local (*loc*) or distributed (*distr*).

Integration schema and its design/derivation. The *process* of obtaining the integration schema can either be manual (*man*), i.e., it is designed, or it can be derived semi-automatically (*s_aut*), e.g., requiring users to select between alternatives, or automatically (*aut*) without any manual intervention. A variety of information can be used as *Input* for designing or deriving the schema, which is depicted by the different locations of the *Design* and *Derive* steps in Figure 2. The schema can be designed using *schema*, or instance (*inst*) information from the sources. Matchings (*match*) or mappings (*map*) can also be used as input. Even when all the available information is taken into account to derive the integration schema automatically, there may still be some uncertainty as to whether the schema models the conceptual world appropriately. This degree of *uncertainty* associated with the schema can be represented as a *score*, which can be, e.g., probabilities or weights derived from inputs. The resulting schema can simply be a union schema, in which source-specific concepts are imported directly into the integration schema, or a schema that merges (e.g. [8]) the source schemas with the aim of capturing more specific the semantic aspects that relate them. The different *types* of resulting schemas are denoted as *union* and *merge* in Table 1, respectively. Integration schemas can also vary in their *scope*. To be able to model a wide variety of data from a variety of domains, generic models (*gen*), such as resource-object-value triples, can be used. In contrast to those, domain-specific models (*dom_sp*) are used in other proposals.

Matchings and their identification. Matchings can vary with respect to their *endpoints*: they can either be correspondences between the source schemas (*src-src*) or between source schemas and the integration schema (*src-int*). The *process* of identifying the matchings can either be manual (*man*), semi-automatic (*s_aut*) or automatic (*aut*). When matchings are identified automatically, some *uncertainty* is often associated with them. This can be represented by the use of *scores*. The identification process may require a variety of different *inputs*, e.g., the *schemas* to be matched, instances that conform to the schemas (*inst*), and training data (*train*), e.g., when machine learning techniques are applied.

Mappings and their identification. Unlike matchings, we take mappings as always been expressed between the sources schemas and the integration schema, not between source schemas. The *process* to derive the mappings can either be manual (*man*), semi-automatic (*s_aut*) or automatic (*aut*). Similar to matchings, the mappings can be associated with a degree of *uncertainty* about their validity, which can be represented by *scores*. The *inputs* to the derivation process may include the *schemas* to be mapped, instances that conform to the schemas (*inst*), matchings (*match*), and training data (*train*), for example, when machine learning techniques are used.

Resulting data resource. The resulting data resources over which queries are expressed can vary with respect to their materialisation (*Materialis.*): they can either be virtual (*virt*), partially materialised (*p_mat*) or fully materialised (*f_mat*). *Uncertainty* that is associated with the content can be denoted by *scores*. During the creation of the integrated database, duplicates (*dupl*) and conflicts (*confl*) can either be reconciled (*Reconciliation*) or be allowed to coexist.

2.2 Usage Phase: Search/Query and Their Evaluation

Searches and queries can be specified (*Specification*) as a workload in advance (*in_adv*) of data integration taking place, or they can be specified after the integration, at runtime (*run*). Specifying queries in advance provides the potential for optimising the integration specifically for a particular workload. Different *types* of searches/queries can be supported by the dataspace: exploratory searches, e.g, browsing (*browse*), which are useful either if the user is unfamiliar with the integration schema, or if there is no integration schema. Other types include keyword search (*key*), select- (*S*), project- (*P*), join- (*J*), and aggregation (*aggr*) queries. An aim for a dataspace is to provide some kind of search at all times [1]. Query *evaluation* can either be complete (*compl*) or partial (*part*), e.g., using top-k evaluation approaches or approaches that are able to deal with the unavailability of data sources [1]. If multiple sources are queried, the results have to be combined (*Combine results*), which may be done by forming the *union* or merging (*merge*) the results, which may include the reconciliation of duplicates and/or conflicts. To indicate the *uncertainty* associated with the results of a query, they may be annotated with *scores* of some form, or they may be *ranked*.

2.3 Maintenance and Improvement Phase

The maintenance phase deals with the fact that the underlying data sources are autonomous [1], and the improvement phase aims to provide tighter integration over time [1]. The steps in both phases are comparable to the steps involved in the initialisation phase, however, additional inputs may need to be considered. Examples include user feedback, as well as previous matchings and mappings that may need to be updated after changes in the underlying schemas.

Despite the general awareness that a DSMS needs to be able to cope with evolving data sources and needs to improve over time, only limited results have been reported to date, making it hard to consolidate the efforts into coherent dimensions. In the following we suggest a set of dimensions, that may be used to characterise future research efforts (see also Table 2).

Table 2. Properties of the maintenance and improvement phase of existing data integration and dataspace proposals

Dimension	DB2 II [9]	Aladin [10]	SEMEX [11,12]	iMeMex[5], iTrails[13]	PayGo[6]	UDI[3]	Roomba [4]	Quarry [14]
Maintenance								
Changes	add/ src_inst		add/ src_inst	add				
Reuse			match/map/ int_sch		match/map/ int_sch			
Improvement								
Approach					alg_match/ exp_user			exp_user
Stage_feedb								match
Stage_impr								match

Maintenance: For effective maintenance, a DSMS needs to be able to cope with a number of different *changes*, including adding (*add*) and removing (*rem*) of resources. A DSMS also needs to be able to cope with changes in the underlying sources, e.g. changes to the instances (*src_inst*) or the schemas (*src_sch*), as well as changes to the integration schema (*int_sch*). Ideally, a DSMS should require little or no manual effort to respond to those changes. It may also be beneficial to *Reuse* the results of previous integration tasks, e.g., previous matchings (*match*), mappings (*map*), integration schemas (*int_sch*), or even user feedback (*feedb*) when responding to source changes.

Improvement: Improvement may be achieved in a number of ways (*Approach*), including the use of different or additional approaches to those used during initialisation for deriving matchings (*a_match*), mappings (*a_map*), or the integration schema (*a_int*). Furthermore, user feedback can be utilised, which could be implicit (*imp_user*) or explicit (*exp_user*). In cases where user feedback is considered, this could be requested about a number of different stages (*Stage_feedb*). This includes requesting feedback on the matchings (*match*), mappings (*map*), integration schema(s) (*int_sch*), reformulated queries (*ref_query*), query results (*res*) or the ranking of the results (*res_ran*). The feedback obtained may not only be used to revise the stage about which it was acquired, but it may also be propagated for improvement at other stages (Stage_impr). The values for this property are the same as for *Stage_feedb*.

3 Data Integration Proposals

For the purpose of comparison, this section uses the framework to characterise the data integration facilities of DB2 [9], as an example of a classical data integration approach; values of the dimensions are provided in Tables 1 and 2.

DB2 [9] follows a database federation approach. It provides uniform access to heterogeneous data sources through a relational database that acts as mediation middleware. The integration schema, which could be a *union* schema, or a *merged* schema defined by views which need to be written *manually*. Data sources are accessed by wrappers, some of which are provided by DB2 and some of which may have to be written by the user. A wrapper supports full SQL and translates (sub)queries of relevance to a source so that they are understood by the external source. Due to the *virtual* nature of the resulting data resource, changes in the underlying data sources may be responded to with limited manual effort. In summary, DB2 relies on largely manual integration, but can provide tight semantic integration and powerful query facilities in return.

4 Dataspace Proposals

This section describes a selection of dataspace proposals. An overview of their properties can be found in Tables 1 and 2.

ALADIN [10] supports semi-automatic data integration in the life sciences, with the aim of easing the addition of new data sources. To achieve this, ALADIN makes use of assumptions that apply to this domain, i.e., that each database tends to be centered around one primary concept with additional annotation of that concept, and that databases tend to be heavily cross-referenced using fairly stable identifiers. ALADIN uses a *union* integration schema, and predominantly *instance-based* domain-specific approaches, e.g., utilising cross-referencing to discover relationships between attributes in entities. The resulting links are comparable to *matchings*. *Duplicates* are discovered during *materialisation* of the data resource. Links and duplicate information are utilised for *exploratory* and *keyword* searches and may help life scientist to discover previously unknown relationships. To summarise, ALADIN provides fairly loose integration and mainly exploratory search facilities that are tailored to the life sciences domain.

SEMEX [11,12] integrates personal information. A domain model, which essentially can be seen as a *merged* integration schema, is provided *manually* upfront, but may be extended manually if required. Data sources are accessed using wrappers, some provided, but some may have to be written manually. The schemas of the data sources are *matched* and *mapped automatically* to the domain model, using a bespoke mapping algorithm that utilises heuristics and *reuses* experience from previous matching/mapping tasks. As part of the *materialisation* of the resulting data resource, *duplicate* references are reconciled, making use of domain knowledge, e.g., exploiting knowledge of the components of email addresses. SEMEX provides support for *adding* new data sources and *changes* in the underlying data, e.g., people moving jobs and changing their email address or phone number, which require domain knowledge to be resolved, e.g., to realise that it is still the same person despite the change to the contact details. SEMEX, therefore, can be seen as a domain-specific dataspace proposal that relies on domain knowledge to match schemas to the given integration schema and reconcile references automatically.

iMeMeX [5] is a proposal for a dataspace that manages personal information; in essence, data from different sources such as email or documents are accessed from a graph data model over which path-based queries can be evaluated. iMeMeX provides low-cost data integration by initially providing a *union* integration schema over diverse data resources, and supports incremental refinement through the *manual* provision of path-based queries known as iTrails [13]. These trail definitions may be associated with a *score* that indicates the *uncertainty* of the author that the values returned by an iTrail is correct. As such, iMeMeX can be seen as a light weight dataspace proposal, in which uniform data representation allows queries over diverse resources, but without automation to support tasks such as the management of relationships between sources.

PayGo [6] aims to model web resources. The schemas of all sources are integrated to form a *union* schema. The source schemas are then matched *automatically* using a schema matching approach that utilises results from the matching of large numbers of schemas [15]. Given the similarity of the schemas determined by matching, the schemas are then clustered. *Keyword* searches are reformulated into structured queries, which are compared to the schema clusters to identify the relevant data sources. The sources are ranked based on the similarity of their schemas, and the results obtained from the sources are *ranked* accordingly. PayGo [6] advocates the *improvement* of the semantic integration over time by utilising techniques that automatically suggest relationships or incorporate user feedback; however, no details are provided as to how this is done. In summary, PayGo can be seen as a large-scale, multi-domain dataspace proposal that offers limited integration and provides keyword-based search facilities.

UDI [3,16] is a dataspace proposal for integration of a large number of domain independent data sources automatically. In contrast to the proposals introduced so far, which either start with a manually defined integration schema or use the union of all source schemas as integration schema, UDI aims to derive a *merged* integration schema *automatically*, consolidating schema and instance references. As this is a hard task, various simplifying assumptions are made: the source schemas are limited to relational schemas with a single relation, and for the purpose of managing uncertainty, the sources are assumed to be independent. Source schemas are matched *automatically* using existing schema matching techniques [17]. Using the result of the matching and information on which attributes co-occur in the sources, attributes in the source schemas are clustered. Depending on the *scores* from the matching algorithms, matchings are deemed to be certain or uncertain. Using this information, multiple mediated schemas are constructed, which are later consolidated into a single *merged* integration schema that is presented to the user. Mappings between the source schemas and the mediated schemas are derived from the matchings and have *uncertainty* measures associated with them. Query results are *ranked* based on the scores associated with the mappings used. In essence, UDI can be seen as a proposal for automatic bootstrapping of a dataspace, which takes the uncertainty resulting from automation into account, but makes simplifying assumptions that may limit its applicability.

Even though the majority of proposals acknowledge the necessity to improve a dataspace over time, Roomba [4] is the first proposal that places a significant emphasis on the *improvement* phase. It aims to improve the degree of semantic integration by asking users for *feedback* on *matches* and *mappings* between schemas and instances. It addresses the problem of choosing which matches should be confirmed by the user, as it is impossible for a user to confirm all uncertain matches. Matches are chosen based on their utility with respect to a *query* workload that is provided *in advance*. To demonstrate the applicability of the approach, a *generic* triple store has been used and *instance-based matching* using string similarity is applied to obtain the matches.

Quarry [14] also uses a *generic* triple store as its resulting data source, into which the data is *materialised*. Using a *union* schema, the data from the data sources coexists without any semantic integration in the form of matchings or mappings. So called signature tables, which contain the properties for each source, are introduced and it is suggested that signature tables with similar properties could be combined. Quarry provides an API for browsing the integrated data and for posing select and project queries.

5 Discussion

To make dataspaces valuable, steps that have been proven to be difficult and/or time consuming and that, to date, have predominantly been carried out manually or semi-automatically, need to be automated. Such steps include designing an integration schema, deriving the matchings between schemas, and deriving mappings between the source and integration schemas, which can then be used for query evaluation. Automating those steps results in uncertainty associated with the matchings, the mappings and the integration schema. As a result, techniques need to be developed that either reduce the uncertainty or take it into account, for example during query evaluation. Further uncertainty may be introduced during reformulation of keyword queries into structured queries, and vice versa, and during reference reconciliation either at the time the target data resource is materialised or when the results from subqueries issued to multiple sources are combined.

Since the dataspace vision has only recently emerged, and since it opens a broad research space covering multiple areas including, among others, data integration, information retrieval, and uncertainty management, existing proposals vary greatly over the dimensions presented in this paper. However, the following general observations can be made, with reference to Tables 1 and 2:

Union schemas are more common than merged schemas. In most proposals to date, the integration schema used by the user to formulate the query is obtained by unioning the source schemas. This choice is driven by the fact that the creation of a merged schema often raises conflicts, the resolution of which requires human intervention. Although there has been some work describing [18] and automating the collection [4] of uncertain information in merged schemas, early dataspace proposals provide limited support for managing the uncertainty associated with merged schemas, and in most cases steer clear of the issue altogether. Furthermore, classical work on data integration typically does not address uncertainty, for the simple reason that this has been dealt with manually at the bootstrapping phase.

Improvement as a means for dealing with or reducing the impact of uncertainty. Although incremental improvement was presented in the dataspace vision as the means for reducing uncertainty, only two of the proposals described in Section 4 consider this issue. This may be explained by the fact that early proposals for dataspace are initially addressing challenges at the bootstrapping phase, rather than looking for means that facilitate the improvement of an existing dataspace.

Dataspaces with short life times are missing. Several applications may generate the need for a dataspace with a short life time. A well known example is that of mashups. The improvement phase of these applications is relatively small compared with dataspace with longer lifetimes. As such, these applications may require dealing with uncertainty to a greater level up-front at bootstrapping time.

6 Conclusions

Dataspaces represent a vision for incremental refinement in data integration, in which the effort devoted to refining a dataspace can be balanced against the cost of obtaining higher quality integration. Comprehensive support for pay-as-you-go data integration might be expected to support different forms of refinement, where both the type and quantity of feedback sought are matched to the specific requirements of an application, user community or individual. Early proposals, however, provide rather limited exploration of the space of possibilities for incremental improvement. A common approach prioritises reduced start-up costs, typically by supporting a *union* integration schema; such an approach provides syntactic consistency, but the extent to which the resulting dataspace can be said to "integrate" the participating sources is strictly limited.

Although there is a considerable body of work outside dataspaces to support activities such as schema matching or merging, early dataspace proposals have made fairly limited use of such techniques. For example, although there has been work on the automated construction of a global model that accommodates uncertainty in matchings [3], this work makes strong simplifying assumptions in terms of the complexity of the models to be integrated. Furthermore, there are no comparable results on automated refinement. As such, although there is certainly a role for different flavours of DSMS, this survey suggests that there is considerable scope for further research into dataspaces, and that proposals to date fall short in different ways of fulfilling the potential of pay-as-you-go data integration.

References

1. Franklin, M., Halevy, A., Maier, D.: From databases to dataspaces: a new abstraction for information management. SIGMOD Record 34(4), 27–33 (2005)
2. Halevy, A., Franklin, M., Maier, D.: Principles of dataspace systems. In: PODS 2006, pp. 1–9. ACM, New York (2006)

3. Das Sarma, A., Dong, X., Halevy, A.: Bootstrapping pay-as-you-go data integration systems. In: SIGMOD 2008, pp. 861–874. ACM, New York (2008)
4. Jeffery, S.R., Franklin, M.J., Halevy, A.Y.: Pay-as-you-go user feedback for dataspace systems. In: SIGMOD 2008, pp. 847–860. ACM, New York (2008)
5. Dittrich, J.P., Salles, M.A.V.: idm: A unified and versatile data model for personal dataspace management. In: VLDB 2006, pp. 367–378. ACM, New York (2006)
6. Madhavan, J., Cohen, S., Dong, X.L., Halevy, A.Y., Jeffery, S.R., Ko, D., Yu, C.: Web-scale data integration: You can afford to pay as you go. In: CIDR 2007, pp. 342–350 (2007)
7. Miller, R.J., Hernández, M.A., Haas, L.M., Yan, L., Ho, C.T.H., Fagin, R., Popa, L.: The clio project: managing heterogeneity. SIGMOD Record 30(1), 78–83 (2001)
8. Pottinger, R., Bernstein, P.A.: Schema merging and mapping creation for relational sources. In: EDBT 2008, pp. 73–84 (2008)
9. Haas, L., Lin, E., Roth, M.: Data integration through database federation. IBM Systems Journal 41(4), 578–596 (2002)
10. Leser, U., Naumann, F. (almost) hands-off information integration for the life sciences. In: CIDR 2005, pp. 131–143 (2005)
11. Dong, X., Halevy, A.Y.: A platform for personal information management and integration. In: CIDR 2005, pp. 119–130 (2005)
12. Liu, J., Dong, X., Halevy, A.: Answering structured queries on unstructured data. In: WebDB 2006, pp. 25–30 (2006)
13. Vaz Salles, M.A., Dittrich, J.P., Karakashian, S.K., Girard, O.R., Blunschi, L.: itrails: Pay-as-you-go information integration in dataspaces. In: VLDB 2007, pp. 663–674. ACM, New York (2007)
14. Howe, B., Maier, D., Rayner, N., Rucker, J.: Quarrying dataspaces: Schemaless profiling of unfamiliar information sources. In: ICDE Workshops, pp. 270–277. IEEE Computer Society Press, Los Alamitos (2008)
15. Madhavan, J., Bernstein, P.A., Doan, A., Halevy, A.: Corpus-based shema matching. In: ICDE 2005, pp. 57–68 (2005)
16. Dong, X., Halevy, A.Y., Yu, C.: Data integration with uncertainty. In: VLDB 2007, pp. 687–698 (2007)
17. Rahm, E., Bernstein, P.A.: A survey of approaches to automatic schema matching. VLDB Journal: Very Large Data Bases 10(4), 334–350 (2001)
18. Magnani, M., Rizopoulos, N., McBrien, P., Montesi, D.: Schema integration based on uncertain semantic mappings. In: Delcambre, L.M.L., Kop, C., Mayr, H.C., Mylopoulos, J., Pastor, Ó. (eds.) ER 2005. LNCS, vol. 3716, pp. 31–46. Springer, Heidelberg (2005)

The Use of the Binary-Relational Model in Industry: A Practical Approach

Victor González-Castro[1], Lachlan M. MacKinnon[2], and María del Pilar Angeles[3]

[1] School of Mathematical and Computer Sciences, Heriot-Watt University, Edinburgh, EH14 4AS, Scotland
victor@macs.hw.ac.uk

[2] School of Computing and Creative Technologies, University of Abertay Dundee, Dundee, DD1 1HG, Scotland
l.mackinnon@abertay.ac.uk

[3] Facultad de Ingeniería, Universidad Nacional Autónoma de México, Ciudad Universitaria, CP04510, México
pilarang@unam.mx

Abstract. In recent years there has been a growing interest in the research community in the utilisation of alternative data models that abandon the relational record storage and manipulation structure. The authors have already reported experimental considerations of the behaviour of n-ary Relational, Binary-Relational, Associative and Transrelational models within the context of Data Warehousing [1], [2], [3] to address issues of storage efficiency and combinatorial explosion through data repetition. In this paper we present the results obtained during the industrial usage of Binary-Relational model based DBMS within a reference architectural configuration. These industrial results are similar to the ones obtained during the experimental stage of this research at the University laboratory [4] where improvements on query speed, data load and considerable reductions on disk space are achieved. These industrial tests considered a wide set of industries: Manufacturing, Government, Retail, Telecommunications and Finance.

Keywords: Binary-relational, Alternative Data Models, Industrial Application, Disk Compression, Column-oriented DBMS.

1 Introduction

The present research is focused on solving a number of issues regarding data storage within Relational Data Warehouses such as excessive allocation of disk space, high query processing times, long periods of Extraction, Transformation and Loading data processing (ETL), among others [4].

Alternative data models have been benchmarked (n-ary Relational, Binary-Relational, Associative and Transrelational) within the context of Data Warehousing [3], [4]. The global evaluation has shown that the Binary-Relational model has obtained better results in almost all the aspects defined on the extended version of the TPC-H benchmark [4]. Therefore the use of such a data model within an Alternative

A.P. Sexton (Ed.): BNCOD 2009, LNCS 5588, pp. 67–77, 2009.

Data Warehouse Reference Architectural Configuration (ADW) has been proposed in [9]. The main intention of this paper is to present the results obtained by implementing the proposed configuration in industrial situations and the advantages of using it regarding the above mentioned issues. The ADW is a different implementation of an infrastructure for real life data warehouses utilising already existing elements that have not been used together before.

2 Industrial Tests

After the results obtained from the laboratory [4] using the synthetic TPC-H data set and where the Binary-Relational model emerged as the best model for Data Warehousing environments, it was appealing to move forward and validate the conclusions obtained in the laboratory by using ADW within industrial environments. The Binary-Relational model instantiation used was implemented under the Sybase IQ product [6] as this is a mature product commercially available. Sybase Inc. refers to its product as a column-oriented DBMS.

The results obtained by Sybase IQ are similar to those achieved by MonetDB [5], which was the Binary-Relational model instantiation used in the laboratory.

The presented Alternative Data Warehouse Reference Architectural Configuration (ADW) has been tested in industry under real life scenarios and the results obtained are presented in this section. The industrial implementations have been made for five organisations in different industries –manufacturing, government, retail, telecommunications and finance. Owing to confidentiality, the names of the organisations have been removed and in some cases the queries have been renamed to keep the data anonymous.

2.1 Tested Organizations Profiles

Manufacturing Industry Company
This is a company with operations in more than 50 countries, but only data for 7 countries have been loaded in this test. The main problems were the high query response time and the disk space required to store data. They were using a traditional n-ary-Relational DBMS: Microsoft SQL Server. Figure 1 shows the ADW configuration made for the manufacturing industry company. The names of the commercial products used in this particular configuration have been included; as ADW is a reference architectural configuration, it can be materialized by different products in a particular configuration. In this case, an ETL approach has been followed as it was not the intention to change all the ETL programs. ETL phases were materialized by Cognos' Decision Stream Product [10].

Government Agency
The main problems were long query response time plus the disk space required to store data. They were using a traditional n-ary-Relational DBMS: Oracle. Figure 2 represents the ADW configuration for the government agency; in this case an ELT approach has been followed as the agency wants to load directly from Oracle and avoid generating intermediate flat files. The names of the commercial products used

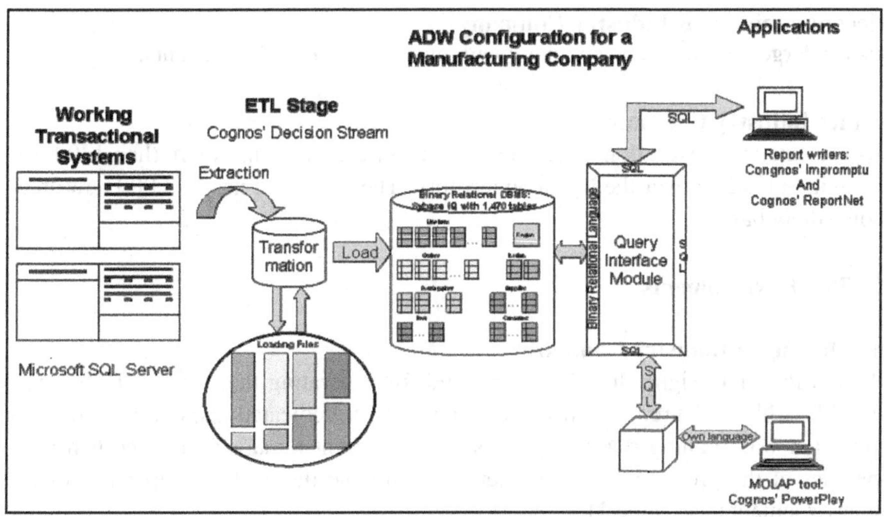

Fig. 1. ADW Instantiation for a Manufacturing Company

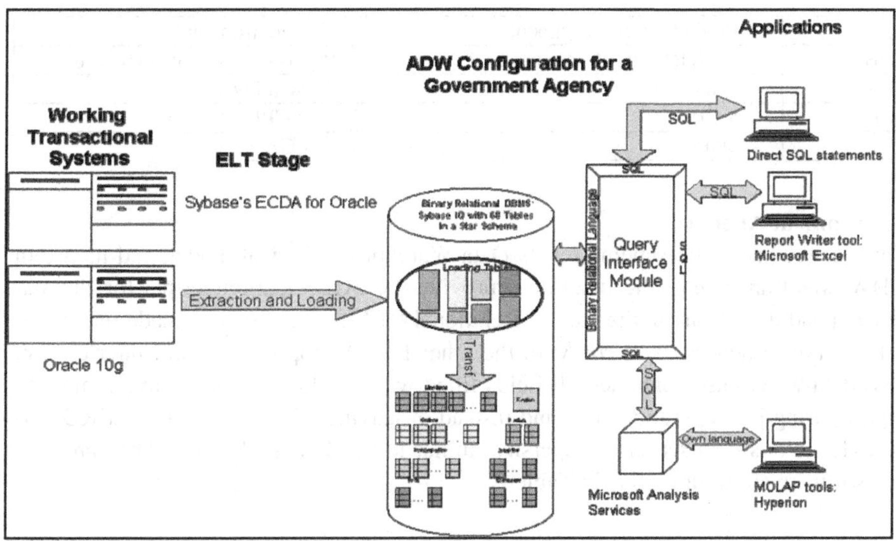

Fig. 2. Government Agency's ADW configuration

to materialise the ADW elements have been listed on the same figure. Microsoft Analysis Services are used to generate the cube structures required to be used by the OLAP tool.

Retail Industry Company

This is a retail company with more than 130 stores. The main problems of this company were the long queries response times plus the lack of disk space for growth.

Telecommunications Industry Company
This is a large mobile phone company with more than 100 million clients.

Finance Industry Company
This is a financial institution oriented to consumer credits. It has more than 130 business units spread all over the Mexican territory. They generate nearly 3,000 new credits on a daily basis.

2.2 Test Environments

Manufacturing Industry Company
Sybase's Power Designer tool has been used for generating the scripts which create the 1,470 tables required. All tests were made with dedicated servers and no other application was running during the tests. An important remark is that the hardware capacities of the production environment were double those of the capacities of the test environment (see Table 1).

Table 1. Manufacturing Company Hardware Configuration

	Production Environment	Test environment
DBMS	n-ary-Relational: Microsoft SQL Server	Binary-Relational: Sybase IQ
OS	Windows	Windows
CPUs	8 CPUs	4 CPUs
RAM	8GB	4GB

Government Agency
This agency had a star schema for its Data Warehouse's logical model, and its use in ADW was transparent; the institution provide the SQL creation scripts for the star schema and it was run on the Binary-Relational DBMS. The test was made on a dedicated server. Each test was run with the other DBMS stopped to avoid interference. This data warehouse contained all fields filled with a value in order to avoid missing tuples during join operations [7], and instead of leaving NULLs, they use ZERO values. Hence this data set has no sparsity similar to the TPC-H data set. The hardware and software configuration is in Table 2.

Table 2. Government Agency's Test Configuration

	Test Environment
n-ary-Relational DBMS	Oracle
Binary-Relational DBMS	Sybase IQ
OS	AIX
CPUs	IBM pSeries670 with 8 Power CPUs at 1.1 GHz
RAM	16GB

Retail Industry Company

This retail company is using an n-ary-Relational instantiation (Microsoft SQL Server). The Data Warehouse consists of 25 tables which include Inventory, Departments, Suppliers, Items, and summary tables with some pre-calculated results.

Telecommunications Industry Company

In this case, as the volumes were very high, the company wanted to test on the same production machine to have realistic numbers. The characteristics of the machine used are presented in Table 3.

Table 3. Telecommunications Company's Hardware

	Production Environment	Test Environment
DBMS	Oracle 10G	Sybase IQ 12.5
OS	Sun Solaris 5.9	Sun Solaris 5.9
CPUs	6 CPUs	6 CPUs
RAM	12 GB	12 GB

Finance Industry Company

They were using the n-ary-Relational database from Sybase "Sybase ASE". As they were having performance problems, the suggestion to use the same vendor Binary-Relational DBMS "Sybase IQ" was made. After seeing the results, the company changed its Data Warehouse to the Binary-Relational model based DBMS, and its production environment is running on Sybase IQ. The characteristics of the machine used are presented in Table 4.

Table 4. Finance Company Hardware Configuration

	Old Production Environment	New Production Environment
DBMS	Sybase ASE 12.5.3	Sybase IQ 12.6
OS	HP-UX 11.23 64 bits	Linux Red Hat 3 32 bits
CPUs	4 CPUs Itanium 64 bits	2 CPUs
RAM	12 GB	4 GB

The main problem with the n-ary Relational DBMS was a slow query response time. They need to create a lot of summary tables in order to try to remedy the slow query performance with the implications of more night processing time, more additional disk storage and increasing the database administration complexity. However the most important issue from the business point of view was the difficulty in undertaking complex business analyses because of the slow system performance.

From the testing and production shown in previous tables, it is important to remark that there were important reductions regarding hardware capacities and therefore saving on hardware acquisition.

2.3 Comparison of Extraction, Transformation and Loading Processes

The results obtained regarding the ETL processes for the manufacturing, government, retail and telecommunication companies are shown in Table 5.

Table 5. ETL Results

	Traditional DW Architecture (hh:mm)	ADW (hh:mm)	**Savings**
Manufacturing	08:41	04:46	**45%**
Government	27:00	12:20	**54%**
Retail	04:35	00:23	**92%**
Telecommunications	02:55	02:23	**18%**

In the particular case of the government agency savings of 54% were achieved by ADW with loading time of 12 hours and 20 minutes, while the traditional configuration makes it in 27 hours. It represents 15 hours' less processing time for this agency.

In the case of the retail company, we require to generated and load flat files into each DBMS using their bulk loading utilities. The results of the ETL processes are as follows: The Binary-Relational instantiation required 23 minutes to load. It contrasts with the 4 hours and 35 minutes required by the n-ary-Relational instantiation. In approximate terms the Binary-Relational instantiation required 8% of the time required to load the same data; this time savings that will benefit their nightly processing.

For the Telecommunications company the loading of one day of transactions was made (187,183, 140 rows); the Binary-Relational instantiation load the data set in 2 hours and 23 minutes and the n-ary-Relational made it in 2 hours and 55 minutes. It represents savings of 18% which is low comparing to the rest of the cases. This low percentage might be due to the loading files were compressed and the DBMS had to wait for the Operating System to uncompress the files.

Comparing the results of the ETL processes among different industries, we can confirm the assertion made in [4], where it was established that Extraction and Transformation times are constant, but Loading times are reduced when using a Binary-Relational model based DBMS in ADW as it can start one loading thread process for each Binary-Relation (column) of the table; thus each row is loaded in parallel. As a conclusion we can state that by using the ADW important reductions in loading time will represent performance improvement. However, such improvements will depend on the characteristics of each company.

2.4 Comparison of Disk Space Utilisation

The Database space required for both traditional and alternative configurations was measured for all the industries and the results are summarized in Table 6.

Table 6. Disk Space Results

	Traditional DW Architecture (GB)	ADW (GB)	**Savings**
Manufacturing	148.00	38.00	**74%**
Government	97.00	12.00	**88%**
Retail	22.74	6.62	**71%**
Telecommunications	37.22	23.36	**37%**
Finance	13.22	8.06	**39%**

In the case of the government institution, the ADW can manage the same amount of information in 12% of the uncompressed disk space required by the Traditional DW, considering data+ indexes+ temporal space, or in other words it is achieving savings of 88%. If considering data space only, savings of 50% are achieved by ADW.

The indexes in the Traditional DW occupied more than the data itself (47GB of indexes vs. 24GB of data). Those indexes were built in order to try to remedy the query performance problem of this agency. Similar patterns exist in many other organizations.

Even when using the compression techniques offered by the n-ary-Relational DBMS [9], the Binary-Relational DBMS has achieved 76% savings when compared against the compressed version of the n-ary-Relational DBMS (52 GB vs. 12GB).

The n-ary-Relational DBMS required 5.5 hours to compress data. This time must be added to the 27 hours required to load time, giving then 32.5 processing hours before data is available for querying on the traditional DW; while in ADW, data are ready for querying after 12 hours of processing to have data ready for querying.

The input raw data was 8,993 MB for the Retail Company, and again here the presumption that the data will grow once loaded into the Database has been challenged, as the Binary-Relational DBMS decreases the data size achieving savings of over 70%. The n-ary Relational used 22.74 GB and the Binary-Relational used 6.62 GB of disk space.

The Binary-Relational instantiation showed their benefits in space reduction for the Telecommunications Company, it achieved savings of 37%. The space occupied by the Binary-Relational instantiation was 23.36 GB and the n-ary-Relational used 37.22 GB.

The Finance Company obtained disk savings of 40% approximately.

The implementation of the alternative configuration represented savings on disk space for all industries, when comparing the disk space utilisation they had with the traditional DW architecture, but it depends on the nature of the data. The synthetic TPC-H data set achieved 32% disk reduction but with this set of companies, the reductions were higher with up to 88% less disk space than the Traditional DW architectural configuration.

2.5 Query Response Time

One of the main benefits found during the laboratory test was the improvement in query response time [4]. In Table 7, the query response time has been measured for the different industrial cases.

Table 7. Query Response Times

	Traditional DW Arch. (min:sec)	ADW (min:sec)	**Savings (%)**
Manufacturing (24 queries)	29:48	08:39	**70.9%**
Government (256 queries)	404:54	07:06	**98.2%**
Retail (2 queries)	00:46	00:05	**89.1%**
Telecommunications (5 queries)	37:51	07:53	**79.1%**
Finance (queries)	16:49	00:01	**99.9%**

The manufacturing company uses Cognos as its analysis tool. A set of relevant queries were executed, the test included direct SQL reports using Cognos, the other set of queries was the full invoicing analysis to the business. Another full business analysis was run; it is the "Accounts Receivable". For ADW its total query workload represents savings of 70.9% in query execution time. Another important fact is that the applications did not require any modification; they were just redirected from the Microsoft SQL Server ODBC to the Sybase IQ ODBC, and the reports ran successfully.

A set of 128 queries used in production for the government agency was provided and benchmarked in both architectures. These queries are SQL sentences which are sent directly to the DBMS. In one of the runs, the queries were limited to a specific geographical entity (filtered) and in the second run this condition was removed so the queries were run at national level, for such reason, the number of queries was 256, and the results show 98.2% savings on query times, in other words, instead of running in 7 hours, the answers were ready in 7 minutes. The worst query in ADW was much better than the best query on the Traditional DW Architecture. The queries were run in both architectures without any modification.

In the retail company, the analytical reports had been written as Database stored procedures using the Microsoft SQL Server proprietary language. Once they were adjusted to the Sybase IQ Database stored procedures language, they ran successfully. As this was an initial test, only the two most relevant reports were ported to the Binary-Relational instantiation. The queries against the Binary-Relational instantiation achieved savings of 89%.

The Telecommunications company query set achieved savings of 80% when compared to the traditional configuration.

For the Finance company, some of the more representative business queries were run. It is remarkable that the total time has been reduced from 1,009 seconds (16.49 minutes) to just 1 second in the Binary-Relational model instantiation using less hardware power. In other words, it achieved savings of 99.9%, considering the query execution time.

2.6 Government Agency Additional Tests

Schema Modification

This agency was also interested in benchmarking the difficulty of modifying its logical schema within ADW, which had better performance than the Traditional DW. In two of the operations Error messages were obtained when executed on the n-ary-Relational DBMS. The first Error appeared after 7 hours of processing, it was caused by the lack of space on the *Log* area of the n-ary-Relational instantiation; while the Binary-Relational instantiation successfully added the new column in only 1.5 minutes. There second error displayed by the n-ary-Relational DBMS was because as it was using the compression feature [8] and in order to eliminate a column, it is necessary to uncompress the database and then proceed to the column elimination. This is an important drawback of following the data compression approach suggested in [8], strictly speaking, the disk space size is not reduced as the space must remain available for certain operations.

Cube Construction

This agency uses different exploitation tools; one of these requires building cubes to provide OLAP facilities. Measurements of the cube construction were made and the results are given in Table 8. ADW reduces the time required to build cubes. The Binary-Relational instantiation built the cube in nearly 5 hours; in contrast the n-ary-Relational instantiation was stopped after 19 hours being running and it had processed only 7 million rows from a total of 103 million rows. If a linear extrapolation is made, the n-ary-Relational DBMS will require around 280 hours (11.6 days) to complete the cube construction. An existing problem on this agency's production environment is that tables have been built for each geographical entity in order to construct the cubes; in the case of ADW, all rows were stored on a single table and it built the cube in few hours. This is another contribution achieved with the proposed ADW Architectural configuration: simplifying the database administration tasks owing to the need of having fewer tables to be maintained.

Table 8. Government Agency's Cube Build

Task	Traditional DW (hh:mm:ss)	ADW (hh:mm:ss)
Cube Build on Microsoft Analysis Services	7,000,000 rows in 19:00:00	+103 million rows in 04:52:00

Database Administration Tasks

ADW requires fewer database administration tasks, no additional indexes are required, nor statistics computation as can be seen in Table 9. The total maintenance period has been reduced from nearly one day to 10 minutes, giving this agency considerable extra time to do strategic analysis.

Table 9. Database Administration Tasks

Task	n-ary-Relational	Binary-Relational	Savings
Database Backup time (min)	18	10	45%
Indexes building time (min)	120	Not Required	100%
Statistics Computation (min)	960	Not Required	100%
Database Compression (min)	336	Not Required	100%
TOTAL (minutes)	1,434	10	99.3%
(hours)	23.9		

Traditional Relational DBMS utilized histograms stored in a separated metadata in order to permit efficient estimation of query result sizes and access plan costs. As a consequence, maintenance of such histograms is required according to the volatility of data within the database. In contrast, the Binary-Relational DBMS stores the data value and its corresponding frequency. Therefore, the histogram is implicit in the data. The frequency of the value and the value itself are updated at the same time.

3 Conclusions

Manufacturing Company

Even with half of the computer hardware power, ADW has demonstrated their benefits in a manufacturing company. In terms of hardware, it will represent computer cost savings of 50% in CPUs and memory, plus 74% savings in disk space. From the business point of view, the query executing time obtained savings around 70% off comparing to the traditional DW architectural configuration.

Government Agency

ADW has been tested in a Government Agency with excellent results. Government institutions usually manage large amounts of data. The decisions made by these institutions are relevant for the national development plans. Hence it is important for government to have efficient architectures on their Decision Support Systems and Data Warehouses, and here is where ADW can achieve importance for different countries. In Table 10 the main benefits that ADW has provided to the government agency are presented.

Table 10. ADW Benefits for a Government Agency

87% savings in Disk storage costs
55% savings in Load time
98% savings in query response time
The applications did not require any modification in order to use ADW

Retail Company

The Binary-Relational Model has demonstrated its benefits for a Retail industry company; having savings of 70% for disk space; 92% in loading time, and 89% in query execution time.

Telecommunications Company

For this company, the query response times were extremely fast with savings of 80%. Other benefits are 37% disk space savings, and the applications did not required major adjustments, as they were direct SQL statements.

Finance Company

In terms of computer hardware it was reduced to one third of the computer power in terms of memory, and half in terms of CPUs, even then considerable performance improvements were achieved. This company made savings of 66% in RAM memory costs, 50% savings on CPU costs and 40% savings in disk costs. From the business point of view, the users are able to run complex analyses and getting the answers within the right times.

General Conclusions

ADW has been tested in broad group of industries (Manufacturing, Government, Telecommunications, Retail and Finance), demonstrating its feasibility to perform well in real environments. These industrial results demonstrate that organizations can benefit from the use of ADW and solve many of the problems they are actually facing. The following benefits are achieved:

• Reduction in the time required to Load data into the Data Warehouse.

• The query execution time is dramatically improved. Business can now execute analyses that they cannot execute on a traditional Data warehouse architecture.

• Therefore organisations can save money by acquiring less hardware.

• Electricity savings are achieved by having fewer disks to be powered and at the same time reducing the cooling kWhs needed to cool the computer centre by having fewer disks generating heat, and even savings when buying a smaller cooler machine.

• Savings in secondary storage (tapes) in the same proportion by having fewer disks to be backed up; therefore reductions in the size of the room to keep the historical tapes can be also considered.

• Maintenance task reduction; it includes: (a) Small backup size. (b) Less backup time. (c) Less restore time.

• No extra indexes are required; having an impact in two aspects: (a) fewer structures to be maintained (CPU time); and (b) less disk space.

• No need to execute statistics maintenance; owing to the frequency of data value is stored as part of the data (see section 2.6). Such advantage positively impacts the CPU time and reduces the batch processing time.

References

1. Gonzalez-Castro, V., MacKinnon, L.: Using Alternative Data Models in the context of Data Warehousing. In: Petratos, P., Michalopoulos, D. (eds.) 1st International Conference on Computer Science and Information Systems by the Athens Institute of Education and Research ATINER, Athens, Greece, pp. 83–100 (2005), ISBN-960-88672-3-1
2. Gonzalez-Castro, V., MacKinnon, L.: Data Density of Alternative Data Models and its Benefits in Data Warehousing Environments. In: British National Conference on Databases BNCOD 22 Proceedings, Sunderland, England U.K., vol. 2, pp. 21–24 (2005)
3. Gonzalez-Castro, V., MacKinnon, L.M., Marwick, D.H.: An Experimental Consideration of the Use of the Transrelational^{TM}Model for Data Warehousing. In: Bell, D.A., Hong, J. (eds.) BNCOD 2006. LNCS, vol. 4042, pp. 47–58. Springer, Heidelberg (2006)
4. Gonzalez-Castro, V., MacKinnon, L., Marwick, D.: A Performance Evaluation of Vertical and Horizontal Data Models in Data Warehousing. In: Olivares Ceja, J., Guzman Arenas, A. (eds.) Research in Computing Science, vol. 22, pp. 67–78 (November 2006), Special issue: Data Mining and Information Systems. Instituto Politécnico Nacional, Centro de Investigación en Computación, México, ISSN: 1870-4069
5. MonetDB web site, `http://monetdb.cwi.nl/` (updated 20-March-2009)
6. Sybase Inc. Migrating from Sybase Adaptive Server Enterprise to SybaseIQ White paper USA (2005)
7. Codd, E.F.: The Relational Model for database Management Version 2. Addison Wesley Publishing Company, Reading (1990)
8. Poss, M., Potapov, D.: Data Compression in Oracle. In: Proceedings of the 29th VLDB conference, Berlin Germany (2003)
9. Gonzalez-Castro, V., MacKinnon, L.M., Angeles Maria del, P.: An Alternative Data Warehouse Reference Architectural Configuration. In: Proceedings of the 26th British National Conference on Databases, BNCOD 2009, Birmingham, UK (to be published)
10. Cognos Inc. website, `http://www.cognos.com/` (updated 20-March-2009)

Hyperset Approach to Semi-structured Databases

Computation of Bisimulation in the Distributed Case

Richard Molyneux[1] and Vladimir Sazonov[2]

[1] Department of Computer Science, University of Liverpool, Liverpool,
molyneux@liv.ac.uk

[2] Department of Computer Science, University of Liverpool, Liverpool,
sazonov@liv.ac.uk

Abstract. We will briefly describe the recently implemented hyperset approach to semi-structured or Web-like and possibly distributed databases with the query system available online at `http://www.csc.liv.ac.uk/~molyneux/t/`. As this approach is crucially based on the bisimulation relation, the main stress in this paper is on its computation in the distributed case by using a so called bisimulation engine and local approximations of the global bisimulation relation.

1 Introduction

Unlike the more widely known traditional view on semi-structured data (SSD) as labelled graphs or as text based XML documents [1], the hyperset approach discussed in this paper assumes an abstract data model where such data is considered as set of sets of sets, etc. All elements of such sets, which are themselves sets, are assumed to have labels. Labels are the proper carriers of the atomic information, whereas sets just organise this information. Moreover, non-well-founded sets or hypersets allowing cyclic, as well as nested structure adopted in this approach are, in fact, a known extension of classical set theory.

Hyperset data model and graphs. The set theoretical view is, in fact, closely related with the graph view on semi-structured data by the evident and natural correspondence between labelled set equations like $s = \{l_1 : s_1, \ldots, l_n : s_n\}$, with s and s_i set names, and "fork" graph fragments determined by labelled directed edges outgoing from s: $s \xrightarrow{l_1} s_1, \ldots, s \xrightarrow{l_n} s_n$. This way a system of such (labelled) set equations can generate arbitrary directed graph (possibly with cycles), and vice versa. All sets and graphs are assumed to be finite. Another way to view such graphs is thinking on set names as analogies of URLs (Web addresses) and on the labels l_i as marking hyperlinks $s \xrightarrow{l_i} s_i$ between these URLs. This analogy also leads to the term Web-like database (WDB) [12] for any such system of set equations (and corresponding graph). "Graph" is essentially synonymous to the "web", and "semi-structured" usually means "structured as

A.P. Sexton (Ed.): BNCOD 2009, LNCS 5588, pp. 78–90, 2009.

a graph". Semi-structured databases arose also essentially due to Web [1] as the possibility of easy accessing various data having no unique and uniform structure. Moreover, the analogy with the WWW can be further supported by allowing system of set equations to be distributed amongst many WDB files, not necessarily in one computer, where a set name s participating in one file as $\ldots = \{\ldots, l : s, \ldots\}$ can possibly refer to other file in which s is described by appropriate set equation $s = \{l_1 : s_1, \ldots, l_n : s_n\}$. This assumes distinguishing between simple set names and full set names involving the URL of appropriate WDB file. The idea of distributed WDB seems inevitable and reasonable in the age of Internet (and Intranets) as it is unpractical and illogical to have all set equations of a big WDB to be contained in one file and in one computer.

Hypersets and relational databases. Before going further in describing hyperset approach, it makes sense to note the analogy of a set $s = \{l_1 : s_1, \ldots, l_n : s_n\}$ with a relational record where it is assumed the restriction that the fields or attributes $l_1, \ldots, l_n$ are all different. Thus, a set of such records with the same list of attributes l_i constitute a relation. Then, the possible distributed character of such data mentioned above would correspond, for example, to a heterogeneous system of relational databases as a very special case of arbitrary semi-structured hyperset data in the form of a system of such equations. Even if the same could be represented in terms of a (distributed) graph, the set theoretic approach (assuming considering arbitrary sets s of sets s_i, etc.) is closer to the spirit of relational (or nested relational) databases having essentially the same logical background which cannot be said so straightforwardly concerning the purely graph theoretical view. Also, the general hyperset query language Δ discussed below would potentially allow to query arbitrary heterogeneous distributed relational databases in a quite natural and logically coherent way. In fact, the great success of relational databases arose due to their highly transparent logical and set theoretical nature, and we develop the same set theoretical paradigm having, as we believe, good theoretical and practical potential.

Equality between hypersets. The concept of equality makes the main and crucial difference between graph nodes and hypersets. Hyperset view assumes that two set names (graph nodes) can denote the same abstract hyperset as the order of elements and repetitions—occasional *redundancies* in the data—should not play a role in the abstract meaning of these data. For example, given set equations

```
BibDB = {book:b,paper:p}
    b = {author:"Jones",title:"Databases"}
    p = {author:"Jones",title:"Databases"},
```

we should conclude that `b = p`.[1] Mathematically, this means that two set names (graph nodes) which are *bisimilar* or *informationally equivalent* denote the same

[1] It might be either intended or not by the designer of this database that the same publication is labelled both as a `book` and as a `paper`. Of course, this WDB can also be redesigned by replacing `book` and `paper` with `publication` and by adding elements `type:"book"` to `b` and `type:"paper"` to `p` thereby distinguishing `b` and `p`.

hypersets (see also [3,4]). That is, possible redundancies and order in set equations should be ignored (or eliminated). In this sense the hyperset approach is a kind of graph approach where graph nodes are considered up to bisimulation. Anyway, the main style of thought implied here is rather a set theoretical one, particularly in writing queries to such databases. The corresponding set theoretic query language Δ will be briefly described, which was recently implemented [15] and now is fully functioning at least as a demo query system available online [7].

On the other hand, pure graph approach to SSDB such as [2] assumes that different graph nodes are considered as different "objects" or "object identities". Also in XML the predefined order of data is considered as essential (see e.g. [5] on querying XML and tree data). In the literature we find only one approach via the query language UnQL [6] which is most close to the hyperset one where queries to graph databases are also required to be bisimulation invariant. However, it is argued in [18], where UnQL is imitated by our query language Δ, that the former is still more graph rather than set theoretic approach because the graphs considered in UnQL have so called input and output nodes (which made the embedding of UnQL into Δ rather complicated and not very natural as these multiple inputs and outputs conflict with the hyperset view). Anyway, Δ has quite sufficient expressive power for this imitation. Also the mere bisimulation invariance used both in UnQL and Δ is only a half-story. An essential theoretical advantage of pure set theoretic query language Δ and its versions consists in precise characterisation of their expressive power in terms of PTIME [14,17] and (N/D)LOGSPACE [11,13].

As the bisimulation relation is crucial to this approach, and because its computation is in general sufficiently complicated, especially in the case of a distributed WDB, it is therefore highly important to find a practical way of dealing with bisimulation, particularly in the extremal distributed case. We suggest to compute the (global) bisimulation relation by using its local approximations computed locally. The computation of the global bisimulation is being done in background time by means of the so called centralised bisimulation engine, which requests local approximations of the bisimulation relation to compute the global bisimulation relation. This work to support computation of the bisimulation relation may be a permanent process as the (global) WDB can be locally updated and this should be repeatedly taken into account. Meanwhile, if a user runs a query q involving the set theoretic equality $x = y$, the query system asks the centralised bisimulation engine whether it already knows the answer to the question "x=y?". If not, the query system tries to compute it itself. But as soon as the engine will get the answer, it will send it to the query system which probably has not computed it yet. In this way, query evaluation involving equality becomes more realistic in the distributed case.

Note that there are known some approaches to efficient computing bisimulation relation such as [8,9,10] which should also be taken into account, although they do not consider the distributed case as we do. Of course, the distributed feature is not an innate one or belonging exclusively to the hyperset approach. But being a natural and in a sense an extreme possibility in general this makes bisimulation relation—a really inherent to this approach—particularly challenging implementation task. We intend to show (and this paper and experiments in [15] are only a first step) that potentially even in this distributed case hyperset approach could be sufficiently realistic.

2 Hyperset Query Language Δ

The abstract syntax of the hyperset query language Δ is as follows:

$$\begin{aligned}\langle\Delta\text{-term}\rangle ::= {} & \langle\text{set variable or constant}\rangle \mid \emptyset \mid \{l_1 : a_1, \ldots, l_n, a_n\} \mid \textstyle\bigcup a \mid \mathsf{TC}(a) \mid \\ & \{l : t(x,l) \mid l : x \in a \,\&\, \varphi(x,l)\} \mid \mathsf{Rec}\, p.\{l : x \in a \mid \varphi(x,l,p)\} \mid \mathsf{Dec}(a,b) \\ \langle\Delta\text{-formula}\rangle ::= {} & a = b \mid l_1 = l_2 \mid l_1 < l_2 \mid l_1 \, R \, l_2 \mid l : a \in b \mid \varphi \,\&\, \psi \mid \varphi \vee \psi \mid \neg\varphi \mid \\ & \forall l : x \in a.\varphi(x,l) \mid \exists l : x \in a.\varphi(x,l)\end{aligned}$$

Here we denote: $a, b, \ldots$ as (set valued) Δ-terms; $x, y, z, \ldots$ as set variables; l, l_i as label values or variables (depending on the context); $l : t(x,l)$ is any l-labelled Δ-term t possibly involving the label variable l and the set variable x; and φ, ψ as (boolean valued) Δ-formulas. Note that labels l_i participating in the Δ-term $\{l_1 : a_1, \ldots, l_n : a_n\}$ need not be unique, that is, multiple occurrences of labels are allowed. This means that we consider arbitrary sets of labelled elements rather than records or tuples of a relational table where l_i serve as names of fields (columns). Label and set variables l, x, p of quantifiers, collect, and recursion constructs (see the descriptions below) should not appear free in the bounding term a (denoting a finite set). Otherwise, these operators may become actually unbounded and thus, in general, non-computable in finite time.

More details on the meaning of the above constructs and on the implemented version of Δ can be found e.g. in [18,16,15,7]. Many constructs of Δ should be quite evident. Here we only mention briefly that

- $\bigcup$ means *union* of a set of sets,
- TC means *transitive closure* of a set (elements, elements of elements, etc.),
- the *collection* operator $\{l : t(x,l) \mid l : x \in a \,\&\, \varphi(x,l)\}$ denotes the set of all labelled elements $l : t(x,l)$ such that $l : x \in a \,\&\, \varphi(x,l)$ holds,
- the *recursion* operator $\mathsf{Rec}\, p.\{l : x \in a \mid \varphi(x,l,p)\}$ defines iteratively a set π satisfying the identity $\pi = \pi \cup \{l : x \in a \mid \varphi(x,l,\pi)\}$,
- the *decoration* operator $\mathsf{Dec}(g,v)$ denotes a unique hyperset corresponding to the vertex v of a graph g. Here g is any set of labelled ordered pairs $l : \{fst : x, snd : y\}$ understood as graph edges $x \xrightarrow{l} y$, and v is one of these x, y participating in such pairs. Note that x, y, v are considered here as arbitrary hypersets. Dec can also be naturally called *plan performance operator*. Given a graphical plan consisting of g and v, it constructs a unique hyperset $\mathsf{Dec}(g,v)$ according to this plan.
- $<$ and R denote alphabetic ordering on labels and, respectively, substring relation.

2.1 Rough Description of the Operational Semantics of Δ [18]

Operational semantics of Δ is required to implement this language. The working query system is described in detail in [16,15,7] where examples of queries can be found and, in fact, run on the implementation available online.

Consider any set or boolean query q in Δ which involves no free variables and whose participating set names (constants) are taken from the given WDB system of set equations. Resolving q consists in the following two macro steps:

- **Extending** this system by the new equation $res = q$ with res a fresh (i.e. unused in WDB) set or boolean name, and
- **Simplifying** the extended system $\mathrm{WDB}_0 = \mathrm{WDB} + (res = q)$ by some (quite natural) reduction steps $\mathrm{WDB}_0 \rhd \mathrm{WDB}_1 \rhd \ldots \rhd \mathrm{WDB}_{res}$ until it will contain only flat bracket expressions as the right-hand sides of the equations or the truth values *true* or *false* (if the left-hand side is boolean name).

After simplification is complete, these set equations will contain no complex set or boolean queries (like q above). In fact, the resulting version WDB_{res} of WDB will consist (alongside the old equations of the original WDB) of new set equations (new, possibly auxiliary set names equated to flat bracket expressions) and boolean equations (boolean names equated to boolean values, *true* or *false*). We cannot go here into further details, except saying that in the case of a set query q the simplified version $res = \{\ldots\}$ of the equation $res = q$ will give the resulting value for q. We only will consider below the case of equality query $x = y$ (with the value *true* or *false*) in terms of the bisimulation relation $x \approx y$.

3 Bisimulation

Assume WDB is represented as a system of set equations $\bar{x} = \bar{b}(\bar{x})$ where $\bar{x}$ is a list of set names $x_1, \ldots, x_k$ and $\bar{b}(\bar{x})$ is the corresponding list of bracket expressions (for simplicity, "flat" ones). The set of all set names $x_1, \ldots, x_k$ of the given WDB is also denoted as *SNames*. Visually equivalent representation can be done in the form of labelled directed graph, where labelled edges $x_i \xrightarrow{label} x_j$ correspond to the set memberships $label{:}x_j \in x_i$ meaning that the equation for x_i has the form $x_i = \{\ldots, label : x_j, \ldots\}$. In this case we also call x_j a child of x_i. Note that this particular, *concrete* usage of the membership symbol $\in$ as relation between set names or graph nodes is non-traditional but very close to the traditional set theoretic membership relation between *abstract* (hyper)sets, hence we decided not to introduce a new kind of membership symbol here. For the simplicity of our description below labels are ignored as they would not affect essentially the nature of our considerations.

3.1 Hyperset Equality and the Problem of Efficiency

One of the key points of our approach is the interpretation of WDB-graph nodes as set names $x_1, \ldots, x_k$ where different nodes x_i and x_j can, in principle, denote the same (hyper)set, $x_i = x_j$. This particular notion of equality between nodes is defined by the bisimulation relation denoted also as $x_i \approx x_j$ (to emphasise that set names can be syntactically different, but denote the same set) which can be computed by the appropriate recursive comparison of child nodes or set names. Thus, in outline, to check bisimulation of two nodes we need to check bisimulation between some children, grandchildren, and so on, of the given nodes, i.e. many nodes could be involved. If the WDB is distributed amongst many WDB files and remote sites then downloading the relevant WDB files might be necessary in this process and will take significant time. (There is also the analogous problem with the related transitive closure operator TC whose efficient implementation in the distributed case can require similar considerations.) So, in practice the equality relation for hypersets seems intractable, although theoretically it takes polynomial time with respect to the size of WDB. Nevertheless, we consider that the hyperset approach to WDB based on bisimulation relation is worth implementing because it suggests a very clear and mathematically well-understood view on semi-structured data and the querying of such data. Thus, the crucial question is whether the problem of bisimulation can be resolved in any reasonable and practical way. One approach related with the possibly distributed nature of WDB and showing that the situation is manageable in principle is outlined below.

3.2 Computing Bisimulation Relation $\approx$ over WDB

Bisimulation relation can be computed (and thereby defined) by deriving negative ($\not\approx$) bisimulation facts by means of the following recursive rule:

$$x \not\approx y : -\ \exists x' \in x \forall y' \in y(x' \not\approx y') \vee \exists y' \in y \forall x' \in x(x' \not\approx y') \qquad (1)$$

where initial negative facts can be obtained by the partial case of this tule:

$$x \not\approx y : -\ (x = \emptyset \ \&\ y \neq \emptyset) \vee (y = \emptyset \ \&\ x \neq \emptyset).$$

This means that any set described as empty one is nonbisimilar to any set described as non-empty in the WDB. The evident dual rule to (1) can also be used for deriving positive bisimulation facts. However in principle this is unnecessary as such facts will be obtained, anyway, at the moment of stabilisation in the derivation process by using only (1) as there are only finitely many of WDB set names in *SNames*.

Equivalently, $\not\approx$ is the least relation satisfying (1), and its positive version $\approx$ is the largest relation satisfying

$$x \approx y \Rightarrow \forall x' \in x \exists y' \in y(x' \approx y') \ \&\ \forall y' \in y \exists x' \in x(x' \approx y'). \qquad (2)$$

It is well-known known that bisimulation $\approx$ is an equivalence relation which is completely coherent with hyperset theory as it is fully described in [3,4] for the pure case, and this fact extends easily to the labelled case. It is by this reason that the bisimulation relation $\approx$ between set names can be considered as an equality relation $=$ between corresponding abstract hypersets.

3.3 Local Approximations of $\approx$

Now, let a non-empty set $L \subseteq SNames$ of "local" vertices (set names) in a graph WDB (a system of set equations) be given, where $SNames$ is the set of all WDB vertices (set names). Let us also denote by $L' \supseteq L$ the set of all "almost local" set names participating in the set equations for each set name in L both at left and right-hand sides. Considering the graph as a WDB distributed among many *sites*, L plays the role of (local) set names defined by set equations in some (local) WDB files of one of these sites. Then $L' \setminus L$ consists of non-local set names which, however, participate in the local WDB files, have defining equations in other (possibly remote) sites of the given WDB. Non-local (full) set names can be recognised by their URLs as different from the URL of the given site.

We will consider *derivation rules* of the form $xRy : - \ldots R \ldots$ for two more relations $\approx_{-}^{L}$ and $\approx_{+}^{L}$ over $SNames$:

$$\approx_{-}^{L} \subseteq \approx \subseteq \approx_{+}^{L} \quad \text{or, rather, their negations} \quad \not\approx_{+}^{L} \subseteq \not\approx \subseteq \not\approx_{-}^{L} \tag{3}$$

defined formally on the whole WDB graph (however, we will be mainly interested in the behaviour of $\approx_{-}^{L}$ and $\approx_{+}^{L}$ on L). We will usually omit the superscript L as we currently deal mainly with one L, so no ambiguity can arise.

3.4 Defining the Local Upper Approximation $\approx_{+}^{L}$ of $\approx$

Let us define the relation $\not\approx_{+} \subseteq SNames^2$ by derivation rule

$$x \not\approx_{+} y : - x, y \in L \;\&\; [\exists x' \in x \forall y' \in y (x' \not\approx_{+} y') \vee \ldots]. \tag{4}$$

Here and below "..." represents the evident symmetrical disjunct (or conjunct). Thus the premise (i.e. the right-hand side) of (4) is a *restriction* of that of (1). It follows by induction on the length of derivation of the $\not\approx_{+}$-facts that,

$$\not\approx_{+} \subseteq \not\approx, \quad \approx \subseteq \approx_{+}, \tag{5}$$

$$x \not\approx_{+} y \Rightarrow x, y \in L, \tag{6}$$

Let us also consider another, "more local" version of the rule (4):

$$x \not\approx_{+} y : - x, y \in L \;\&\; [\exists x' \in x \forall y' \in y (x', y' \in L \;\&\; x' \not\approx_{+} y') \vee \ldots]. \tag{7}$$

It defines the same relation $\not\approx_{+}$ because in both cases (6) holds implying that the right-hand side of (7) is equivalent to the right-hand side of (4). The advantage of (4) is its formal simplicity whereas that of (7) is its "local" computational meaning. From the point of view of distributed WDB with L one of its local sets of vertices/set names (corresponding to one of the sites of the distributed WDB), we can derive $x \not\approx_{+} y$ for local x, y via (7) by looking at the content of local WDB files only. Indeed, participating URLs (full set names) $x' \in x$ and $y' \in y$, although likely non-local names ($\in L' \setminus L$), occur in the locally stored WDB files with local URLs x and $y \in L$. However, despite the possibility that x' and y' can be in general non-local, we will need to use in (7) the facts of the kind $x' \not\approx_{+} y'$ derived on the previous steps for local $x', y' \in L$ only. Therefore,

Note 1 (Local computability of $x \not\approx_+ y$). For deriving the facts $x \not\approx_+ y$ for $x, y \in L$ by means of the rule (4) or (7) we will need to use the previously derived facts $x' \not\approx_+ y'$ for set names x', y' from L only, and additionally we will need to use set names from a wider set L' (available, in fact, also locally)[2]. In this sense, the derivation of all facts $x \not\approx_+ y$ for $x, y \in L$ can be done locally and does not require downloading of any external WDB files. (In particular, facts of the form $x \not\approx_+ y$ or $x \approx_+ y$ for set names x or y in $L' \setminus L$ present no interest in such derivations.)

The upper approximation $\approx_+$ (on the whole WDB graph) can be equivalently characterised as the largest relation satisfying any of the following (equivalent) implications for all graph vertices x, y:

$$x \approx_+ y \Rightarrow x \notin L \vee y \notin L \vee [\forall x' \in x \exists y' \in y(x' \approx_+ y') \mathbin{\&} \ldots]$$
$$x \approx_+ y \mathbin{\&} x, y \in L \Rightarrow [\forall x' \in x \exists y' \in y(x' \approx_+ y') \mathbin{\&} \ldots] \tag{8}$$

It is easy to show that the set of relations $R \subseteq \mathit{SNames}^2$ satisfying (8) (in place of $\approx_+$) **(i)** contains the identity relation $=$, **(ii)** is closed under unions (thus the largest $\approx_+$ does exist), and **(iii)** is closed under taking inverse. Evidently, any ordinary (global) bisimulation relation $R \subseteq \mathit{SNames}^2$ (that is, a relation satisfying (2)) satisfies (8) as well. For any $R \subseteq L^2$ the converse also holds: if R satisfies (8) then it is actually a global bisimulation relation (and $R \subseteq \approx$). It is also easy to check that **(iv)** relations $R \subseteq L^2$ satisfying (8) are closed under compositions. It follows from **(i)** and **(iii)** that $\approx_+$ is reflexive and symmetric. Over L, the relation $\approx_+$ (that is the restriction $\approx_+ \restriction L$) is also transitive due to **(iv)**. Therefore, $\approx_+$ is an *equivalence relation* on L. (In general, $\approx_+$ cannot be an equivalence relation on the whole graph due to (6) if $L \neq \mathit{SNames}$.) Moreover, any $x \notin L$ is $\approx_+$ to all vertices (including itself).

3.5 Defining the Local Lower Approximation $\approx_-^L$ of $\approx$

Consider the derivation rule for the relation $\not\approx_- \subseteq \mathit{SNames}^2$:

$$x \not\approx_- y :- [(x \notin L \vee y \notin L) \mathbin{\&} x \neq y] \vee$$
$$[\exists x' \in x \forall y' \in y(x' \not\approx_- y') \vee \ldots] \tag{9}$$

which can also be equivalently replaced by two rules:

$$x \not\approx_- y :- (x \notin L \vee y \notin L) \mathbin{\&} x \neq y - \text{“a priori knowledge”}, \tag{10}$$
$$x \not\approx_- y :- \exists x' \in x \forall y' \in y(x' \not\approx_- y') \vee \ldots .$$

[2] This is the case when $y = \emptyset$ but there exists according to (7) an x' in x which can be possibly in $L' \setminus L$ (or similarly for $x = \emptyset$). When $y = \emptyset$ then, of course, there are no suitable witnesses $y' \in y$ for which $x' \not\approx_+ y'$ hold. Therefore, only the existence of some x' in x plays a role here.

Thus, in contrast to (4), this is a *relaxation* or an *extension* of the rule (1) for $\not\approx$. It follows that

$$\not\approx\,\subseteq\,\not\approx_- \;(\approx_-\,\subseteq\,\approx).$$

It is also evident that

$$\text{any } x \notin L \text{ is } \not\approx_- \text{ to all vertices different from } x,$$
$$x \approx_- y \;\&\; x \neq y \Rightarrow (x, y \in L).$$

The latter means that $\approx_-$ (which is an equivalence relation on *SNames* and hence on L as it is shown below) is non-trivial only on the local set names. Again, like for $\not\approx_+$, we can conclude from the above considerations that,

Note 2 (Local computability of $x \not\approx_- y$). We can compute the restriction of $\not\approx_-$ on L locally: to derive $x \not\approx_- y$ for $x, y \in L$ with $x \neq y$ (taking into account reflexivity of $\approx_-$) by (9) we need to use only $x', y' \in L'$ (by $x' \in x$ and $y' \in y$) and already derived facts $x' \not\approx_- y'$ for $x', y' \in L, x \neq y$, as well as the facts $x' \not\approx_- y'$ for x' or $y' \in L' \setminus L$, $x' \neq y'$ following from the "a priori knowledge" (10).

The lower approximation $\approx_-$ can be equivalently characterised as the largest relation satisfying

$$x \approx_- y \Rightarrow (x, y \in L \vee x = y) \;\&\; (\forall x' \in x \exists y' \in y (x' \approx_- y') \;\&\; \ldots).$$

Evidently, $=$ (substituted for $\approx_-$) satisfies this implication. Relations R satisfying this implication are also closed under unions and taking inverse and compositions. It follows that $\approx_-$ is reflexive, symmetric and transitive, and therefore an *equivalence relation over the whole WDB graph*, and hence *on its local part L*.

Thus, both approximations $\approx_+^L$ and $\approx_-^L$ to $\approx$ are computable "locally". Each of them is defined in a trivial way outside of L, and the computation requires only knowledge at most on the L'-part of the graph. In fact, only edges from L to L' are needed, everything being available locally.

3.6 Using Local Approximations to Aid Computation of the Global Bisimulation

Now, assume that the set *SNames* of all set names (nodes) of a WDB is disjointly divided into a family of local sets L_i, for each "local" site $i \in I$ with local approximations $\approx_+^{L_i}$ and $\approx_-^{L_i}$ to the global bisimulation relation $\approx$ computed locally. Now the problem is how to compute the global bisimulation relation $\approx$ with the help of many its local approximations $\approx_+^{L_i}$ and $\approx_-^{L_i}$ in all sites $i \in I$.

Granularity of sites. However, for simplicity of implementation and testing the above idea and also because this is reasonable in itself we will redefine the scope of i to a smaller granularity. Instead of taking i to be a site, consisting of many WDB files, we will consider that each i itself is a name of a single WDB file

file$_i$. More precisely, i is considered as the URL of any such a file. This will not change the main idea of implementation of the bisimulation Oracle on the basis of using local information for each i. That is, we reconsider our understanding of the term local – from being *local to a site* to *local to a file*. Then L_i is just the set of all (full versions of) set names defined in file i (left-hand sides of all set equations in this file). Evidently, so defined sets L_i are disjoint and cover the class *SNames* of all (full) set names from the WDB considered.

Then the relations $\approx_+^{L_i}$ and $\approx_-^{L_i}$ should be automatically computed and maintained as the current local approximations for each WDB file i each time this file is updated. In principle a suitable tool is necessary for editting (and maintaining) WDB, which would save a WDB file i and thereby generate and save the approximation relations $\approx_+^{L_i}$ and $\approx_-^{L_i}$ automatically.

In general, we can reasonably use even more levels of locality distributing the workload between many servers of various levels acting in parallel.

Local approximations giving rise to global bisimulation facts. It evidently follows from (3) that

- *each positive local fact of the form* $x \approx_-^{L_i} y$ *gives rise to the fact* $x \approx y$, *and*
- *each negative local fact of the form* $x \not\approx_+^{L_i} y$ *gives rise to the fact* $x \not\approx y$.

Let $\approx^{L_i}$ (without subscripts + or −) denote the set of positive and negative facts for set names in L_i on the global bisimulation relation $\approx$ obtained by these two clauses. This set of facts $\approx^{L_i}$ is called the *local simple approximation set* to $\approx$ for the file (or site) i. Let the *local Oracle* LO_i just answer *"Yes"* ("$x \approx y$"), *"No"* ("$x \not\approx y$") or *"Unknown"* to questions $x \overset{?}{\approx} y$ for $x, y \in L_i$ according to $\approx^{L_i}$.

In the case of i considered as a site (rather than a file), LO_i can have delays when answering *"Yes"* ("$x \approx y$") or *"No"* ("$x \not\approx y$") because LO_i should rather compute $\approx^{L_i}$ itself and find out in $\approx^{L_i}$ answers to the questions asked which takes time. But, if i is understood just as a file saved together with all the necessary information on local approximations at the time of its creation then LO_i can submit the required answer and, additionally, all the other facts it knows at once (to save time on possible future communications).

Therefore, a centralised Internet server (for the given distributed WDB) working as the (global) Oracle or *Bisimulation Engine*, which derives positive and negative ($\approx$ and $\not\approx$) global bisimulation facts, can do this by a natural algorithm based on the derivation rule (1), additionally asking (when required) various local Oracles LO_i concerning $\approx^{L_i}$. That is, the algorithm based on (1) and extended to exploit local simple approximations $\approx^{L_i}$ should, in the case of the currently considered question $x \overset{?}{\approx} y$ with $x, y \in L_i$ from the same site/WDB file i[3], additionally ask the oracle LO_i whether it already knows the answer (as described in the above two items). If the answer is known, the algorithm should just use it (as it was in fact derived in a local site). Otherwise (if LO_i does not know the answer or x, y do not belong to one L_i – that is, they are "remote"

[3] $x, y \in L_i$ iff the full versions of set names x, y have the same URL i.

one from another), the global Oracle should work according to (1) by downloading the necessary set equations, making derivation steps, asking the local Oracles again, etc. Thus, local approximations serve as auxiliary local Oracles LO_i helping the global Oracle.

4 Bisimulation Engine (The Oracle)

The idea is to have a centralised service providing answers to bisimulation question which would improve query performance (for those queries exploiting set equality). This service could be named *Bisimulation Engine* or just the *Oracle.* The goal of such bisimulation engine would consist in:

- **Answering bisimulation queries** asked by (any of possible copies[4] of) Δ-query system via appropriate protocol.
- **Computing bisimulation** by deriving bisimulation facts in background time, and strategically prioritising bisimulation questions posed by the Δ-query systems by temporary changing the fashion of the background time work in favour of resolving these particular questions.
- **Exploiting local approximations** $\approx_{-}^{L_i}$, $\approx_{+}^{L_i}$ and $\approx^{L_i}$ corresponding to WDB servers/files i of a lower level of locality to assist in the computation of bisimulation.
- **Maintaining cache of set equations** downloaded in the previous steps. These set equations may later prove to be useful in deriving new bisimulation facts, saving time on downloading of already known equations.

Moreover, it is reasonable to make the query system adopt its own "lazy" prioritisation strategy while working on a query q. This strategy consists of sending bisimulation subqueries of q to the Oracle but not attempting to resolve them in the case of the Oracle's answer "Unknown". Instead of such attempts, the query system could try to resolve other subqueries of the given query q until the resolution of the bisimulation question sent to the Oracle is absolutely necessary. The hope is that before this moment the bisimulation engine will have already given a definite answer.

However these useful prioritisation features have not yet been implemented. Moreover, currently we have only a simplified imitation of bisimulation engine which resolves all possible bisimulation questions for the given WDB in some predefined standard order without any prioritisation and answers these questions in a definite way when it has derived the required information. The Oracle, while doing its main job in background time, should only remember all the pairs (client, question) for questions asked by clients and send the definite answer to the corresponding client when it is ready.

More detailed algorithm of such a Bisimulation Engine and some encouraging experiments and artificial examples of a WDBs for these experiments were described in detail in [15] which we briefly present below. They show the benefit

[4] There could be many users running each their own copy of the Δ-query system.

both of background work of this engine and of using local approximations to the bisimulation relation on the WDB.

Determining the benefit of background work by the bisimulation engine on query performance. For 51 set names distributed over 10 WDB files, connected in chains and an isomorphic copy of the same it was shown that querying $x \approx x'$ for the root nodes with a delay d (after the Bisimulation Engine started working) has performance time with exponential decay depending on d. Without the Bisimulation Engine execution of $x \approx x'$ would take about 20 seconds, but with using it after 5 seconds of delay it takes about 8 seconds, and after 20 seconds of delay it takes 10 milliseconds. As Bisimulation Engine is assumed to work permanently, the delay time should not count in the overall performance.

Determining the benefit of exploiting local approximations by the Bisimulation Engine on query performance. For two isomorphic chains $x_1 \to x_2 \to \ldots \to x_n$ and $x'_1 \to x'_2 \to \ldots \to x'_n$ the bisimulation $x_1 \approx x'_1$ takes 112 and 84 minutes for $n = 70$ with the Bisimulation Engine not exploiting local approximations and, respectively, without using the Bisimulation Engine at all, and it takes 40 seconds with using it and local approximations to these two chains. The dependence on n was also experimentally measured. Here we assume $d = 0$, so the difference $112 - 84 = 34$ min is the additional expense of $\sim 70^2$ communications with the Bisimulation Engine instead of getting benefit from two files with $\approx^L$ and $\approx^{L'}$.

Determining the benefits of background work by the bisimulation engine exploiting local approximations. For three chain files $x \to x_1 \to x_2 \to \ldots \to x_{20}$, $y_1 \to y_2 \to \ldots \to y_{20}$, $z_1 \to z_2 \to \ldots \to z_{20}$ and two additional external edges $x \to y_1$ and $x \to z_1$ + an isomorphic copy of the same checking $x \approx x'$ shows that the execution time sharply falls in from about 17 minutes of pure querying to 180 milliseconds (and then to 10 ms) for the delay $d = 5$ seconds.

Of course, these are rather oversimplified experiments. We also did not take into account other works on efficient computation of bisimulation for non-distributed case [8,9,10]. But it is evident that the general qualitative picture should remain the same. Also more realistic large scaled experiments are required.

Note that the current implementation of the hyperset language Δ [7] does not use yet any bisimulation engine. These experiments were implemented separately and only to demonstrate some potential benefits of using such an engine.

5 Conclusion

While the hyperset approach to semi-structured databases is very natural theoretically and even practically as example queries in [7,15,16] show, developing some efficient way of dealing with hyperset equality (bisimulation) is required. We have demonstrated one such approach based on local approximations to the global bisimulation relation and on the idea of background time computation somewhat similar to that of Web search engines. The experiments presented in [15] and above show that this idea is promising, however further experiments and improvements should be done to make this approach more realistic. Another important topic to be considered is the impact of (possibly) dynamic updates of the WDB on the whole process, and also whether and under which conditions local updates can have small consequences. More general, the current implementation of the query language Δ [7] already working as a demo version and the language itself should be further improved/extended to make it more efficient and user friendly.

References

1. Abiteboul, S., Buneman, P., Suciu, D.: Data on the Web — From Relations to Semistructured Data and XML. Morgan Kaufmann Publishers, San Francisco (2000)
2. Abiteboul, S., Quass, D., McHugh, J., Widom, J., Wiener, J.L.: The Lorel query language for semistructured data. International Journal on Digital Libraries 1(1), 68–88 (1997)
3. Aczel, P.: Non-Well-Founded Sets. CSLI, Stanford (1988)
4. Barwise, J., Moss, L.: Vicious circles: on the mathematics of non-well-founded phenomena. Center for the Study of Language and Information (1996)
5. Benedikt, M., Libkin, L., Neven, F.: Logical definability and query languages over ranked and unranked trees. ACM Transactions on Computational Logic (TOCL) 8(2), 1–62 (2007)
6. Buneman, P., Fernández, M., Suciu, D.: UnQL: a query language and algebra for semistructured data based on structural recursion. The VLDB Journal 9(1), 76–110 (2000)
7. Delta-WDB Site (2008), `http://www.csc.liv.ac.uk/~molyneux/t/`
8. Dovier, A., Piazza, C., Policriti, A.: An efficient algorithm for computing bisimulation equivalence. Theoretical Computer Science 311(1-3), 221–256 (2004)
9. Fernandez, J.-C.: An implementation of an efficient algorithm for bisimulation equivalence. Science of Computer Programming 13(1), 219–236 (1989)
10. Jancar, P., Moller, F.: Techniques for decidability and undecidability of bisimilarity. In: Baeten, J.C.M., Mauw, S. (eds.) CONCUR 1999. LNCS, vol. 1664, pp. 30–45. Springer, Heidelberg (1999)
11. Leontjev, A., Sazonov, V.: Δ: Set-theoretic query language capturing logspace. Annals of Mathematics and Artificial Intelligence 33(2-4), 309–345 (2001)
12. Lisitsa, A., Sazonov, V.: Bounded hyperset theory and web-like data bases. In: Proceedings of the Kurt Goedel Colloquium, vol. 1234, pp. 178–188 (1997)
13. Lisitsa, A., Sazonov, V.: Δ-languages for sets and LOGSPACE computable graph transformers. Theoretical Computer Science 175(1), 183–222 (1997)
14. Lisitsa, A., Sazonov, V.: Linear ordering on graphs, anti-founded sets and polynomial time computability. Theoretical Computer Science 224(1–2), 173–213 (1999)
15. Molyneux, R.: Hyperset Approach to Semi-structured Databases and the Experimental Implementation of the Query Language Delta. PhD thesis, University of Liverpool, Liverpool, England (2008)
16. Molyneux, R., Sazonov, V.: Hyperset/web-like databases and the experimental implementation of the query language delta - current state of affairs. In: ICSOFT 2007 Proceedings of the Second International Conference on Software and Data Technologies, vol. 3, pp. 29–37, INSTICC (2007)
17. Sazonov, V.: Hereditarily-finite sets, data bases and polynomial-time computability. Theoretical Computer Science 119(1), 187–214 (1993)
18. Sazonov, V.: Querying hyperset / web-like databases. Logic Journal of the IGPL 14(5), 785–814 (2006)

Multi-Join Continuous Query Optimization: Covering the Spectrum of Linear, Acyclic, and Cyclic Queries

Venkatesh Raghavan, Yali Zhu, Elke A. Rundensteiner, and Daniel Dougherty

Department of Computer Science, Worcester Polytechnic Institute,
Worcester, MA 01609, USA
{venky,yaliz,rundenst,dd}@cs.wpi.edu

Abstract. Traditional optimization algorithms that guarantee optimal plans have exponential time complexity and are thus not viable in streaming contexts. Continuous query optimizers commonly adopt heuristic techniques such as Adaptive Greedy to attain polynomial-time execution. However, these techniques are known to produce optimal plans only for linear and star shaped join queries. Motivated by the prevalence of acyclic, cyclic and even complete query shapes in stream applications, we conduct an extensive experimental study of the behavior of the state-of-the-art algorithms. This study has revealed that heuristic-based techniques tend to generate sub-standard plans even for simple acyclic join queries. For general acyclic join queries we extend the classical IK approach to the streaming context to define an algorithm *TreeOpt* that is guaranteed to find an optimal plan in polynomial time. For the case of cyclic queries, for which finding optimal plans is known to be NP-complete, we present an algorithm *FAB* which improves other heuristic-based techniques by (i) increasing the likelihood of finding an optimal plan and (ii) improving the effectiveness of finding a near-optimal plan when an optimal plan cannot be found in polynomial time. To handle the entire spectrum of query shapes from acyclic to cyclic we propose a *Q-Aware* approach that selects the optimization algorithm used for generating the join order, based on the shape of the query.

1 Introduction

1.1 Continuous Query Plan Generation

In traditional, static, databases, query optimization techniques can be classified as either techniques that generate optimal query plans [1, 2, 3, 4], or heuristic based algorithms [5, 6, 7], which produce a good plan in polynomial time. In recent years there has been a growing interest in continuous stream processing [8, 9, 10, 11]. Continuous query processing differs from its static counterpart in several aspects. First, the incoming streaming data is unbounded and the query life span is potentially infinite. Therefore, if state-intensive query operations such as joins are not ordered correctly they risk consuming all resources. Second,

A.P. Sexton (Ed.): BNCOD 2009, LNCS 5588, pp. 91–106, 2009.

live-stream applications such as fire detection and stock market tickers are time-sensitive, and older tuples are of less importance. In such applications, the query execution must keep up with incoming tuples and produce real-time results at an optimal output rate [12]. Third, data stream statistics typically utilized to generate the execution query plan such as input rates, join selectivity and data distributions, will change over time. This may eventually make the current query plan unacceptable at some point of the query life span. This volatile nature of the streaming environment makes re-optimization a necessity. We conclude that an effective continuous query optimizer must have (1) polynomial time complexity, (2) the ability to generate optimal or at least near-optimal plans, (3) provide migration strategies to move the existing plan states to the newly generated plan. Existing approaches either have exponential complexity [1,3,4] or fail to produce an optimal plan. In streaming databases, the typical approach [13,14,10,9] is to use a *forward greedy heuristic* [3]. Although it was shown in [15] that a greedy approach can perform within a constant factor of optimal, the scope of this analysis was restricted to ordering unary filters over a single stream. Migration strategies proposed by [16] to safely transfer the current suboptimal query plan to the re-optimized query plan can be employed.

1.2 Spectrum of Linear, Acyclic and Cyclic Queries

In any streaming domain many flavors of join graphs from cyclic to acyclic may coexist. As motivation we examine the processing of sensor readings obtained from sensors placed in the mobile home (Figure 1.a) conducted by National

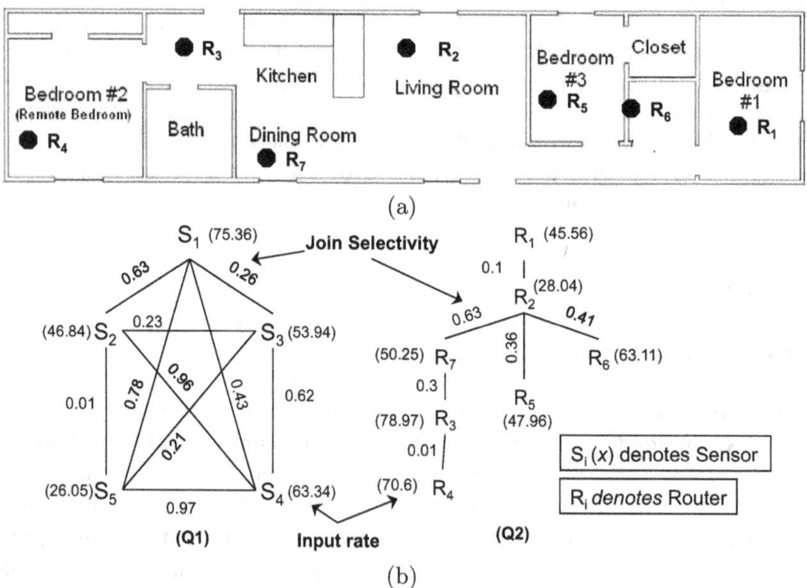

Fig. 1. a) NIST Fire Lab - Mobile Home Test Arena, b) Join Graphs for $Q1$ and $Q2$

Institute of Standards and Technology (NIST) [17]. Each sensor generates a reading feed S_i made up of tuples that contain the sensor identifier *sid*, the time stamp of the reading and the actual reading. In addition, each room contains a router (R_i) that generates a summarizing stream of all the sensors in its respective room. Fire engineers interested in determining false positive alarms or rogue sensors can submit the following query:

`Query 1 (Q1): A smoke sensor in bedroom #1 has detected an abnormality. Monitor and compare its behavior with all sensors within the same room.`

To evaluate $Q1$ we must compare the readings of each sensor against the readings of all the sensors in bedroom #1 to identify abnormalities. Such user queries are represented as complete (cyclic) join graphs, as in Figure 1.b. In the same domain, user queries can also correspond to acyclic join graphs. For example, first responders must identify the fire and smoke path in an arena.

`Query 2 (Q2): Find the direction and the speed of the smoke cloud caused by the fire generated in bedroom #1.`

The spread of fire and smoke is guided by access paths such as doors and vents. Such queries can be answered by querying the summarized streams from the routers placed in each room. Query $Q2$ is therefore guided by the building layout, which in Figure 1.b is an acyclic join graph.

1.3 Our Contributions

The optimization of cyclic join queries is known to be NP-complete [5]. Even in the case of acyclic join queries, the state-of-the-art techniques [15, 18, 19] used in streaming database systems do not guarantee optimality. In addition, these techniques are insensitive to the shape of user query and so represent a "one-algorithm fits-all" policy. There is a need for a comprehensive approach to handle the entire spectrum of join queries, from acyclic to cyclic in the most effective manner.

In this effort, we begin by studying the performance of the commonly adopted heuristic-based techniques [15, 18, 19]. These techniques are known to produce sub-standard query plans for general acyclic and cyclic queries and may prove fatal for time-critical applications. More specifically, when handling acyclic queries our experiments demonstrate several cases when these techniques are shown to produce sub-optimal plans that are 5 fold more expensive than their optimal counterparts. In response, we tackle this shortcoming by extending the classical *IK algorithm* [5] in the streaming context (Section 4). The resulting *TreeOpt* approach while still featuring polynomial time complexity now also guarantees to produce optimal join plans for acyclic continuous join queries, such as $Q2$.

Subsequently, we focus on query plans represented by cyclic join graphs. Our experiments reveal that when handling cycles, in several cases the popular heuristic-based techniques generate plans that are 15 fold more expensive than their optimal counterpart. Unfortunately, even the adaptation of TreeOpt for cyclic queries is not guaranteed to generate optimal plans. Since the optimization problem in this setting is NP-complete we refine our goals as follows. We ask that our optimizer (i) be polynomial in time complexity, (ii) be able to increase the probability of finding optimal plans, and (iii) in scenarios where an optimal solution cannot be found, the technique should decrease the ratio of how expensive the generated plan is in comparison to the optimal plan. Towards this end, we introduce our *Forward and Backward Greedy* (*FAB*) algorithm that utilizes our *global impact ordering* technique (see Section 5). This can be applied in parallel with the traditional *forward greedy* approach [3]. Through our experimental evaluation we show that our FAB algorithm has a much higher likelihood of finding an optimal plan than state-of-the-art techniques. In scenarios when an optimal plan cannot be found, FAB is shown to generate plans that are closer to the optimal than those generated by current approaches. Finally, we put the above techniques together into a system that is query shape aware while still having a polynomial time complexity, called the *Q-Aware* approach (see Section 6). This technique is equipped to generate the best possible (optimal or a good) plans guided by the shape of the query.

2 Background

2.1 Continuous Multi-Join Processing

The common practice for executing multi-join queries over windowed streams is by a multi-way join operator called mjoin [18], a single multi-way operator that takes as input the continuous streams from all join participants. See Figure 2.a. Two benefits of mjoin are that the order in which the inputs tuples from each stream are joined with remaining streams can be dynamic, and intermediate tuples are not longer stored, saving space. To illustrate, in Figure 2.b new tuples from S_1 (ΔS_1 for short) are first inserted into the state of S_1 (denoted as ST_{S_1}), then used to probe the state of S_4 (ST_{S_4}), and the resulting join tuples then go on to probe ST_{S_2}, etc. This ordering is called S_1's pipeline. The *join graph* (JG), as in Figure 1.b, represents a multi-join query along with statistical information such as input rates and selectivities. In this work, we assume independent join selectivities and time window constraints [20, 21].

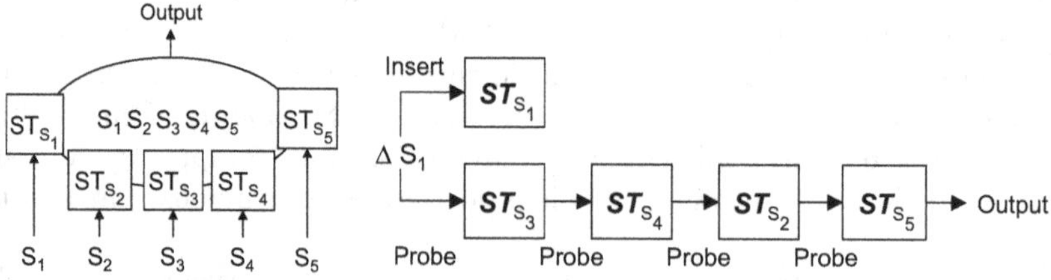

Fig. 2. a) 5-way Mjoin operator b) Sample join order to process tuples from S_1 (ΔS_1)

Table 1. Terms Used In The Cost Model

Term	Meaning
n	Number of participant streams
ΔS_i	ith update stream
$\bowtie_{S_i:1}, ..., \bowtie_{S_i:n-1}$	Join order (pipeline) for ΔS_i
$\bowtie_{S_i:j}$ or $S_i : j$	jth stream to join in ΔS_i's pipeline
C_{insert}	Cost of inserting a tuple into a state
C_{delete}	Cost of deleting a tuple into a state
C_{join}	Cost of joining a pair of tuples
$\lambda_{S_i:j}$	Average input rate of stream $S_i : j$
$\sigma_{S_i:j}$	Selectivity for joining stream $S_i : j$
W	Sliding time-based window constraint
$\lvert ST_{S_i} \rvert$	Number of tuples in state of stream S_i

2.2 Cost Analysis

The cost model is based on the commonly used per-unit-time cost metrics [20]. The estimated processing cost of an mjoin is the cumulative sum of the costs to process updates from each of the n input streams.

Without loss of generality, we estimate the per-unit-time CPU cost of processing the update from stream S_1. It is the sum of the costs incurred in inserting the new tuples ($insert(S_1)$) to its corresponding state (ST_{S_1}), to purge tuples ($purge(S_1)$) that fall outside the time-frame (W) and to probe the states of the participating streams ($probe(\bowtie_{S_1:1}, \bowtie_{S_1:2}, ..., \bowtie_{S_1:n-1})$).

$$CPU_{S_1} = insert(S_1) + purge(S_1) + probe(\bowtie_{S_1:1}, \bowtie_{S_1:2}, ..., \bowtie_{S_1:n-1}) \quad (1)$$

The cost for inserting new tuples from stream S_1 into the state ST_{S_1} is $\lambda_{S_1} * C_{insert}$ where C_{insert} is the cost to insert one tuple. Tuples whose time-stamp is less than ($time_{current}$ - W) are purged. Under the uniform arrival rate assumption, the number of tuples that will need to be purged is equivalent to λ_{S_1}. If the cost for deleting a single tuple is given by C_{delete}, then the purging cost for stream S_1 is $\lambda_{S_1} * C_{delete}$. Updates from stream S_1 are joined with the remaining join participants in a particular order as specified by its pipeline $\bowtie_{S_1:1}, \bowtie_{S_1:2}, \ldots, \bowtie_{S_1:n-1}$. The cost of joining every new tuple from S_1 with $\bowtie_{S_1:1}$ ($= S_3$) is $\lambda_{S_1} * (\lambda_{\bowtie_{S_1:1}} * W) * C_{join}$. This is due to the fact that under a constant arrival rate at most ($\lambda_{\bowtie_{S_1:1}} * W$) tuples exist in the state of $\lambda_{S_1:1}$ stream which will join with the new updates from S_1. Now, the resulting tuples $(\lambda_{S_1} * (\lambda_{\bowtie_{S_1,1}} * W) * \sigma_{\bowtie_{S_1:1}})$ probe the state of $\bowtie_{S_1:2}$ and so on. Thus the total update processing cost is:

$$\begin{aligned} CPU_{S_1} =& \lambda_{S_1} * C_{insert} + \lambda_{S_1} C_{delete} + \lambda_{S_1} * (\lambda_{\bowtie_{S_1:1}} * W) * C_{join} \\ &+ (\lambda_{S_1} * [\lambda_{\bowtie S_1:1} * W] * \sigma_{\bowtie_{S_1:1}}) * (\lambda_{\bowtie_{S_1:2}} * W) * C_{join} + \ldots.. \\ =& \lambda_{S_1} [C_{insert} + C_{delete} + (\sum_{i=1}^{n-1} [\prod_{j=1}^{i} \lambda_{\bowtie_{S_1:j}} * \sigma_{\bowtie_{S_1:j-1}}] * W^i * C_{join})]; \end{aligned} \quad (2)$$

where $\sigma_{i:0} = 1$.

It follows that the CPU cost for any n-way mjoin is:

$$CPU_{mjoin} = \sum_{k=1}^{n}(\lambda_{S_k}[C_{insert} + C_{delete} + \sum_{i=1}^{n-1}[\prod_{j=1}^{i} \lambda_{\bowtie S_{k:j}} \sigma_{\bowtie S_{k:j-1}}]W^i C_{join})]) \quad (3)$$

where k is the number of streams, while i and j are the index over the pipeline.

3 Assessment of Popular Optimization Algorithms

3.1 Dynamic Programming Techniques for Ordering MJoin

The classical bottom-up dynamic programming algorithm due to Selinger [1] is guaranteed to find an optimal ordering of *n-way* joins over relational tables, and can easily be adopted for mjoin ordering (DMJoin). The aim of the algorithm to find the join order that produces the least number of intermediate tuples. This is consistent with the CPU cost model described in Equation 3. To illustrate, consider our query $Q1$ (as depicted in Figure 1.b) and the processing of the new tuples from stream S_1. In the first iteration, the algorithm computes the cost of all join subsets with S_1 and one other stream. For example, $\{S_1, S_2\}$, $\{S_1, S_3\}$, etc., are each considered. Next, the algorithm considers all subsets having k = *3* streams and S_1 being one of those streams. In the kth iteration $\binom{n-1}{k-1}$ join pairs need to be maintained. For each of these subsets there can be $k-1$ ways of ordering. Several extensions to this core approach have been proposed [3, 4] along with better pruning techniques [22, 2] have been designed. However, their exponential time complexity makes them not viable for streaming databases.

3.2 Forward Greedy Algorithm

In the streaming context, it is a common practice to adopt some variations of the forward greedy algorithm [3] to order mjoins [19,18,15], here called *F-Greedy*. In each iteration, the candidate that generates the smallest number of intermediate tuples is selected as the next stream to be joined.

For $Q1$ (Figure 1.b), in the first iteration the algorithm computes the cost incurred by new tuples from S_1 joining with the remaining streams. For example,

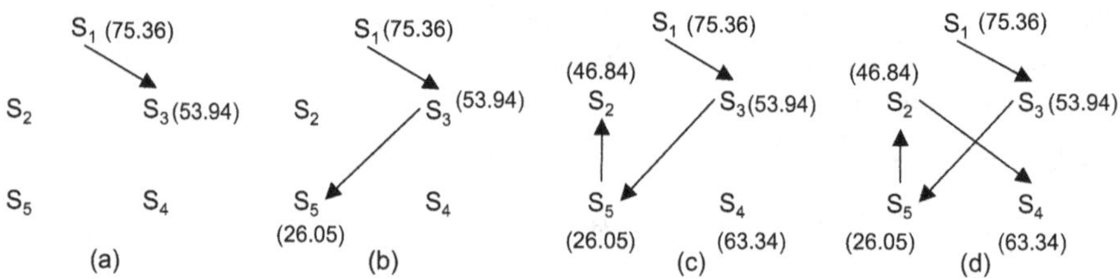

Fig. 3. Forward Greedy Algorithm

stream S_3 generates $75.36 * 53.94 * 0.26 \approx 1057$ tuples/sec, while with stream S_5 is $75.36 * 26.05 * 0.78 \approx 1531$ tuples/sec. Since S_3 produces the smallest number of intermediate results, it is chosen for joining first S_1's pipeline and so on. F-Greedy returns the plan $S_1 \bowtie S_3 \bowtie S_5 \bowtie S_2 \bowtie S_4$ which generates ≈ 7212 intermediate tuples per sec. For comparison, the optimal plan for processing input tuples from S_1 is $S_1 \bowtie S_5 \bowtie S_2 \bowtie S_3 \bowtie S_4$ which generates ≈ 3629 intermediate tuples per second. Therefore, F-Greedy plan is 2 fold more expensive than the optimal plan generated by DMJoin.

Time complexity to generate a query plan that processes the new tuples from an input stream is $\mathcal{O}(n^2)$. Therefore, the time complexity for ordering an n-way mjoin operator is $\mathcal{O}(n^3)$.

3.3 Experimental Evaluation

Environment. Our experiments were conducted on a 3.0 GHz 4 Intel machine with 1 GB memory. We executed the generated plans on the CAPE continuous query engine [10] to verify the cost model and record the execution time. All algorithms where implemented in Java 1.5.

Experimental Setup. In line with state-of-the-art techniques [23, 24] we use synthesized data to control the key characteristics of the streams namely input rate and selectivites between streams. More importantly, to assure statistical significance of our results and scopes of applicability we work with wide variations of settings as elaborated below. The setup included varying the number N of streams from 3 to 20. For each N, we randomly generate: 1) the input rate for each stream [1–100] tuples/sec, and 2) the join selectivities among the streams. For each N, we repeat this setup process 500 times. Therefore we have a total of (20-3+1) * 500 = 9000 different parameter settings.

Objectives. First, we compare time needed by the algorithm to generate a plan. Second, we measure the effectiveness measured as % of runs[1] of each algorithm to produce an optimal plan. Third, we compare the plan produced by heuristic-based algorithm against the optimal plan returned by DMJoin. Lastly, we observe the effectiveness of generating optimal plans as well as how expensive non-optimal plans can be for a diversity of join graph shapes.

Evaluation of Popular Algorithms. We begin by comparing the time needed to generate a plan by F-Greedy vs. DMJoin. The plan generation time for each distinct N is the average time over 500 distinct runs. As it is well known, Figure 4.a re-affirms the exponential nature of DMJoin. Next, we study the capability of F-Greedy to generate optimal plans for different query plan shapes. We achieve this by comparing the cost of plans generated by F-Greedy to those generated by DMJoin. As in Figure 4.b F-Greedy generates optimal plans for linear and star-shaped join queries. However, for general acyclic and cyclic join queries, F-Greedy generates substandard plans for many settings. Next, to provide a deeper understanding of the behavior of F-Greedy when applied to general

[1] *A run is an unique assignment of input rates and selectivities in a join graph.*

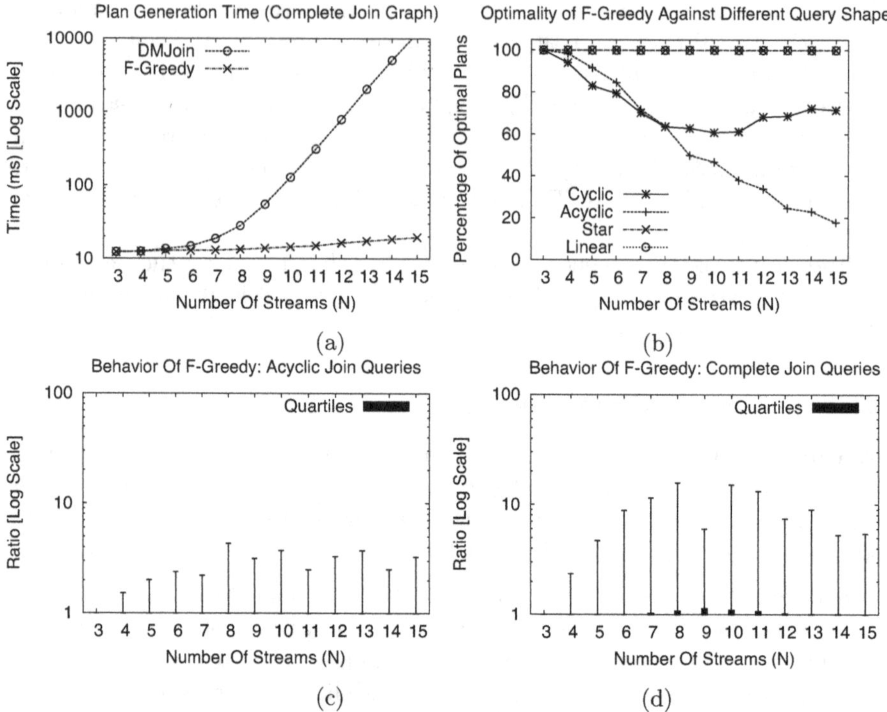

Fig. 4. F-Greedy vs. DMJoin: (a) Plan Generation Time; (b) % of Optimal Plans Generated by F-Greedy; (c) General Acyclic Join Graph; (d) Complete Join Graph

acyclic and cyclic join queries, we compute the ratio of the average number of intermediate results generated by plans produced by heuristic-based algorithm against those generated by optimal plan. A ratio = 1 is ideal since it reflects that F-Greedy plan generates the same number of intermediate results and therefore have similar CPU costs. Figures 4.c and 4.d confirms that F-Greedy could potentially generate plans that are many fold more expensive than the optimal plan, thereby triggering re-optimization sooner.

4 The TreeOpt Algorithm

In the previous section, F-Greedy is shown to generate substandard plans even for acyclic join queries. We now extend the *classical IK algorithm* [5] to solve the optimal ordering of acyclic queries, called *TreeOpt*. [5] has been largely ignored in the streaming context. To illustrate the main intuition of this approach let us consider acyclic join graphs such as $Q2$ (Figure 1.b) and the processing of new tuples from stream S_i. The join graph can now be viewed as a directed tree with S_i as its root. The aim is to transform this tree into a chain (representing the join order) with S_i as its root. However, a directed tree can be composed of vertexes linked as a chain or as wedges (two or more children). If this directed tree is a chain, then the chain hierarchy dictates the join order. When this directed tree is not a chain, then TreeOpt starts the optimization from its leaves. Each vertex (S_r) is assigned a rank (defined later). If a parent vertex S_q has a greater rank than that of its child vertex S_r, then the two vertexes are merged into one with the name S_qS_r and their rank is calculated. The merging of unordered vertexes ensures that the structural information of the query is not lost. For example, S_q is always joined before S_r. To transform a wedge into a chain, we merge all the children of a vertex into a single chain arranged in ascending order by their respective rank.

Next, we show that the cost model, as in Section 2.2, satisfies the Adjacent Sequence Interchange (ASI) property of [5] and thereby is guaranteed to generate an optimal plan for acyclic join graphs. Consider a given acyclic join graph JG and a join sequence $\zeta = (\bowtie_{S_1:0}, ..., \bowtie_{S_1:n-1})$ starting from input S_1. By Equation 2 the total CPU cost of this sequence is: $CPU_{S_1} = \lambda_{S_1} [C_{insert} + C_{delete} + (\sum_{i=1}^{n-1} [\prod_{j=1}^{i} \lambda_{\bowtie_{S_1:j}} * \sigma_{\bowtie_{S_1:j-1}}] * W^i * C_{join})$. The terms $\lambda_{S_1}(C_{insert}+C_{delete})$ and C_{join} are order-independent, and are therefore ignored. The order-dependent part of CPU_{S_1}, that is, $\sum_{i=1}^{n-1} [\prod_{j=1}^{i} \lambda_{\bowtie_{S_1:j}} * \sigma_{\bowtie_{S_1:j-1}}] * W^i$ can be defined recursively as below (similar to [5]):

$C(\Lambda) = 0$	Null sequence Λ.
$C(S_1) = 0$	Starting input stream.
$C(\bowtie_{S_1:i}) = \lambda_{\bowtie_{S_1:i}} \sigma_{\bowtie_{S_1:i}} W$	Single input $S_i (i > 1)$.
$C(\zeta_1\zeta_2) = C(\zeta_1) + T(\zeta_1)C(\zeta_2)$	Sub-sequences ζ_1 and $\zeta_2$in join sequence ζ.
where T(*) is defined by:	
$T(\Lambda) = 1$	Null sequence Λ.
$T(S_1) = \lambda_{S_1}$	Starting input stream.
$T(\bowtie_{S_1:i}) = \sigma_{\bowtie_{S_1:i}} \lambda_{\bowtie_{S_1:i}} W$	Single input $S_i (i > 1)$.
$T(\zeta_1) = \prod_{k=i}^{j} (\sigma_{\bowtie_{S_1:k}} \lambda_{\bowtie_{S_1:k}} W)$	Subsequence $\zeta_1 = (\bowtie_{S_1:i}, \ldots, \bowtie_{S_1:j-1})$.

Each node S_q is marked by the rank, $rank(S_q) = (T(S_q) - 1)/C(S_q)$, where $C(S_q)$ and $T(S_q)$ are defined as above. This modified cost model satisfies the Adjacent Sequence Interchange (ASI) property [5] and therefore is guaranteed to produce an *optimal join order* for acyclic graphs.

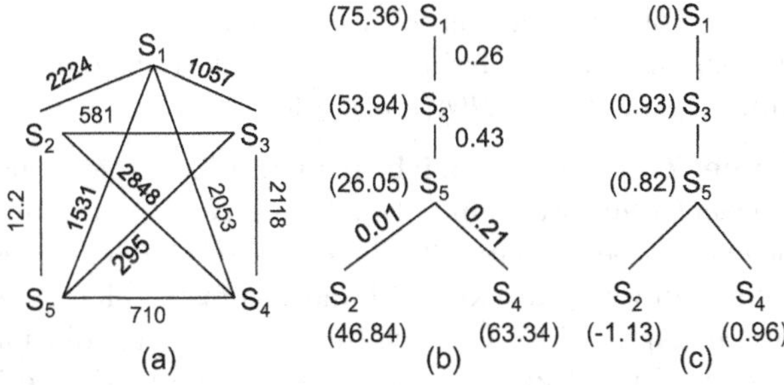

Fig. 5. a) Weighted Join Graph, b) Minimum Spanning Tree c) A Rooted Tree

Handling Cycles in TreeOpt: For the problem of optimizing cyclic join graphs the strategy is to first transform the graph into some acyclic graph and then apply TreeOpt. The aim now is to construct a good acyclic graph. Note that when ordering the pipeline of a given stream S_i, the goal is to reduce the number of intermediate tuples. Therefore, we propose to generate a *minimal spanning tree* (MST), where the weight of an edge connecting two vertexes S_i and S_j is computed as $\lambda_{S_i} * \lambda_{S_j} \sigma_{S_i S_j}$. In the static database context, [25] proposed a similar heuristic accounting for the cost of disk accesses. *TreeOpt* is then applied to produce an optimal ordering for this MST.

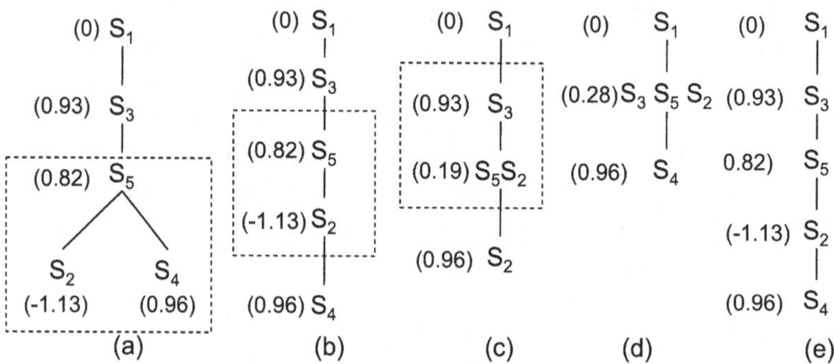

Fig. 6. Finding the Optimal Join Ordering

Example: For the weighted cyclic join graph in Figure 5.a we generate the MST shown in Figure 5.b and then compute the *rank* for each node S_q. We traverse the rooted tree (Figure 6.a) bottom up. S_5 is the first node with more than one child and we check to see if all of its children's branches are ordered by non-decreasing ranks. We then merge the children's nodes into one sequence by the ascending order of their ranks as in Figure 6.b. The resulting chain is not ordered since $rank(S_2) < rank(S_5)$ and so we merge nodes S_5 and S_2, and recompute the rank for the merged node S_5S_2 (Figure 6.c). As a final step, we expand all combined nodes. The resulting sequence is the optimal join for the MST shown in Figure 5.b.

Time Complexity: For a join graph with n vertexes, it takes $\mathcal{O}(n^2 log(n))$ to find a minimum spanning tree. Ordering a stream takes $\mathcal{O}(nlog(n))$. Therefore, *TreeOpt* has a time complexity of $\mathcal{O}(n^2 log(n))$ for ordering an n-way mjoin.

Evaluation of The *TreeOpt* Algorithm: Figure 7.a depicts the percentage of runs in which *TreeOpt* generates an optimal plan. Note that all lines in Figure 7.a except for the cyclic case overlap fully. That is, *TreeOpt* generates an optimal plan for any acyclic join query as expected and has faster plan generation time than F-Greedy (as in Figure 7.b). Clearly, this is a win-win solution for acyclic join queries. However, for the cyclic queries we observe in Figure 7.b that the ability of *TreeOpt* to generate an optimal plan rapidly goes down to zero. A closer investigation of the distribution of generated plan costs reveals that in

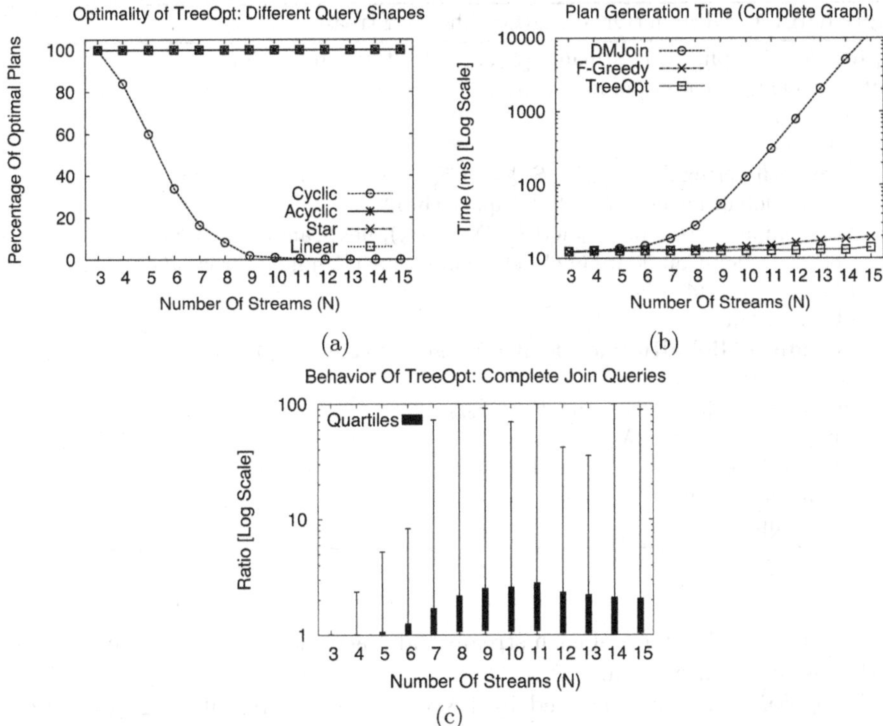

Fig. 7. a) % of Optimal Plans b) Execution Time c) Comparison: TreeOpt vs. DMJoin

most cases the upper-quartile (top 75%) of generated plans are ≈ 2.5 fold more expensive than the optimal plan generated by the DMJoin. Additionally, we note in Figure 7.c that there are many cases when *TreeOpt* performs unacceptably, sometimes 50 fold or worse depending on the MST considered.

Summarizing: *TreeOpt* generates optimal plans for any acyclic join graph in polynomial time, while the widely used F-Greedy [15, 18, 19] often produces sub-standard plans. However, for cyclic join queries, where neither offer any guarantees on optimality, F-Greedy outperforms *TreeOpt*.

5 The FAB Algorithm

We now present our *Forward and Backward greedy* (*FAB*) algorithm. The F-Greedy algorithm incrementally picks the next best stream and is therefore too greedy at the early stages of optimization. In Figure 3, during the first iteration F-Greedy chooses S_3 as it generates the fewest number of join results. However, S_3 has a higher input rate, and its join selectivities with the remaining streams are greater than S_5. This early greedy decision affects the total plan cost by producing more intermediate tuples in comparison to the case when S_2 is chosen. By contrast, our FAB approach uses a *global impact-based ordering*, in which we

Algorithm 1. GImpactOrdering(Graph JG, Stream S_{start})

Input: Join Graph, JG of streams $\{S_1, .., S_n\}$; Start stream, S_{start}
Output: Query Plan *cPlan* for $\triangle S_{start}$

1: **for** $i = n$ to 2 **do**
2:   impact = ∞
3:   **for** each vertex $S_q \in \{S_1, .., S_n\}$ and $S_q \neq S_{start}$ **do**
4:     **if** GlobalImpact(JG, S_q) < impact **then**
5:       impact = GlobalImpact(JG, Node S_q); nextCandidate = S_q
6:   cPlan = nextCandidate $\bowtie$ cPlan; Remove nextCandidate from JG
7: cPlan = $R_{start} \bowtie$ cPlan
8: **return** cPlan
9: **procedure GlobalImpact**(JoinGraph JG, Stream S_{curr})
10: impact = 1;
11: **for** each $S_x \in \{S_1, .., S_n\}$ and $S_x \neq S_{curr}$ **do**
12:   impact = impact * λ_{S_x}
13:   **for** each $S_y \in \{S_1, .., S_n\}$ and $S_y \neq S_{curr}$ and $S_y \neq S_x$ **do**
14:     impact = impact * $\sigma_{S_x S_y}$
15: return impact

explore the global impact of each stream and place the most expensive streams as the last candidate to join.

The *global impact* metric used by FAB is listed in Algorithm 1 (Line 12). The global impact of stream S_q ($S_q \in S$ and $S_q \neq S_{start}$) is the product of the arrival rates of all remaining streams and the join selectivities not related to S_q. The intuition here is that if a stream, S_q, has the least impact, i.e., high input rate and poor selectivities, then the generated plan will be the most expensive. Following this principle, FAB starts by picking the candidate that has the least impact and places it as $\bowtie_{S_{start}:n-1}$. Next, we remove this candidate stream from the join graph and proceed to determine the next-to-last stream to join in the update pipeline of stream S_{start}. This process is done iteratively until all $n-1$ positions in the pipeline are filled. Our FAB approach makes use of the *global impact ordering* in unison with the F-Greedy to generate optimal plans for ordering multi-join queries.

Time complexity of our global impact ordering is $\mathcal{O}(n^2)$. Therefore, to order the pipeline for processing new tuples from a single stream by FAB is $\mathcal{O}(n^2)$; thus, ordering an *n-way* mjoin operator has complexity $\mathcal{O}(n^3)$.

Example: Consider $Q1$ and the processing of tuples from stream S_1. We compute the global impact of all remaining join participants. As in Figure 8, the candidate with the least impact is identified. Here, the impact of S_4 is ≈ 300 while that of S_3 is $\approx$ 11K and is therefore the last stream to be joined in S_1's pipeline i.e., $\bowtie_{S_1:4} = S_4$. Iteratively, the resultant ordering is $S_1 \bowtie S_5 \bowtie S_2 \bowtie S_3 \bowtie S_4$, which generates ≈ 3629 intermediate tuples/sec. This is equivalent to the optimal plan generated by DMJoin. Recall in Section 3.2, F-Greedy returns a sub-optimal plan that generates ≈ 7212 tuples/sec, which is **2** fold more expensive.

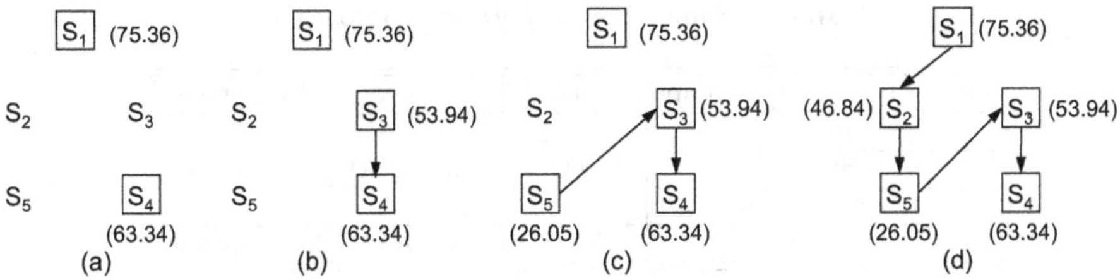

Fig. 8. Finding Optimal Join Ordering Through GImpactOrdering (Algorithm 1)

Evaluation of the FAB Algorithm. In Figure 9.b, the FAB approach has a higher likelihood of generating an optimal for the entire spectrum of queries than F-Greedy (Figure 9.b). However, FAB does not provide optimality guarantees as TreeOpt (Figure 7.b) for acyclic queries.

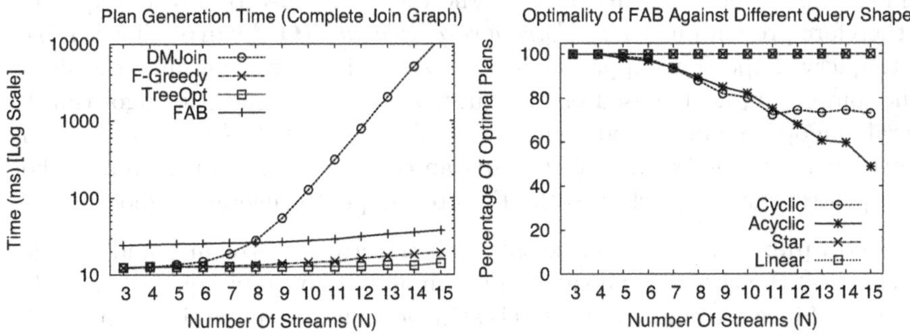

Fig. 9. FAB vs. DMJoin: (a) Plan Generation Time (b) % of Optimal Plans by FAB

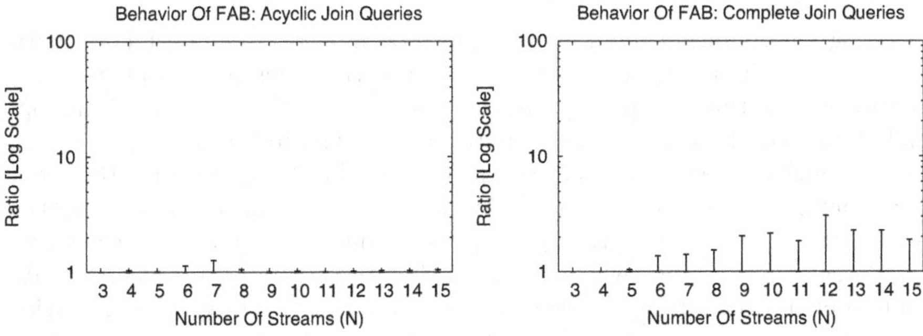

Fig. 10. FAB vs. DMJoin (Plan Cost Comparison): (a) Acyclic (b) Complete

For acyclic queries, FAB produces near-optimal plans with the upper quartile of the runs generating optimal plans (as in Figure 10.a). FAB is shown to generate plans that are at-most 1.25 fold more expensive in cost than those generated by DMJoin. Figure 10.b highlights similar trends in the upper quartile of runs when

Table 2. Summarizing *Q-Aware* Approach

Query (Join Graph) Shape	Algorithm	Complexity	Properties
Linear	TreeOpt	$\mathcal{O}(n^2 log(n))$	Optimal
Star			
General-Acyclic			
Cyclic	FAB	$O(n^3)$	Near-optimal

processing cyclic join queries. The most expensive plan in FAB are cyclic queries, which are at most 2 fold more expensive than those generated by DMJoin.

6 The Q-Aware Approach

Due to the NP-completeness of the query optimization problem no single algorithm can effectively and efficiently handle the entire spectrum of join queries. We therefore present our *query shape aware approach* (**Q-Aware**) that is sensitive to query shape. This approach has two steps. First we determine the shape of the join graph; next, based on the query shape, we choose the algorithm to generate an optimal or a good plan, as described in Table 6. An acyclic query is always ordered using TreeOpt as it is guaranteed to return an optimal plan. For cyclic queries, the approach uses the FAB technique to generate a good plan.

Time complexity: For a join graph with n vertexes (streams) and e edges (selectivities) the time complexity to determine if a given join query has cycles is $\mathcal{O}(n+e)$. Since the number of edges is smaller than $\mathcal{O}(n^3)$, the time complexity of *Q-Aware* $\mathcal{O}(n^3)$ for ordering an n-way mjoin.

7 Conclusion

Streaming environments require their query optimizers to support (1) reoptimization and migration techniques to keep up with fluctuating statistics, (2) have polynomial time complexity, (3) increase the probability of generating optimal plans, and in the worst case scenarios it must aim to generate a good plan, and (4) handle a diversity of query types. Motivated by this, we revisit the problem of continuous query optimization. We begin by experimentally studying the effectiveness of the state-of-the-art continuous query optimization techniques for different query shapes which confirms that these techniques generate substandard plans even for the simple problem of ordering acyclic queries. To tackle this deficiency, we extend the classical *IK algorithm* to the streaming context called *TreeOpt*. TreeOpt is a polynomial-time algorithm, and is superior to the greedy approach for general acyclic queries. For the harder problem of ordering cyclic graphs, *TreeOpt* is not be a viable alternative. Therefore, we introduce our *FAB* algorithm utilizes our *global impact ordering* technique. This approach increases the chances of finding an optimal plan as well as generating a less expensive plan when an optimal plan cannot be found in polynomial time. Lastly, we put forth our *Q-Aware* approach that generates optimal or near-optimal plans for any query shape in guaranteed polynomial time.

Acknowledgement

This work is supported by the National Science Foundation (NSF) under Grant No. IIS-0633930, CRI-0551584 and IIS-0414567.

References

1. Selinger, P.G., Astrahan, M.M., Chamberlin, D.D., Lorie, R.A., Price, T.G.: Access path selection in a relational database management system. In: SIGMOD, pp. 23–34 (1979)
2. Vance, B., Maier, D.: Rapid bushy join-order optimization with cartesian products. In: SIGMOD, pp. 35–46 (1996)
3. Kossmann, D., Stocker, K.: Iterative dynamic programming: a new class of query optimization algorithms. ACM Trans. Database Syst. 25(1), 43–82 (2000)
4. Moerkotte, G., Neumann, T.: Dynamic programming strikes back. In: SIGMOD, pp. 539–552 (2008)
5. Ibaraki, T., Kameda, T.: On the optimal nesting order for computing n-relational joins. ACM Trans. Database Syst. 9(3), 482–502 (1984)
6. Swami, A.N., Iyer, B.R.: A polynomial time algorithm for optimizing join queries. In: ICDE, pp. 345–354 (1993)
7. Ioannidis, Y.E., Kang, Y.C.: Left-deep vs. bushy trees: An analysis of strategy spaces and its implications for query optimization. In: SIGMOD, pp. 168–177 (1991)
8. Abadi, D.J., Ahmad, Y., Balazinska, M., Çetintemel, U., Cherniack, M., Hwang, J.H., Lindner, W., Maskey, A., Rasin, A., Ryvkina, E., Tatbul, N., Xing, Y., Zdonik, S.B.: The design of the borealis stream processing engine. In: CIDR, pp. 277–289 (2005)
9. Ali, M.H., Aref, W.G., Bose, R., Elmagarmid, A.K., Helal, A., Kamel, I., Mokbel, M.F.: Nile-pdt: A phenomenon detection and tracking framework for data stream management systems. In: VLDB, pp. 1295–1298 (2005)
10. Rundensteiner, E.A., Ding, L., Sutherland, T.M., Zhu, Y., Pielech, B., Mehta, N.: Cape: Continuous query engine with heterogeneous-grained adaptivity. In: VLDB, pp. 1353–1356 (2004)
11. Madden, S., Shah, M.A., Hellerstein, J.M., Raman, V.: Continuously adaptive continuous queries over streams. In: SIGMOD, pp. 49–60 (2002)
12. Ayad, A., Naughton, J.F.: Static optimization of conjunctive queries with sliding windows over infinite streams. In: SIGMOD, pp. 419–430 (2004)
13. Arasu, A., Babcock, B., Babu, S., Datar, M., Ito, K., Nishizawa, I., Rosenstein, J., Widom, J.: Stream: The stanford stream data manager. In: SIGMOD Conference, vol. 665 (2003)
14. Chandrasekaran, S., Cooper, O., Deshpande, A., Franklin, M.J., Hellerstein, J.M., Hong, W., Krishnamurthy, S., Madden, S., Raman, V., Reiss, F., Shah, M.A.: Telegraphcq: Continuous dataflow processing for an uncertain world. In: CIDR (2003)

15. Babu, S., Motwani, R., Munagala, K., Nishizawa, I., Widom, J.: Adaptive ordering of pipelined stream filters. In: SIGMOD, pp. 407–418 (2004)
16. Zhu, Y., Rundensteiner, E.A., Heineman, G.T.: Dynamic plan migration for continuous queries over data streams. In: SIGMOD, pp. 431–442 (2004)
17. Bukowski, R., Peacock, R., Averill, J., Cleary, T., Bryner, N., Walton, W., Reneke, P., Kuligowski, E.: Performance of Home Smoke Alarms: Analysis of the Response of Several Available Technologies in Residential Fire Settings. NIST Technical Note 1455, 396 (2000)
18. Viglas, S., Naughton, J.F., Burger, J.: Maximizing the output rate of multi-way join queries over streaming information sources. In: VLDB, pp. 285–296 (2003)
19. Golab, L., Özsu, M.T.: Processing sliding window multi-joins in continuous queries over data streams. In: VLDB, pp. 500–511 (2003)
20. Kang, J., Naughton, J.F., Viglas, S.: Evaluating window joins over unbounded streams. In: ICDE, pp. 341–352 (2003)
21. Hammad, M.A., Franklin, M.J., Aref, W.G., Elmagarmid, A.K.: Scheduling for shared window joins over data streams. In: VLDB, pp. 297–308 (2003)
22. Ganguly, S., Hasan, W., Krishnamurthy, R.: Query optimization for parallel execution. In: SIGMOD, pp. 9–18 (1992)
23. Jain, N., Amini, L., Andrade, H., King, R., Park, Y., Selo, P., Venkatramani, C.: Design, implementation, and evaluation of the linear road benchmark on the stream processing core. In: SIGMOD, pp. 431–442 (2006)
24. Tao, Y., Yiu, M.L., Papadias, D., Hadjieleftheriou, M., Mamoulis, N.: Rpj: Producing fast join results on streams through rate-based optimization. In: SIGMOD, pp. 371–382 (2005)
25. Krishnamurthy, R., Boral, H., Zaniolo, C.: Optimization of nonrecursive queries. In: VLDB, pp. 128–137 (1986)

An XML-Based Model for Supporting Context-Aware Query and Cache Management

Essam Mansour and Hagen Höpfner

International University in Germany
School of Information Technology
Campus 3, D-76646 Bruchsal, Germany
essam.mansour@ieee.org, hoepfner@acm.org

Abstract. Database systems (DBSs) can play an essential role in facilitating the query and cache management in context-aware mobile information systems (CAMIS). Two of the fundamental aspects of such management are update notifications and context-aware query processing. Unfortunately, DBSs does not provide a built-in update notification function and are not aware of the context of their usage. This paper presents an XML model called XReAl (XML-based Relational Algebra) that assists DBSs in extending their capabilities to support context-aware queries and cache management for mobile environments.

1 Introduction

Database systems (DBSs) can play an essential role in facilitating the advanced data management required to modern information systems, such as context-aware mobile information systems (CAMIS). Usually, this advanced data management is provided by adding middle-wares over DBSs. Update notifications [8] and context-aware query processing [9] are part of the fundamental management aspects in CAMIS. Unfortunately, DBSs does not provide a built-in update notification function and are not aware of the context of their usage.

The main focus of this paper is to provide a model, which could be directly integrated into existing DBSs. One of the main requirements for this model is to be realized within DBSs in a way, which assists in extending DBSs capabilities to support cache management and the processing of context-aware queries as built-in DBS functions. Such extension is to reduce the code-complexity and increase the performance of CAMIS due to avoiding the middle-wares.

This paper presents a model called XReAl (XML-based Relational Algebra) that supports context-aware queries and cache management in CAMIS. The XReAl model provides XML representation for the contextual information of the mobile clients, queries issued by these clients and manipulation operations. This XML representation is to be stored as XML documents in modern DBSs that provide XML management support, such as DB2 and Oracle.

The rest of this paper is organized as follows. Section 2 highlights related work. Section 3 outlines the context-aware services and cache management. Sections 4, 5 and 6 present the three sub-models of XReAl. Section 7 presents our DBS-based implementation for XReAl. Section 8 outlines evaluation results. Section 9 concludes the paper.

A.P. Sexton (Ed.): BNCOD 2009, LNCS 5588, pp. 107–119, 2009.

2 Related Work

In caching techniques, query results are stored and indexed with the corresponding query [6]. So, one can analyze whether new queries can be answered completely or partially from the cache [10]. Semantic caches tend to use semantic correlations between different cached items as a cache replacement criteria [1]. However, updating the base table is ignored in these works. Our proposed model supports the cache maintenance as a DBS built-in function.

Different research efforts, such as [4,7], investigated into the topic of XML algebra. Our proposed model, XReAl, is distinguished by providing an XML representation for relational algebra queries. XReAl is to extend relational algebra to support context-aware operators.

3 Query and Cache Management

This section discusses the management requirements for supporting advanced context aware services. These services support mobile service providers to be more receptive to mobile users needs. The management requirements are classified mainly into two categories query and cache requirements.

3.1 Advanced Context Aware Services

In our proposed context aware services, we assume a mobile service provider (*MSP*) is to prepare for their customers contextual information document. We utilized the classification of contexts in the ubiquitous computing environment proposed by Korkea-Aho in [5]. This document divides the contextual information into several contexts, *physical*, *environmental*, *informational*, *personal*, *social*, *application*, and system. For the *physical* context, *MSP* detects the location of a mobile user (*MS*) and the corresponding time of such location. The *environmental* context includes information related to the current location of *MS* or locations of interest, such as user's home and work location. Examples for information related to this context are a traffic jam, parking spots, and weather.

A mobile user might ask *MSP* to consider interesting business quotes in the user's *informational* context. For example, the user might need to know the newspaper's name, date, section, and page number combined with the quote value. The *personal* context records the user's personal plans, such as plans for working days, and hobbies, such as drinking coffee or visiting new food shops. The *social* context includes information about social issues, such as the user's kids, neighbors and social activities. The *application* context includes information concerning used applications, such as email accounts of the user in order to provide notification by received emails. Finally, the *system* context records information concerning systems used by the user, such as heating system and water supply. The users expects a very high level of security and privacy for their contextual information. The mobile users are supposed to issue ad-hoc or preregister queries. These queries might be continuous or non-continuous queries.

A	B
$q : \{\pi \vert \pi^a\}([\sigma]([\rho](R)))$	
$q : \{\pi \vert \pi^a\}([\sigma](\rho(q)))$	$QS \leftarrow \pi_{ShopName,tele,PID}(\sigma_{status='NEW'}(shop))$
$q : \{\pi \vert \pi^a\}([\sigma](cp))$	$QL \leftarrow \pi_{street,ID}(\sigma_{postal\_code=76646}(location))$
$cp : \{[\rho](R) \vert \rho(q)\} \times \{[\rho](S) \vert \rho(q) \vert cp\}$	$Q \leftarrow \pi_{ShopName,tele,street}(\sigma_{PID=ID}(QS \times QL))$
$q : \{\pi \vert \pi^a\}(q\{\cup \vert - \vert \cap\}q)$	

Fig. 1. A) the relational algebra recursive structure; B) A relational algebra of *NCQ*

3.2 Query Representation

In mobile information systems, applications generate queries and send them to the server. Therefore, there is no need to support descriptive query languages, such as SQL. Queries are to be represented in a useful way for storage and retrieval. The relational algebra representation [2] is an efficient way to represent queries over data stored in relational database. However, one can always translate SQL-queries into such expressions.

The query notation used in this paper is the notation of the relational algebra operators [2], such as selection (σ), θ-join ($\bowtie_\theta$), and projection (π). The θ-join is represented in our work as Cartesian product ($\times$) and selection (σ). It is possible to optimize the transformation into query trees in order to improve the support for query indexing by reducing the number of alternatives.

In general and formally, a database query q can have the recursive structure shown in Figure 1.A. A relational algebra query can be interpreted as a tree. However, there are several semantically equivalent relational algebra trees for each query. We, for the purpose of detecting update relevance [8], push selection and join operations inside a join using the algebraic properties for query optimization [2]. Then, we convert the θ-join to Cartesian product ($\times$) and selection (σ) to be in the form shown in Figure 1.A.

The query *NCQ* that retrieves the attributes (ShopName, Tele and *street*) of shops, whose status is *NEW* and postal_code is the postal_code of the user's current position. Assume that the user was in the area, whose postal_code is 76646, when the user requested the result of the query *NCQ*. The tables *shop* and *location* are joined using the attributes *PID* and *ID*. The query Q shown in Figure 1.B is an example of a relational algebra query for the query *NCQ*. The query Q is represented using the recursive structure shown in Figure 1.A.

3.3 Cache Management

Managing redundant data is a must for mobile information systems. However, redundancies might also lead to inconsistencies. Replication techniques use synchronization approaches but tend to strongly restrict the language used for defining replicas. Caches, on the other hand, are created by storing queried data implicitly. In [3,8] we discussed how to calculate the relevancy of server side updates for data cached on mobile devices and how to invalidate outdated caches. Therefore, the server must keep track of the queries posted by the mobile clients. Hence, the DBS must maintain a query index that indexes the caches and the clients. Each server update has to be checked against this index. All clients that hold updated data must be at least informed by such update.

4 Formalizing Contextual Information

This section presents the overall structure of the XREAL *mobile client* sub-model and examples for the sub-model.

4.1 The Structure

The *mobile client* sub-model consists of an identification attribute, called *MCID*, and a sequence of elements (*physical context, environmental context, informational context, personal context, social context, application context,* and *system context*). Figure 2.A shows the XML schema of *mobile client* at an abstract level.

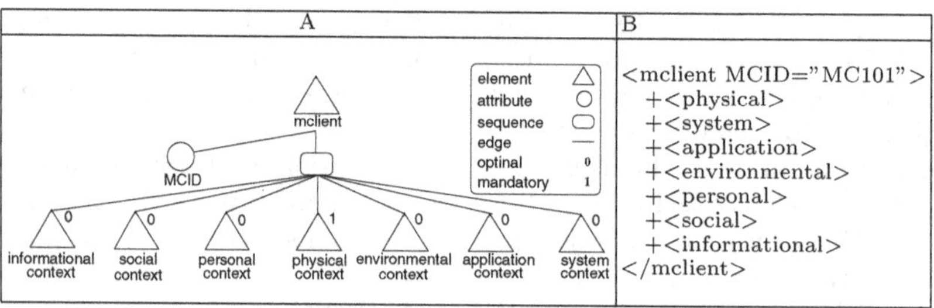

Fig. 2. A) the XREAL query model; B) the XREAL contextual information document

Any *mobile client* is assigned a *MCID* number, to be recognized by the system. *Physical context* provides information related to location and time. The location is a position, elevation, and direction. The position could be represented using a geographical coordinates and/or relative coordinates, such as a street, area and city. The time represents time zone, which could be inferred from the location information. The time zone determines the absolute time, day, week, month, quarter, and year. *Physical context* might help to infer information at a generic level related to *environmental context*, such as weather and light. Other methods are needed to determine an accurate environmental information.

Informational context formalizes information of interest to the mobile client, such as currency rates, stoke quotes and sports scores. *Personal context* specifies information such as health, mood, biographical information, habit and activities. *Social context* formalizes information concerning group activity and social relationships. *Application context* models information, such as email received and websites visited. The *system context* represents information related to systems used by the client and specs of her mobile, such as processor, and memory.

The user of a mobile client might provide personal and social information to be recorded as contextual information related to her mobile client. It is assumed that the minimum level of information is the information of *physical context*. So, the *physical context* element is a mandatory element. However, the other elements are optional. Furthermore, it is assumed that there is a repository of contextual information related to the environment, in which mobile clients are moving, such as parking spots or food shops.

4.2 Examples

Figure 2.B shows the contextual information document specified using XREAL that is generated by a *MSP* for one of its customers as discussed in Section 3. The contextual information document is assigned *MC101* as an ID. The XML language is very flexible in representing variety of contextual information. Figure 3 depicts part of the *physical* and *informational* contexts. Figure 3.A shows a representation for information of the relative position, and Figure 3.B illustrates a representation for business information as a part of informational context.

A	B
<relative_position>	<quote>
<country>Germany</country>	+<value>
<city>Bruchsal</city>	+<newspaper>
<area>south</city>	+<section>
<street>Durlacher<street>	+<date>
<postal_code>76646</postal_code>	+<description>
</relative_position>	</quote>

Fig. 3. A) part of the physical context, B) part of the informational context

5 Formalizing Queries

This section presents the XREAL sub-model for formalizing a relational algebra query based on the recursive structure discussed in Section 3.

5.1 Fundamental Elements

The XREAL model formalizes a relational algebra query as a *query* element that consists of two attributes, *QID* and *MCID*, and a sequence of elements, *relations*, *projection* and *join*. Figure 4.A shows the XML schema of XREAL *query*. The *QID* attribute represents a query identification. The *MCID* attribute represents the identification number of a mobile client that issued the query. A query might access only one relation. Therefore, a *query* element contains at least a *relations* element and *projection* element, and might has a *join* element. The *query* sub-model provides a formalization for queries represented as discussed in Section 3.

The *relations* element is composed of a sequence of at least one *relation* element. The *relation* element consists of an identification attribute, called *RID*, and a sequence of elements, *name, rename, selections* and *rprojection*. The *name* element represents the relation name. The *rename* element denotes the temporally name used to refer to the relation in the query. The *selection* element is composed of a sequence of a *spredicate* element of type *predicateUDT*. The *rprojection* element consists of a sequence of at least one *attribute* element of type *attributeUDT*. The *predicateUDT* type is a complex type that is able to represent simple predicate or composite predicate. The *projection* element is similar to the *rprojection* element, but *projection* represents the original projected attributes used in the query. The *join* element specifies the join predicates used to join together the relations (sub-queries).

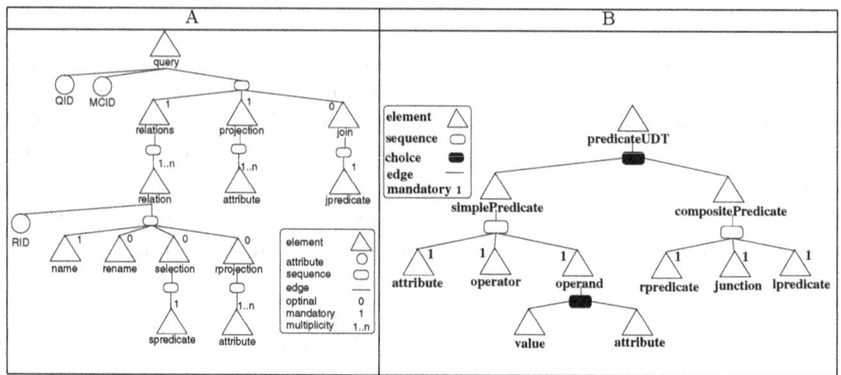

Fig. 4. A) the XML Schema of the query model; B) the predicateUDT Schema

The *query* sub-model is able to represent a query, such as the query Q shown in Figure 1.B. Th query Q consists of two sub-queries QS and QL. The specification of the query Q is to contain the elements *relations*, *projection* and *join*. Each sub-query is to be represented as a *relation* element that has its own *selection* and *rprojection* element. The *projection* element represents the original projection of the query. The *join* element represents the join predicate that joins QS and QL.

5.2 User Defined Data Types

The *query* sub-model has two main user defined data types (UDT), *predicateUDT* and *attributeUDT*. The *predicateUDT* type is a complex type composed of one of the elements *simplePredicate* or *compositePredicate*, as depicted in Figure 4.B. The *simplePredicate* element consists of a sequence of elements, *attribute, operator* and *operand.* The *attribute* element is of type *attributeUDT*. The *operator* element is of type *logicalOperatorUDT*, which is a simple type that restricts the token datatype to the values (*eq, neq, lt, lteq, gt,* and *gteq*). Respectively, they refer to equal, not equal, less than, less than or equal, greater than, and greater than or equal. The *operand* element is composed of one of the elements *value* or *attribute.* The *value* element is to be used with selection predicates. The *attribute* element is to be used with join predicates.

The *compositePredicate* element consists of a sequence of elements, *rpredicate, junction* and *lpredicate.* The *rpredicate* and *lpredicate* elements are of type *predicateUDT*. Consequentially, the *rpredicate* and *lpredicate* elements might consist of simple or composite predicate. The *junction* element is of type *junctionUDT*, which is a simple type that restricts the token datatype to the values (*and* and *or*). The *attributeUDT* type is a complex type composed of an attribute, called *ofRelation*, and a sequence of elements, *name* and *rename.* The *ofRelation* attribute represents a relation ID, to which the attribute belongs. The *name* element denotes the name of the attribute. The *rename* element represents the new name assigned to the attribute in the query.

```
<query QID="QID1" MCID="MC101">
  <relations>
    +<relation RID="RID01">
    +<relation RID="RID02">
  </relations>
  +<projection>
  <join>
    <jpredicate>
      <simplePredicate>
        <attribute ofRelation="RID01">
          <name>PID</name>
        </attribute>
<operator>eq</operator>
<operand>
  <attribute
   ofRelation="RID02">
    <name>ID</name>
  </attribute>
</operand>
</simplePredicate>
</jpredicate>
</join>
</query>
```

Fig. 5. The specification of *NCQ*

```
<relation RID="RID01">
  <name>shop</name>
  <selection>
    <spredicate>
      <simplePredicate>
        <attribute>
          <name>status</name>
        </attribute>
        <operator>eq</operator>
        <operand>
          <value>'NEW'</value>
        </operand>
      </simplePredicate>
    </spredicate>
  </selection>
<projection>
  <attribute>
    <name>ShopName</name>
  </attribute>
  <attribute>
    <name>tele</name>
  </attribute>
  <attribute>
    <name>PID</name>
  </attribute>
</projection>
</relation>
```

Fig. 6. The specification of the relation RID01

5.3 An Example

Figure 5 illustrates an overview of the XReAL specification for the query *NCQ*, whose relational algebra expression is shown in Figure 1.B. This specification consists of a *query* element. The query ID is *QID1* and is issued by a mobile client, whose ID is *MC101*. There are two sub-queries over the relations (*shop* and *location*), which are joined together using one join predicate.

Figure 6 illustrates the XReAL specification of the sub-query *QS*, which queries the relation (*shops*). The ID of the relation is *RID01*. This specification consists of a *relation* element, whose name is *shop*. The selection predicate associated with *shop* checks that the shop's status is equal to *NEW*. There is also a projection operation that picks the attributes (*ShopName,tele* and *PID*).

6 Formalizing Manipulation Operations

The following sub-sections present the structure of *moperation* and examples.

6.1 The Structure

A manipulation operation might be an insert, delete or update operation. Figure 7 shows the XML schema of the *moperation* component, which might consists of one *IStatement, DStatement*, or *UStatement*. The *IStatement* element

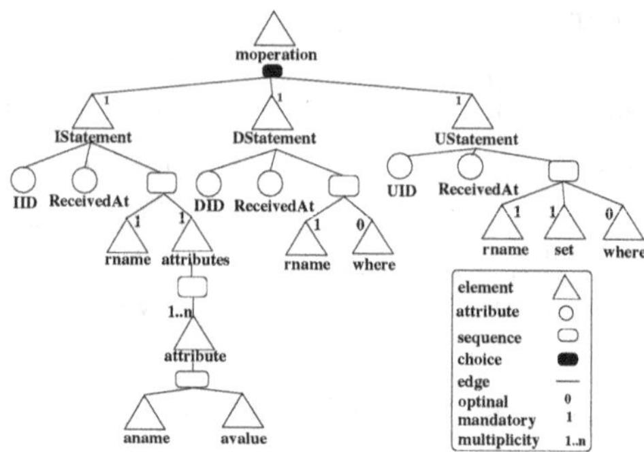

Fig. 7. The XML Schema of the manipulation operations

```
<IStatement IID="I3001"
receivedAt="2008-09-12T11:34:27">
  <rname>shop</rname>
  <attributes>
    <attribute>
      <aname>SID</aname>
      <avalue>9905</avalue>
    </attribute>
    <attribute>
      <aname>
      SHOPNAME</aname>
      <avalue>MEMO</avalue>
    </attribute>
    <attribute>
      <aname>PID¡/aname>
      <avalue>102¡/avalue>
    </attribute>
<attribute>
  <aname>TELE</aname>
  <avalue>111333888</avalue>
</attribute>
<attribute>
  <aname>RATE</aname>
    <avalue>7</avalue>
</attribute>
<attribute>
  <aname>
  STATUS</aname>
  <avalue>NEW</avalue>
</attribute>
</attributes>
</IStatement>
```

Fig. 8. The specification of ***MO1***

consists of attributes, *IID* and *ReceiveAt*, and a sequence of elements, *rname* and *attributes*. The *rname* element represents the name of the manipulated relation. The *attributes* element represents the attributes of the inserted tuple and the corresponding value for each attribute.

The *DStatement* element consists of attributes, *DID* and *ReceiveAt*, and a sequence of elements, *rname* and *where*. The *where* element is of type *predicateUDT*. The *UStatement* element consists of attributes, *UID* and *ReceiveAt*, and a sequence of elements, *rname, set* and *where*. The *where* element is of type *predicateUDT*. The *set* element is of type *simplePredicate* and restricted to use an equal operator only. Assume the update statement modifies only one attribute.

6.2 Examples

It is assumed that the server is to execute several manipulation operations over the *shop* table. The first operation (MO1) inserts a new shop, whose id, name, tele, rate and status are 9905, MEMO, 111333888, 7 and NEW, respectively. This shop is located in Karlsruhe, whose position ID is 102. The second operation (MO2) deletes a shop tuple, whose id is 9903.

```
<DStatement DID="D5001"
receivedAt="2008-09-12T11:34:27">
<rname>shop</rname>
<where>
  <spredicate>
    <simplePredicate>
      <attribute>
        <name>SID</name>
      </attribute>
        <operator>eq</operator>
        <operand>
          <value>9903</value>
        </operand>
      </simplePredicate>
    </spredicate>
</where>
</DStatement>
```

Fig. 9. The specification of ***MO2***

```
<UStatement UID="U7001"
  receivedAt="2008-09-12T11:34:27">
  <rname>shop</rname>
  <set>
    <spredicate>
      <simplePredicate>
        <attribute>
          <name>RATE</name>
        </attribute>
        <operator>eq</operator>
        <operand>
          <value>7</value>
        </operand>
      </simplePredicate>
    </spredicate>
  </set>
<where>
  <spredicate>
    <simplePredicate>
      <attribute>
        <name>PID</name>
      </attribute>
      <operator>eq</operator>
      <operand>
        <value>103</value>
      </operand>
    </simplePredicate>
  </spredicate>
</where>
</UStatement>
```

Fig. 10. The specification of ***MO3***

The third operation (MO3) updates the rate of the shop tuples, which is located at 103, to seven. Figure 8 illustrates the XReAl specification for the insert operation. *IStatement* of the insert operation consists of attributes, *IID* whose value is *I3001* and *receivedAt* that determines the receipt time. There are six elements of type *attribute* that specify the name and value of an attribute, such as *SID* and *9905* for the first attribute of the insert statement. Figure 9 illustrates the XReAl specification for the delete operation. *DStatement* of the delete operation consists of attributes, *DID* whose value is *D5001* and *receivedAt* that determines the receipt time.

There is a *where* element under *DStatement* that formalizes the where clause of the delete statement, which is *SID = 9903*. Figure 10 illustrates the XReAl specification for the update operation . *UStatement* of the update operation consists of attributes, *UID* whose value is *U7001* and *receivedAt* that determines the receipt time. The *set* element formalizes the set clause of the update statement, which is *RATE = 7*. The *where* element of *UStatement* formalizes the where clause of the update operation, which is *PID = 103*.

7 Realizing the XReAl Model within DBSs

Modern DBSs support XML data management by providing an XML data type with storage and retrieval support. The XReAl specifications are represented as well-formed XML documents that could be stored in an attribute of XML type. This XML document could be validated against an XML Schema.

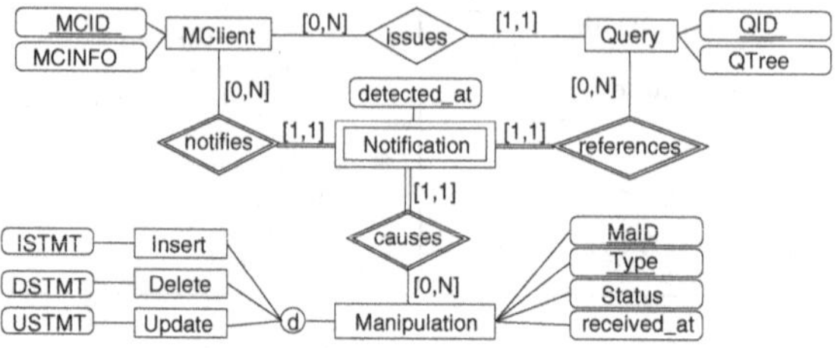

Fig. 11. The ER diagram of the XReAl Repository

7.1 The XReAl Repository Structure

The XReAl repository is based on a relational database schema, in which XML type is supported to store well-formed and validated XML documents. Figure 11 depicts the database schema of the XReAl repository. The schema consists of four fundamental relations, *mclient, query, manipulation*, and *notification.*

The relations, *mclient* and *query*, consist of a primary key attribute (*MCID* and *QID*) and an attribute of XML type (*MCINFO* and *QTree*). Each manipulation operation has an identification number and is classified into three types, *insert, delete*, and *update.* The attributes *MaID* and *Type* store the identification number and the type of the manipulation operation. Both attributes represent the primary key of the relation. Manipulation operations are classified also into two status new (N) or tested (T) operations. The *Status* attribute represents the status of an operation. The time at which the operation is received is to be stored into the *received_at* attribute. The *ISTMT, DSTMT* and *USTMT* attributes are of XML type and store XML documents representing XReAl specification for *insert, delete* or *update* operations respectively. The content of the attributes of XML type is to be validated by the XML schema of the XReAl model.

7.2 Repository Maintenance

The XReAl specifications are to be maintained (modified or queried) as any other XML documents using XQuery scripts. Modern DBSs provide means for maintaining the XML documents. In particular, DB2, which was adopted in this research, supports both the XQuery language and the XQuery update facilities provided by W3C. Moreover SQL/XML language is also supported by DB2. Such language provides support for querying the relational data of the application and the XReAl specifications. That assists in providing a unified management environment within the DBS for *CAMIS.*

8 Evaluation

We have utilized DB2 Express-C 9.5 and the Sun Java 1.6 language to implement XReAl, and built-in functions within DB2 for update notification and the context-aware query processing. This section outlines the fundamental ideas of these functions, and shows our empirical results of the update notification.

8.1 Update Notification

Based on the XREAL model, we have developed DBS-built-in method [8] that detects the relevance of manipulation operations over multi-set semantics of the relational data model. The modified data is specified by a manipulation operation, which is formalized using XREAL. Also, the cached data is specified by a query, which is formalized using XREAL.

The main idea of detecting the relevancy of an operation is to check the intersection between the specifications of the operation and query. A non-empty intersection means that there is cached data affected by the operation. Consequentially, the operation is a relevant update. For more details concerning our method for update notification based on XREAL, the reader is referred to [8].

As shown in Figure 11, a manipulation operation might cause notification(s) to be sent to mobile clients issuing queries, whose cached result intersects with data affected by the manipulation operation. The *notification* relation shown in Figure 11, consists of the attributes, *MCID, QID*, (*MaID, Type*) and *detected_at* that represents the time at which the notification is detected. The tuples of the *notification* relation are to be inserted as a result of testing the intersections between cached and modified data.

8.2 Context-Aware Query Processing

The XREAL specifications of the contextual information is the base for processing any context-aware query. Our main idea is to represent the context-aware semantics using relational algebra operations. The specifications of the contextual information and a query is used to generate an instance of this query according to the current context(s) of the user, who issued this query. This instance is generated by replacing relatives attributes with its corresponding values from the context of the user.

The query *NCQ* shown in Figure 1.B is an example for such process. Figures 5 and 6 show part of the specification of the instance query. Finally, a SQL query is generated from such instance and executed using the DBS, which at the same time manages the relational data of the application, in this case shops database. Our context-aware query processor is in-progress.

We are implementing the processor as built-in DBS function supported with Java-stored procedures. Currently, we are supporting context-aware queries based on location specified using relative position, such as postal_code. More advanced context-aware functions, such as *close to* and *towards*, are to be supported.

8.3 Experimental Results

Our experiments were done on a standard PC running Ubuntu 8.04 Linux (Intel(R) Core(TM)2 Duo CPU @ 2.20 GHz with 2 GB of RAM). Figure 12 illustrates the time consumption for registering queries on the server and for checking the relevance of insert, update and delete operations. As shown in Figure 12, our method is scalable to the number of queries registered in the systems. Moreover, the maximum required time for checking the relevancy of a manipulation operation to 16,384 related queries is approximately 50 seconds.

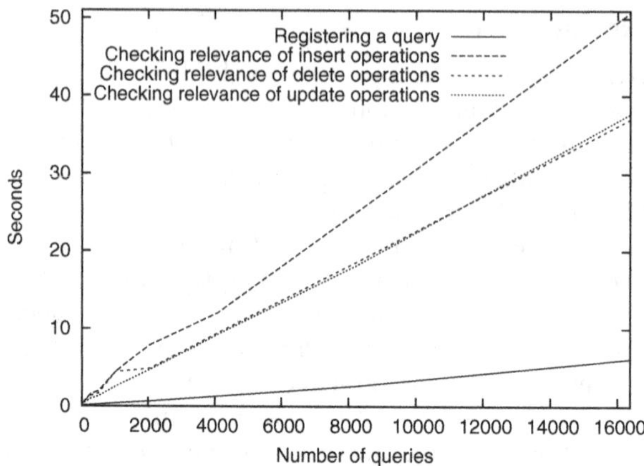

Fig. 12. Evaluation of time consumption

9 Conclusions and Outlook

This paper has presented an XML model called XReAl (XML-based Relational Algebra) that assists DBSs in extending their capabilities to support context-aware queries and cache management for mobile environments. XReAl models the contextual information related to mobile clients, queries issued by these clients and manipulation operations, such as insert, delete and update.

The main advantages of XReAl are inherited from the use of XML, such as: 1) flexibility in exchange and sharing the XReAl specification, 2) high compatibility in representing relational algebra query trees, and 3) seamless integration of XReAl management into DBS. The third point leads to performance improvement due to avoiding several middle-wares introduced to support the advanced management of CAMIS, such as update notifications and context-aware query processing.

The presented work is part of a continuous research project aiming at developing a framework for advanced query and cache management in CAMIS based on DBSs. The development of our proposed context-aware query processor is in-progress. There is a need to extend relational algebra to represent context-aware functions, such as *close to*, *around*, *towards*, and *approaching*.

References

1. Dar, S., Franklin, M.J., Jónsson, B., Srivastava, D., Tan, M.: Semantic Data Caching and Replacement. In: Proc. of 22nd International Conference on Very Large Data Bases (VLDB 1996), September 1996, pp. 330–341. Morgan Kaufmann, San Francisco (1996)
2. Elmasri, R., Shamkant, B.N.: Fundamentals of Database Systems. Addison Wesley, Reading (2007)

3. Höpfner, H.: Relevanz von Änderungen für Datenbestände mobiler Clients. VDM Verlag Dr. Müller, Saarbrücken (2007) (in German)
4. Jagadish, H.V., Lakshmanan, L.V.S., Srivastava, D., Thompson, K.: TAX: A Tree Algebra for XML. In: Ghelli, G., Grahne, G. (eds.) DBPL 2001. LNCS, vol. 2397, pp. 149–164. Springer, Heidelberg (2002)
5. Korkea-Aho, M.: Context-aware applications survey. Technical report, Department of Computer Science, Helsinki University of Technology (2000)
6. Lee, K.C.K., Leong, H.V., Si, A.: Semantic query caching in a mobile environment. ACM SIGMOBILE Mobile Computing and Communications Review 3(2), 28–36 (1999)
7. Magnani, M., Montesi, D.: XML and Relational Data: Towards a Common Model and Algebra. In: IDEAS 2005: Proceedings of the 9th International Database Engineering & Application Symposium, pp. 96–101. IEEE Computer Society Press, Washington (2005)
8. Mansour, E., Höpfner, H.: An Approach for Detecting Relevant Updates to Cached Data Using XML and Active Databases. In: 12th International Conference on Extending Database Technology (EDBT 2009) (2009)
9. Mansour, E., Höpfner, H.: Towards an XML-Based Query and Contextual Information Model in Context-Aware Mobile Information Systems. In: The MDM Workshop ROSOC-M (2009)
10. Ren, Q., Dunham, M.H., Kumar, V.: Semantic Caching and Query Processing. IEEE Trans. on Knowl. and Data Eng. 15(1), 192–210 (2003)

Answering Multiple-Item Queries in Data Broadcast Systems*

Adesola Omotayo, Ken Barker, Moustafa Hammad, Lisa Higham, and Jalal Kawash

Department of Computer Science,
University of Calgary,
Calgary, Alberta, Canada
{adesola.omotayo,kbarker,hammad,higham,jkawash}@ucalgary.ca

Abstract. A lot of research has been done on answering single-item queries, only a few have looked at answering multiple-item queries in data broadcast systems. The few that did, have proposed approaches that are less responsive to changes in the query queue. It is not immediately clear how single-item scheduling algorithms will perform when used in answering pull-based multiple-item queries. This paper investigates the performance of existing single-item scheduling algorithms in answering multiple-item queries in pull-based data broadcast systems. We observed that Longest Wait First, a near-optimal single-item data scheduling algorithm, has been used in environments where users' data access pattern is skewed. This paper also investigates the performance of Longest Wait First under various user access patterns. We propose $\mathcal{Q}$LWF: an online data broadcast scheduling algorithm for answering multiple-item queries in pull-based data broadcast systems. For the purpose of comparison with $\mathcal{Q}$LWF, we adapted existing pull single-item algorithm, push single-item algorithm, and push multiple-item algorithm to answer multiple-item queries in pull environments. Results from extensive sets of experiments show that $\mathcal{Q}$LWF has a superior performance compared with the adapted algorithms.

Keywords: Data Management, Mobile Computing, Data Broadcast System, Data Scheduling, Query Processing.

1 Introduction

The increased need for information and the popularity of portable computing devices (e.g. pocket PCs, smart phones and PDAs) with communication capability have paved the way for broadcast-based multiple-item query answering applications. Examples of such applications include stock trading systems, airline reservation systems, and world sport tickers. An example of a multi-item query in the Summer Olympics domain is: $Q = \{$*field hockey, table tennis, football*$\}$,

* This research is partially supported by the Natural Science and Engineering Research Council (NSERC) and Alberta's Informatics Circle of Research Excellence (iCORE).

A.P. Sexton (Ed.): BNCOD 2009, LNCS 5588, pp. 120–132, 2009.

which represents a query issued from a pocket PC for the results of concurrently running competitions in field hockey, table tennis and football.

There are three techniques for accomplishing data broadcasting, namely, push, pull, and push-pull. In a *push-based* data broadcast system, the server repetitively broadcasts a set of data items according to a pre-determined pattern (*broadcast schedule*). Users only need to tune in and listen to the broadcast channel in order to retrieve data items of interest. In a *pull-based* data broadcast system, users initiate the broadcast process. Users prompt the server for data by making specific requests and the server broadcasts data items based on users' requests. Using the *push-pull* method, items which are of interest to the majority of users are repeatedly broadcast in a push manner. Any item of interest which is not in the broadcast schedule is explicitly requested by the user in a pull manner. Users connect to information sources using portable computing devices with limited battery power. Therefore, it is important that data items are scheduled such that users' wait time is minimized and battery power is conserved.

A data server may schedule items for broadcast in response to multiple-item queries in two ways. First, by using the *Query Selection Method* in which the server selects a query from the queue and sequentially broadcasts all individual data items accessed by the query. Second, by using the *Item Selection Method* in which the server selects a data item for broadcast from amongst those accessed by the queries pending result delivery. The Item Selection Method has an advantage of being more responsive to changes in the query queue state (caused by query arrivals and the broadcast of data items) than the Query Selection Method.

A large body of work exists on the scheduling of data for broadcast when users request single items [1,2,3,4]. The few research works that investigated broadcast scheduling for multiple-item queries [5,6,7,8,9,10] propose approaches that are less responsive to changes in query queue. Taking the Item Selection approach, this paper proposes $\mathcal{Q}$LWF, a broadcast scheduling algorithm for multiple-item queries that outperforms the existing ones in minimizing users' wait time. $\mathcal{Q}$LWF determines the items to broadcast and the order in which they are broadcast in a single step. The key contributions in this paper are:

1. $\mathcal{Q}$LWF, a data scheduling algorithm for answering multiple-item queries in pull data broadcast systems is developed. $\mathcal{Q}$LWF schedules for broadcast a data item meeting the following conditions: a) has a high frequency of occurrence among the queries pending result delivery; b) the total wait time of the queries in which the item occurs is large; and (c) the broadcast of the item will cause a large number of queries to be answered. $\mathcal{Q}$LWF adapts dynamically to changes in query queue.
2. The performance of existing pull single-item scheduling algorithm, push single-item scheduling algorithms, and push multiple-item algorithm in answering multiple-item queries is investigated.
3. The performance of LWF under various user access patterns is investigated and it is established that LWF performs best when there is a high level of correlation among users' data requests.

4. The existing claim that LWF is only near-optimal [11] is validated by showing that QLWF has the same performance as LWF in single-item query environments, but outperforms LWF in multiple-item query environments.
5. Round Robin is experimentally shown to have the best maximum wait time amongst known algorithms.
6. The benefits of QLWF are experimentally evaluated and it is shown that QLWF has a superior performance compared with some existing single-item algorithms that were adapted for answering multiple-item requests.

The balance of this paper is organized as follows: Section 2 presents the performance metrics and surveys related work. Section 3 presents the proposed algorithm and the adaptation of existing single-item algorithms to answer multiple-item queries. Sections 4 and 5 present the methodology of the experiment and performance study. Finally, Section 6 summarizes the paper.

2 Preliminaries

2.1 Performance Metrics

Two metrics are of importance in the evaluation of service quality received by users from data broadcast systems. First, the *average wait time*, which expresses the average elapsed time between the arrival of a user's query at a data server and the broadcast of a response. Second, the *maximum wait time*, which is the maximum amount of time that any user's query stays in the query queue before being answered. The average wait time is a measure of performance from the overall system viewpoint while the maximum wait time is a measure of performance from an individual user's viewpoint. Previous works on data broadcast scheduling [2,3,12,13,4,14] also used average and maximum wait times to measure performance, thereby providing a basis for comparison with our work.

2.2 Related Work

Most of the previous works in the area of pull-based data broadcasting consider processing single-item queries. Wong [15] propose the Longest Wait First (LWF) algorithm, which selects the item with the largest total wait time. LWF has been identified [2,12,15] as the best scheduling algorithm for single-item queries. To understand how a scheduling algorithm designed for answering single-item queries may perform with multiple-item queries, we adapt LWF in Section 3.1 and present the performance results in Section 5. The proposal by Prabhu *et al.* [8] for answering transactional multiple-item requests in pull data broadcast environments is query-selection based, thus less responsive to the query queue.

Very little research in push-pull data broadcast focuses on processing multiple-item queries [10]. Some of the existing works focus on multiple-item queries in push-based data broadcasting environments [5,16,6,7,17,18,9]. The existing works assume that user access patterns do not change, hence may suffer from unresponsiveness when used to answer multiple-item queries on demand. The

Frequency/Size with Overlapping (FSO) method, proposed by Lee *et al.* [7,18], selects for broadcast the query with the highest ratio of frequency to the number of items accessed by the query. We adapt FSO for use in pull-based environments in Section 3.1 and evaluate the performance in Section 5.

3 Data Scheduling Solutions

This section explores the use of some of the existing algorithms and the proposed QLWF in answering multiple-item queries in a pull data broadcast system. Each data item accessed by the pending queries is scored by computing Val as described below. The data item with the maximum score is selected for broadcast.

3.1 Existing Approaches

Frequency/Size with Overlapping (FSO). FSO has been proposed as an algorithm for choosing the multiple-item queries whose individual items are inserted in the broadcast schedule of push-based broadcast systems. FSO starts with an empty broadcast cycle that can grow to a maximum length, say BC. The query whose number of data items does not cause the length of the broadcast cycle to be greater than BC and has a maximum frequency/size value is selected for insertion into the broadcast schedule. The data items inserted into the broadcast schedule are removed from the remaining queries. The process is repeated until the broadcast cycle is filled or until there is no query whose accessed items can completely fit into the remaining broadcast cycle. FSO is not a starvation-free algorithm since an unpopular query may never be answered.

In adapting FSO for use in pull-broadcast environments, the data server selects one query, say Q, to broadcast and does not make another query selection until all the individual data items accessed by Q have been broadcast one at a time.

The query queue of length k is now presented in terms of Q and Z, where

$Q = \{Q_1, \ldots, Q_n\}$ is the set of distinct pending multiple-item queries
Z_i is the multiplicity of Q_i for all $1 \le i \le n$ such that
$k = \sum_{i=1}^{n} Z_i$

For each query Q_i in the query queue:

$$Val_{FSO}(Q_i) = \frac{Z_i}{\mid Q_i \mid} \tag{1}$$

Round Robin. Round Robin is an existing algorithm that has been used in answering single-item queries in push-based data broadcast environments. A data server using the Round Robin algorithm consecutively broadcasts, in a cyclic manner, and with equal priority, each item in the items database. The length of time that a query must wait before being answered is determined at the time of arrival of the query. Round Robin is a starvation-free algorithm.

Longest Wait First (LWF). LWF is a well known algorithm proposed for answering single-item queries in pull-based data broadcast environments. The algorithm selects the data item that has the largest total wait time, that is, the sum of the time that all pending requests for the item have been waiting. If the arrival of requests for each data item has a Poisson distribution, LWF broadcasts each item with frequency roughly proportional to the square root of the mean arrival rate of requests for the item, which is essentially optimal[19,11]. All the experimental comparisons of the most natural algorithms have identified LWF as the best data scheduling algorithm for answering single-item queries [2,12,15]. In fact, LWF is known to have a near-optimal 6-speed $O(1)$-competitive worst case performance [11]. The 6-speed $O(1)$-competitiveness of LWF means that LWF has the property that $\max_i \frac{LWF_6(i)}{OPT_1(i)} \leq 1$, where $LWF_6(i)$ is the average wait time of LWF with a 6-speed processor on input i, and $OPT_1(i)$ is the average wait time of the adversarial scheduler for i with a 1-speed processor. Simply put, LWF is slower than an optimal algorithm by a constant factor of 6.

Suppose the query queue contains multiple-item user-issued queries $Q_1, \ldots, Q_k$ that are yet to be fully answered. Val_{LWF} is now defined as an item scoring function for multiple-item querying in pull-based broadcast environment.

For each data item i in the item database D:

$$\begin{aligned} Val_{LWF}(i) &= \sum_{j=1}^{k} (t - t_j) \times X_j^i \qquad X_j^i = \begin{cases} 1 \text{ if } i \in Q_j \\ 0 \text{ otherwise} \end{cases} \\ &= t \sum_{j=1}^{k} X_j^i + \sum_{j=1}^{k} -X_j^i t_j \end{aligned} \tag{2}$$

where Q_j is a user-issued query, t is the current time and t_j is the arrival time of Q_j, and k is the number of users' queries that are yet to be fully answered. The arrival time of a data item that is being requested in a query is the arrival time of the query itself. Thus, the wait time of an item, say i, requested in Q_j is $t - t_j$. The score of i (i.e. $Val_{LWF}(i)$) is the sum of the wait times of all pending requests for i. LWF is a starvation-free algorithm.

3.2 Our Approach: $\mathcal{Q}$LWF

$\mathcal{Q}$LWF is a new data scheduling algorithm proposed in this paper to answer multiple-item queries. Unlike existing multiple-item selection methods, which are query-selection methods and thus less responsive to changes in the query queue, $\mathcal{Q}$LWF is an item-selection method, which selects one item at a time for broadcast. In selecting items for broadcast, $\mathcal{Q}$LWF considers the relationships between the data items being requested in the pending queries, queries' waiting time, data items popularity and the fractions of queries' completed. This enables $\mathcal{Q}$LWF to achieve lower user wait time than existing algorithms as shown in Section 5.

LWF is augmented with $\mathcal{Q}$, which is a normalized utility of items in the queries. $\mathcal{Q}$ is essentially a weight factor for LWF to give a higher priority to the broadcast of a data item that will cause the largest number of pending queries to be fully (or almost fully) answered. Therefore, $\mathcal{Q}$LWF selects an item for broadcast because the queries in which the item appears have the largest aggregate wait time, the item is popular among the pending queries, or the item will cause a large number of queries to be fully (or almost fully) answered. This enhances LWF as will be demonstrated in Section 5. $\mathcal{Q}$LWF is starvation-free.

Suppose the query queue contains multiple-item user-issued queries $Q_1, \ldots, Q_k$ that are yet to be fully answered. We now define $Val_{\mathcal{Q}LWF}$ as an item scoring function for multiple-item querying in pull-based broadcast environment.

For each data item i in the item database D:

$$Val_{\mathcal{Q}LWF}(i) = Val_{LWF}(i) \times \mathcal{Q}(i) \tag{3}$$

where: $Val_{LWF}(i)$ is defined by Equation 2. If there is no pending requests for i, then

$$\mathcal{Q}(i) = 0 \tag{4}$$

otherwise,

$$\mathcal{Q}(i) = \frac{\sum_{j=1}^{k}\left(\frac{1}{|Q_j|}X_j^i\right)}{\sum_{j=1}^{k}X_j^i} \qquad X_j^i = \begin{cases} 1 \text{ if } i \in Q_j \\ 0 \text{ otherwise} \end{cases} \tag{5}$$

where $\mid Q_j \mid$ is the current number of individual data items accessed by Q_j that have not been broadcast since the arrival of Q_j at the query queue, and k is the number of users' queries that are yet to be answered.

Therefore, in the absence of pending requests for i,

$$Val_{\mathcal{Q}LWF}(i) = 0 \tag{6}$$

and in the presence of at least one pending request for i,

$$\begin{aligned} Val_{\mathcal{Q}LWF}(i) &= \mathcal{Q}(i)t\sum_{j=1}^{k}X_j^i - \mathcal{Q}(i)\sum_{j=1}^{k}X_j^i t_j \\ &= t\sum_{j=1}^{k}\left(\frac{1}{\mid Q_j \mid}X_j^i\right) + \left(\sum_{j=1}^{k} -X_j^i t_j\right)\frac{\sum_{j=1}^{k}\left(\frac{1}{|Q_j|}X_j^i\right)}{\sum_{j=1}^{k}X_j^i} \end{aligned} \tag{7}$$

In environments with single-item queries, $\mid Q_j \mid = 1$ and Equation 7 becomes Equation 2. Thus, $\mathcal{Q}$LWF delivers the same performance as LWF when used in answering single-item queries, but outperforms LWF as shown in Section 5 when used in answering multiple-item queries.

4 Experimental Methodology

4.1 Experimental Setup

The simulated environment consists of a pull-based data broadcast system. The data server maintains a database containing a finite set of equally sized read-only data items (earlier work [20,21] shows how to extend data schedulers to incorporate variably-sized items). The server is modeled to retrieve and transmit one data item in a broadcast message at a time. The length of the transmission is referred to as a *broadcast tick* and ticks are used as the unit of time measure in the experiments. Users are simulated as an infinite population, which is representative of data broadcast systems with a large user base.

Simulation models of FSO, Round Robin, LWF, and QLWF) have been implemented. Our goal is to determine the quality of broadcast decisions made by a data server using the algorithms. Thus, the time taken by users' queries to arrive at the server, and the overhead associated with query handling and decision making at the server, are excluded from our study conformant with past practices [2]. In keeping with earlier studies [7,18,15], it is assumed that users do not cancel a submitted query and do not leave until the query is fully answered. A data item broadcast by the server is assumed to be received by all users with pending requests for that item (i.e., transmission errors are not consider). At the end of each simulation run, the server stopped receiving new queries, all queries pending result delivery were processed normally until all had been answered, and all queries were included in the results.

A set of predefined queries is simulated, which the users might choose to send to the server. The predefined queries were simulated as the most popular queries among the users. However, users that did not choose any of the predefined queries were simulated to send *ad-hoc* queries that they made up to the server. To ensure consistency in our experiments, the predefined query set and users' *ad-hoc* queries we simulated were stored permanently in files as test data. Separate test data files containing 10 million queries each were created for Zipf, Uniform and Potpourri distributions. All other system parameters such as the database size, and number of predefined queries, were taken into consideration in generating the test data. All user queries were read from the test data files during our experiments, thus guaranteeing input data fairness in our evaluation of the broadcast scheduling algorithms. The simulation parameters and their values are shown in Table 1.

The Potpourri distribution is a mix of Zipf distribution, Uniform distribution and random selection. This distribution is used to model queries that exhibit no user interest pattern.

Table 1. Simulation Parameters' Settings

Parameter	Range	Default
Database size	10–1000	1000 data items
Size of the predefined set of queries	10–100	100
Percent of users' queries predefined	0%–100%	80%
# of items accessed by each query	2–10	2–5
Arrival mean of queries/broadcast tick	10–2000	200
Distribution of data items across queries	Zipf, Uniform & Potpourri	Zipf with parameter 2.0

5 Experimental Results

This section provides the results of an extensive evaluation of QLWF, the scheduling algorithm proposed in Section 3.2 and various existing algorithms (FSO, Round Robin, and LWF) described in Section 3.1.

5.1 The Effects of Arrival Rate

We study the effects of query arrival rate on the performance of QLWF in comparison to other (LWF, FSO, and Round Robin) broadcast scheduling algorithms. Average and maximum wait times are used as performance measures and gather the performance results shown in Figures 1a and 1b by varying the query arrival rate. As the arrival rate of queries increases, the average wait times of QLWF, LWF, FSO, and Round Robin increase slightly before finally stabilizing. The stabilization is observed for query arrival rates in the range 500 queries/broadcast tick and higher. This implies that the average wait times of QLWF, LWF, FSO, and Round Robin are independent of query arrival rates under heavy query loads and that these algorithms are scalable to a large population of users. Further, QLWF has the lowest average wait time among the tested algorithms with the performance of QLWF being an average of 8.4%, 37.5%, and 95.1% better than those of LWF, FSO, and Round Robin, respectively.

Figure 1b shows the changes in query arrival rate do not affect the maximum wait time experienced by Round Robin as the value remains at 999 broadcast ticks (which is the database size minus one). This is because between two broadcasts of an item, Round Robin schedules all other items in the database for broadcast exactly once. At a low query load of 10 queries/broadcast tick, QLWF has the lowest maximum wait time while FSO has the highest. However, the maximum wait times of QLWF, LWF and FSO increase as query arrival rate increases, but finally converge under a query load of 2000 queries/broadcast tick. The convergence shows that the number of requests (which is the common attribute to the three algorithms) is a dominant factor in scheduling decision making while the arrival times of queries (which is the differentiating attribute between QLWF&LWF and FSO) is less important under high loads. Overall, Round Robin offers the best maximum wait time, followed by QLWF which outperforms FSO and LWF by an average of 20.87% and 2.14% respectively.

5.2 The Effects of Data Access Pattern

We also investigate QLWF, LWF, FSO, and Round Robin's sensitivity to the skewness in the frequency distribution of the data items accessed by users' queries. Figures 1c and 1d show the behaviors of the algorithms as users' data access pattern is varied. All four algorithms give their best performance in environments with highly skewed data access pattern. The average wait time increases across all four algorithms as the skewness in the data access pattern decreases. However, LWF performs better than FSO and Round Robin in environments with skewness of request patterns ranging from low to high, and in

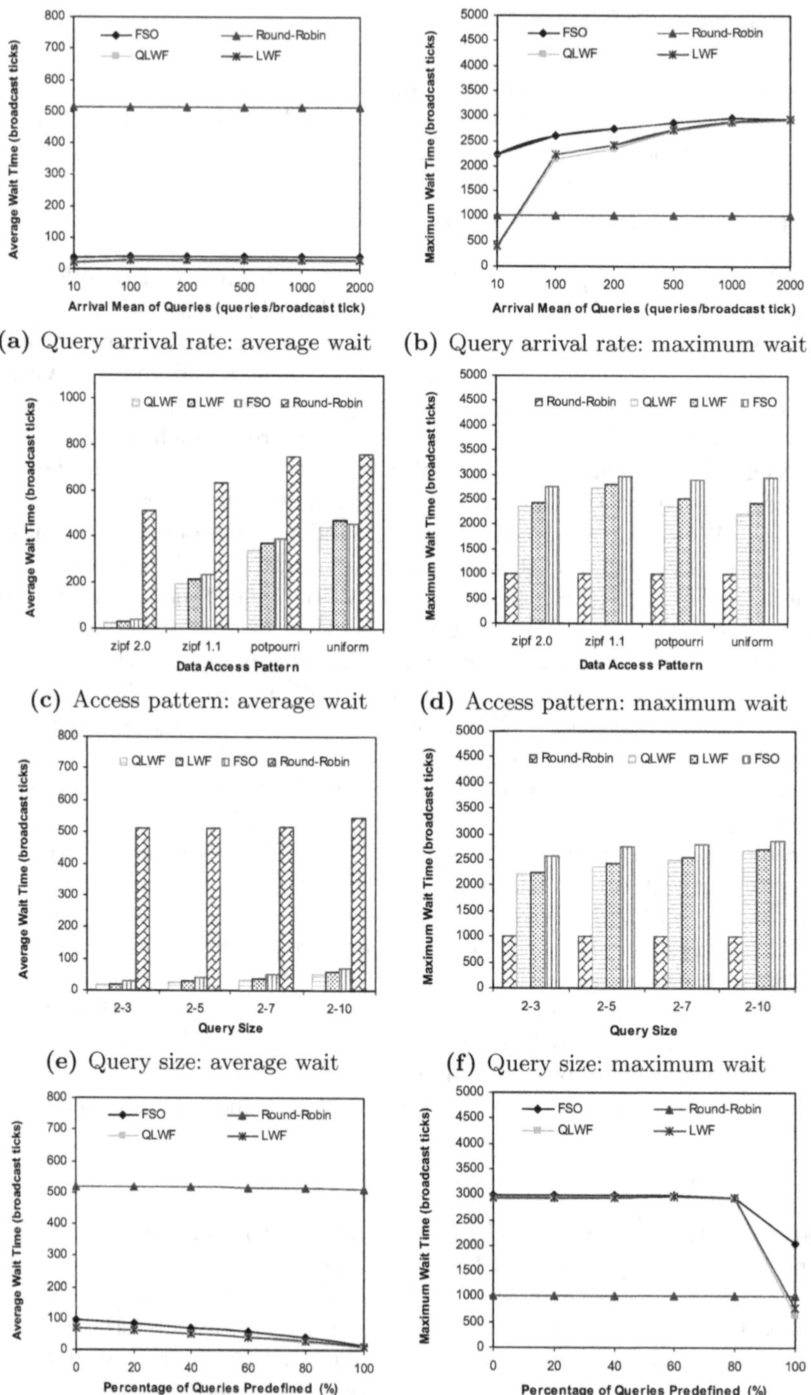

(a) Query arrival rate: average wait (b) Query arrival rate: maximum wait

(c) Access pattern: average wait (d) Access pattern: maximum wait

(e) Query size: average wait (f) Query size: maximum wait

(g) Predefined queries: average wait (h) Predefined queries: maximum wait

Fig. 1. Performance analysis of schedulers while varying query arrival rate, data access pattern, query size, and percentage of queries predefined

environments without a definite request pattern. This is a significant finding because all other studies of LWF [2,20] have been conducted in environments with a high skewness in data request patterns and the performance of LWF in other environments have not been previously examined. Overall, QLWF provides the lowest average wait time, which on the average is 7.55%, 16.97%, and 64.96% lower than those for LWF, FSO, and Round Robin, respectively.

Figure 1d shows the longest length of time that a user query waited before being answered. For the same reason given in Section 5.1, the maximum wait time of Round Robin is the same no matter the distribution of the accessed data. However, the maximum wait time for QLWF is shortest when data is uniformly accessed while it is shortest for LWF and FSO when the data access pattern is highly skewed. The longest maximum wait time for QLWF, LWF, and FSO occurs when data access is slightly skewed. Round Robin has the shortest maximum wait time followed by QLWF which is 5.15% and 16.08% better than LWF and FSO.

5.3 The Effects of Query Size

We consider the impacts of the number of individual data items accessed by multiple-item queries on the average and maximum wait times. Figure 1e shows that users' average wait time under QLWF, LWF, FSO, and Round Robin scheduling increases as the query size increases. This is expected as a query is not considered answered until all the items accessed by the query have been broadcast at least once since the arrival of the query at the server. Our proposed algorithm, QLWF has the best average wait time, which is 6.84%, 33.16%, and 93.76% lower than LWF, FSO, and Round Robin, respectively.

The maximum wait times are shown in Figure 1f. Again, the maximum wait time for Round Robin is the shortest among the tested algorithms and is not affected by the query size. Amongst the other three algorithms, QLWF demonstrates the shortest maximum wait time, which is 1.88% and 11.46% shorter than those for LWF and FSO, respectively.

5.4 The Effects of Percent of Users' Queries Predefined

The predefined queries provide an opportunity to streamline users' choices in the actual data items and the number of data items accessed in a single query. We evaluate the effects of predefined queries on users' average and maximum wait times. Figure 1g shows that the average wait time reduces as the percentage of users' queries that are predefined increases. At 0% query predefinition, all users' queries sent to the data server are *ad-hoc* and consist of any combination of 2-5 items from the database. However, at 100% query predefinition, all of the users' queries are selected from the 100 predefined queries. Therefore, the chance of overlap amongst users' data access becomes higher as the percentage of query predefinition increases. This means that at a high percentage of predefinition, the broadcast of an item has the potential to completely answer (or bring towards being completely answered) many users' queries faster. QLWF has the lowest average wait time which is 5.18%, 30.51%, and 91.75% lower than LWF, FSO, and Round Robin, respectively.

The striking feature of Figure 1h is the sharp drop in the maximum wait times of the algorithms between 80% and 100% query predefinitions. This indicates that a maximum wait time that is better than that of Round Robin is achievable if users' queries access only a small fraction of the items in the database and if there is a high degree of overlap in users' data access. The maximum wait time for Round Robin remains the same irrespective of the percentage of users' requests that are predefined. $\mathcal{Q}$LWF has a maximum wait time that is 3.32% and 12.74% lower than LWF and FSO, respectively.

5.5 The Effects of Database Size

We examine the effects of changes in database size on the average and maximum wait times of $\mathcal{Q}$LWF, LWF, FSO, and Round Robin. Figure 2a shows that the average wait time for Round Robin greatly increases as the database size increases while those for $\mathcal{Q}$LWF, LWF, and FSO only increase slightly. These results show that the average wait time for Round Robin is very sensitive to database size. However, at a size of 1500 items or more, database size ceases to impact the average wait times of $\mathcal{Q}$LWF, LWF, and FSO, implying that these algorithms are scalable with respect to database size. The lowest average wait time is given by $\mathcal{Q}$LWF and it is 7.30% 32.31%, and 79.43% lower than LWF, FSO, and Round Robin, respectively (see the chart inset within Figure 2a).

The maximum wait times of all the algorithms vary with changes in the database size as shown in Figure 2b. The maximum wait time for Round Robin, which is the lowest amongst the algorithms, matches the size of the database (Section 5.1 gives the reason). A surprising finding is the wide margin in the maximum wait time of FSO and the other algorithms for an 100-item database. This shows that FSO is more sensitive to database sizes. However, the maximum

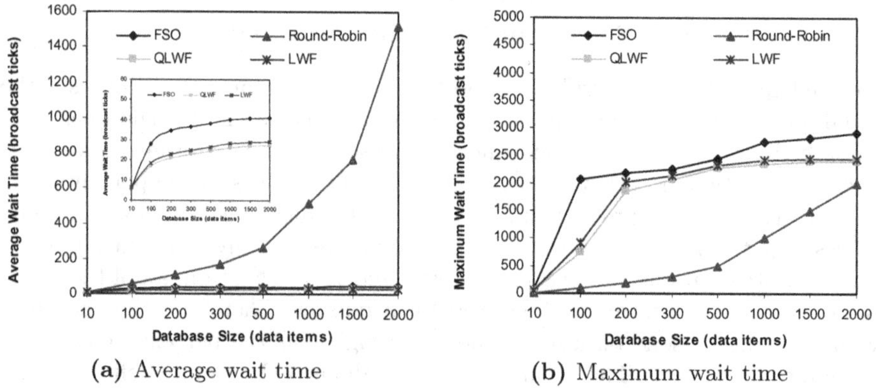

(a) Average wait time (b) Maximum wait time

Fig. 2. Performance analysis of schedulers while varying the database size

wait times for $\mathcal{Q}$LWF and LWF level off at high database sizes. $\mathcal{Q}$LWF has a maximum wait time which is 7.39% and 23.24% lower than LWF and FSO.

6 Conclusions

This paper focuses on the challenges of scheduling data items for broadcast in response to users' multiple-item queries. We propose $\mathcal{Q}$LWF, a new data scheduling algorithm. $\mathcal{Q}$LWF augments an existing algorithm LWF with $\mathcal{Q}$, which is a normalized weight factor to give a higher priority to the broadcast of a data item that will cause the largest number of pending queries to be fully (or almost fully) answered. Essentially, $\mathcal{Q}$LWF selects an item for broadcast because the queries in which the item appears have the largest aggregate wait time, the item is popular among the queries, or the item will cause a large number of queries to be fully (or almost fully) answered.

As with other algorithms, it is established that the average wait time performance of LWF under various user access patterns is best when there is a high level of correlation among users' data requests. The performance of existing single-item scheduling algorithms adapted to answer multiple-item queries relative to one another follows the same pattern for single-item queries.

$\mathcal{Q}$LWF is compared with, LWF (a single-item pull algorithm), FSO (a multiple-item push algorithm), and Round Robin (a single-item push algorithm). All our experimental results show that $\mathcal{Q}$LWF performs better than LWF, FSO, and Round Robin in terms of users' average wait time. This result is important because none of the algorithms proposed in the literature has been shown to outperform LWF. Our results also show that Round Robin provides the lowest maximum wait time amongst all the tested algorithm. $\mathcal{Q}$LWF is an important contribution because it proves that there is still an opportunity for more research in data broadcast scheduling algorithms beyond what has been achieved with LWF. The lower user wait time achieved by $\mathcal{Q}$LWF has a significant implication for power constrained hand-held devices. The users of these devices rely on them for up-to-date critical information from broadcast servers and the ability to operate the devices for a longer duration is desirable.

References

1. Acharya, S., Franklin, M., Zdonik, S.: Balancing push and pull for data broadcast. In: Proceedings of the ACM SIGMOD Conference, May 1997, pp. 183–194 (1997)
2. Aksoy, D., Franklin, M.: RxW: A scheduling approach for large-scale on-demand data broadcast. IEEE/ACM Transactions On Networking 7(6), 846–860 (1999)
3. Bartal, Y., Muthukrishnan, S.: Minimizing maximum response time in scheduling broadcasts. In: Proceedings of the 11th Annual ACM-SIAM Symposium on Discrete Algorithms, January 2000, pp. 558–559 (2000)
4. Kalyanasundaram, B., Pruhs, K., Velauthapillai, M.: Scheduling broadcasts in wireless networks. In: Proceedings of the 8th Annual European Symposium on Algorithms, September 2000, pp. 290–301 (2000)

5. Chang, Y.I., Hsieh, W.H.: An efficient scheduling method for query-set-based broadcasting in mobile environments. In: Proceedings of the 24th International Conference on Distributed Computing Systems Workshops, March 2004, pp. 478–483 (2004)
6. Chung, Y.D., Kim, M.H.: Effective data placement for wireless broadcast. Distributed and Parallel Databases 9(2), 133–150 (2001)
7. Lee, G., Lo, S.C.: Broadcast data allocation for efficient access of multiple data items in mobile environments. Mobile Networks and Applications 8(4), 365–375 (2003)
8. Prabhu, N., Kumar, V.: Data scheduling for multi-item and transactional requests in on-demand broadcast. In: Proceedings of the 6th International Conference on Mobile Data Management, May 2005, pp. 48–56 (2005)
9. Wu, G.M.: An efficient data placement for query-set-based broadcasting in mobile environments. Computer Communications 30(5), 1075–1081 (2007)
10. Yuen, J.C.H., Chan, E., yiu Lam, K.: Adaptive data broadcast strategy for transactions with multiple data requests in mobile computing environments. In: Proceedings of the 6th International Conference on Real-Time Computing Systems and Applications, December 1999, pp. 37–44 (1999)
11. Edmonds, J., Pruhs, K.: A maiden analysis of longest wait first. ACM Transactions on Algorithms 1(1), 14–32 (2005)
12. Dykeman, H., Ammar, M., Wong, J.: Scheduling algorithms for videotex systems under broadcast delivery. In: Proceedings of the IEEE International Conference on Communications, June 1986, vol. 3, pp. 1847–1851 (1986)
13. Fernandez, J., Ramamritham, K.: Adaptive dissemination of data in time-critical asymmetric communication environments. Mobile Networks and Applications 9(5), 491–505 (2004)
14. Liu, Y., Wong, J.: Deadline-based scheduling in support of real-time data delivery. Computer Networks (accepted for publication, 2007)
15. Wong, J.: Broadcast delivery. Proceedings of IEEE 76(12), 1566–1577 (1988)
16. Chung, Y.D., Kim, M.H.: QEM: A scheduling method for wireless broadcast data. In: Proceedings of the 6th International Conference on Database Systems for Advanced Applications, April 1999, pp. 135–142 (1999)
17. Liberatore, V.: Multicast scheduling for list requests. In: Proceedings of the 21st Annual Joint Conference of the IEEE Computer and Communications Societies, vol. 2, pp. 1129–1137 (2002)
18. Lee, G., Yeh, M., Lo, S., Chen, A.L.: A strategy for efficient access of multiple data items in mobile environments. In: Proceedings of the 3rd International Conference on Mobile Data Management, January 2002, pp. 71–78 (2002)
19. Ammar, M.H., Wong, J.W.: The design of teletext broadcast cycles. Performormance Evaluation 5(4), 235–242 (1985)
20. Su, C., Tassiulas, L.: Broadcast scheduling for information distribution. In: Proceedings of IEEE INFOCOM, vol. 1, pp. 109–117 (1997)
21. Vaidya, N.H., Hameed, S.: Data broadcast in asymmetric wireless environments. In: Proceedings of the 1st International Workshop on Satellite-based Information Services (November 1996)

A Study of a Positive Fragment of Path Queries: Expressiveness, Normal Form, and Minimization

Yuqing Wu[1], Dirk Van Gucht[1], Marc Gyssens[2], and Jan Paredaens[3]

[1] Indiana University, USA
[2] Hasselt University & Transnational University of Limburg, Belgium
[3] University of Antwerp, Belgium

Abstract. We study the expressiveness of a positive fragment of path queries, denoted Path$^+$, on node-labeled trees documents. The expressiveness of Path$^+$ is studied from two angles. First, we establish that Path$^+$ is equivalent in expressive power to a particular sub-fragment as well as to the class of tree queries, a sub-class of the first-order conjunctive queries defined over label, parent-child, and child-parent predicates. The translation algorithm from tree queries to Path$^+$ yields a normal form for Path$^+$ queries. Using this normal form, we can decompose a Path$^+$ query into sub-queries that can be expressed in a very small sub-fragment of Path$^+$ for which efficient evaluation strategies are available. Second, we characterize the expressiveness of Path$^+$ in terms of its ability to resolve nodes in a document. This result is used to show that each tree query can be translated to a unique, equivalent, and minimal tree query. The combination of these results yields an effective strategy to evaluate a large class of path queries on documents.

1 Introduction

XQuery [5] is a language to express queries on XML documents (i.e., node-labeled trees). In this paper, we study the expressiveness of an algebraic path query language, denoted Path$^+$, which is equivalent to a sub-language of XQuery, and wherein each query associates with each document a binary relation on its nodes. Each pair (m, n) in such a relation can be interpreted as the unique, shortest path from m to n in the queried document. Hence, whenever we talk in the paper about a path in a document, we represent it by the pair of its start- and end-node.

```
for $i in doc(...)//a/b
  for $j in $i/c/*/d[e]
    for $k in $j/*/f
return ($i, $k)
intersect
for $i in doc(...)//a/b
  for $j in $i/c/a/d
    for $k in $j/c/f
return ($i, $k)
```

Consider the XQuery query on the right. We can express such queries in an algebraic path query language which we denote as the Path$^+$ algebra. The Path$^+$ algebra allows $\emptyset$ formation, label examination, parent/child navigation, composition, first and second projections, and intersection. More precisely, the expressions of Path$^+$ are

$$E ::= \emptyset \mid \varepsilon \mid \hat{\ell} \mid \downarrow \mid \uparrow \mid E;E \mid \Pi_1(E) \mid \Pi_2(E) \mid E \cap E$$

A.P. Sexton (Ed.): BNCOD 2009, LNCS 5588, pp. 133–145, 2009.

where the primitives $\emptyset$, ε, $\hat{\ell}$, $\downarrow$, $\uparrow$ respectively return the empty set of path, the paths of length 0, the labeled paths of length 0, the parent-child paths, and the child-parent paths, and the operators ; , Π_1, Π_2, and $\cap$ denote composition, first projection, second projection, and intersection of sets of paths. Path$^+$ is fully capable of expressing the XQuery query above in an algebraic form as

$$\Pi_2(\hat{a};\downarrow);\hat{b};\downarrow;\hat{c};\downarrow;\downarrow;\hat{d};\Pi_1(\downarrow;\hat{e});\downarrow;\downarrow;\hat{f}\cap\Pi_2(\hat{a};\downarrow);\hat{b};\downarrow;\hat{c};\downarrow;\hat{a};\downarrow;\hat{d};\Pi_1(\downarrow;\hat{e});\downarrow;\hat{c};\downarrow;\hat{f}$$

XPath is a language for navigation in XML documents [6] and is always evaluated in the node-set semantics. Researchers have introduced clean algebraic and logical abstractions in order to study this language formally. Literature on the formal aspects of XPath has become very extensive, which involves full XPath as well as its fragments [8,3,14,13,10]. Research on XPath and its sub-languages has been focusing on the expressiveness [10] and the efficient evaluation of these languages [8,12]. Tree queries, also called pattern trees, are also natural to XML. They have been studied ever since XML and query languages on XML were introduced. Such studies cover areas from the minimization of tree queries [2,15] to the efficient evaluation of pattern trees [1,11,4].

However, XQuery, with its FLWR statement and nested variable bindings, is capable of combining the results of multiple XPath queries. This language feature requires that the path expressions in XQuery be evaluated in the path semantics. In this paper, we study the expressiveness of Path$^+$ from two angles.
(1) We establish that Path$^+$ is equivalent in expressive power to a particular sub-fragment of this language as well as to the class of tree queries, a sub-class of the first-order conjunctive queries defined over label, parent-child, and child-parent predicates. The translation algorithm from tree queries to Path$^+$ expressions yields a normal form for Path$^+$ expressions. Using this normal form, we can decompose a Path$^+$ query into sub-queries that can be expressed in a very small sub-fragment of Path$^+$ for which efficient evaluation strategies are available.
(2) We characterize the expressiveness of Path$^+$ in terms of its ability to resolve pairs of nodes in a document. We show pairs of nodes cannot be resolved if and only if the paths from the root of the documents to these nodes have equal length and corresponding nodes on these paths are bisimilar. This result is then used to show that each tree query can be translated to a unique, equivalent, and minimal tree query.

We conclude the paper by showing that Path$^+$ queries can be regarded as the canonical building blocks for more general path queries, such as those involving union, set difference, ancestor, and descendant operations. As such, Path$^+$ can be viewed to path queries, as SPJ queries are viewed to relational algebra queries.

2 Preliminaries

In this section, we give the definition of documents, a positive fragment of path queries, and the query language of tree queries.[1]

[1] Throughout the paper, we assume an infinitely enumerable set $\mathcal{L}$ of *labels*.

Definition 1. *A* document D *is a labeled tree* (V, Ed, λ)*, with* V *the set of nodes,* $Ed \subseteq V \times V$ *the set of edges, and* $\lambda : V \to \mathcal{L}$ *a node-labeling function.*

For two arbitrary nodes m and n in a document D, there is a unique, shortest path from m to n if we ignore the orientation of the edges. The unique node on this path that is an ancestor of both m and n will henceforth be denoted $\mathrm{top}(m, n)$.

Example 1. Figure 1 shows an example of a document that will be used throughout the paper. Notice that, in this document, $\mathrm{top}(n_8, n_{12}) = n_4$.

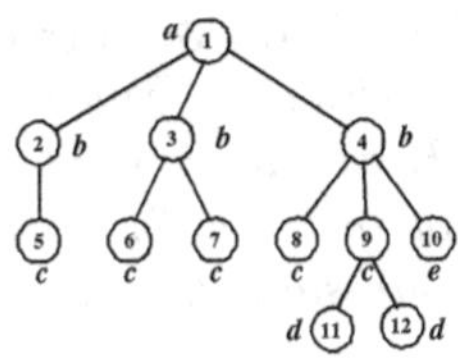

Fig. 1. An Example Document

2.1 The Positive Path Algebra

Here we give the formal definition of the positive path algebra, denoted Path^+, and its semantics.

Definition 2. *Path*$^+$ *is an algebra which consists of the primitives* $\emptyset$, ε, $\hat{\ell}$ $(\ell \in \mathcal{L})$, $\downarrow$, *and* $\uparrow$*, together with the operations composition* $(E_1; E_2)$*, first projection* $(\Pi_1(E))$*, second projection* $(\Pi_2(E))$ *and intersection* $(E_1 \cap E_2)$*.* (E, E_1, *and* E_2 *represent Path*$^+$ *expressions.)*

Given a document $D = (V, Ed, \lambda)$, the *semantics* of a Path^+ expression is a binary relation over V, defined on the right. By restricting the operators allowed in expressions, several sub-algebras of Path^+ can be defined. The following is of special interest to us: $\mathrm{Path}^+(\Pi_1, \Pi_2)$ is the sub-algebra of Path^+ where, besides the primitives and the composition operation, only the first and second projections are allowed. In addition, we will consider the algebra $\mathrm{DPath}^+(\Pi_1)$, where, besides the primitives $\emptyset$, ε, $\hat{\ell}$, $\downarrow$, and the composition operations, only the first projection is allowed. Thus, in $\mathrm{DPath}^+(\Pi_1)$ expressions, the primitive $\uparrow$ and the second projection are not allowed.

$$
\begin{aligned}
\emptyset(D) &= \emptyset;\\
\varepsilon(D) &= \{(n, n) \mid n \in V\};\\
\hat{\ell}(D) &= \{(n, n) \mid n \in V \text{ and } \lambda(n) = \ell\};\\
\downarrow(D) &= Ed;\\
\uparrow(D) &= Ed^{-1};\\
E_1; E_2(D) &= \pi_{1,4}(\sigma_{2=3}(E_1(D) \times E_2(D)));\\
\Pi_1(E)(D) &= \{(n, n) \mid \exists m : (n, m) \in E(D)\};\\
\Pi_2(E)(D) &= \{(n, n) \mid \exists m : (m, n) \in E(D)\};\\
E_1 \cap E_2(D) &= E_1(D) \cap E_2(D);
\end{aligned}
$$

Example 2. The following is an example of a Path^+ expression:

$$\Pi_1(\downarrow); \Pi_2(\hat{d}; \uparrow; \hat{c}); \Pi_2(\hat{a}; \downarrow; \hat{c}); \uparrow; \Pi_2(\Pi_1((\downarrow; \hat{a}; \downarrow) \cap (\downarrow; \downarrow; \hat{c})); \downarrow); \downarrow; \hat{c}; \Pi_1(\hat{c}; \downarrow; \hat{d}); \downarrow .$$

The semantics of this expression given the document in Figure 1 is the following set of pairs of nodes of that document: $\{(n_2, n_{12}), (n_2, n_{13}), (n_4, n_{15}), (n_4, n_{16}), (n_4, n_{17}), (n_4, n_{18})\}$. The above expression is equivalent to the much simpler $\mathrm{Path}^+(\Pi_1, \Pi_2)$ expression $\Pi_1(\downarrow; \hat{d}); \hat{c}; \uparrow; \Pi_2(\downarrow); \hat{a}; \downarrow; \Pi_1(\downarrow; \hat{d}); \hat{c}; \downarrow$. Note that the sub-expressions $\Pi_1(\downarrow; \hat{d})$ and $\hat{a}; \downarrow; \Pi_1(\downarrow; \hat{d}); \hat{c}; \downarrow$ are in $\mathrm{DPath}^+(\Pi_1)$.

2.2 Tree Queries

Here we define the tree query language, denoted **T**, and its semantics.

Definition 3. *A* tree query *is a 3-tuple* (T, s, d)*, with* T *a labeled tree, and* s *and* d *nodes of* T*, called the* source *and* destination *nodes. The nodes of* T *are either labeled with a symbol of* $\mathcal{L}$ *or with a* wildcard *denoted "*", which is assumed not to be in* $\mathcal{L}$*. To the set of all tree queries, we add* $\emptyset$*. The resulting set of expressions is denoted* $\boldsymbol{T}$.

Two symbols of $\mathcal{L} \cup \{*\}$ are called *compatible* if they are either equal or one of them is a wildcard. For two compatible symbols ℓ_1 and ℓ_2, we define $\ell_1 + \ell_2$ to be ℓ_1 if ℓ_1 is not a wildcard, and ℓ_2 otherwise. Let $P = ((V', Ed', \lambda'), s, d)$ be a tree query, and let $D = (V, Ed, \lambda)$ be a document. A *containment mapping* of P in D is a mapping $h : V' \to V$ such that

1. $\forall m', n' \in V'((m', n') \in Ed' \to (h(m'), h(n')) \in Ed$; and
2. $\forall m' \in V'(\lambda'(m') \in \mathcal{L} \to \lambda(h(m')) = \lambda'(m')$.

Observe that a containment mapping is in fact a homomorphism with respect to the parent-child and label predicates if the tree query does not contain wildcards.

We can now define the semantics of a tree query.

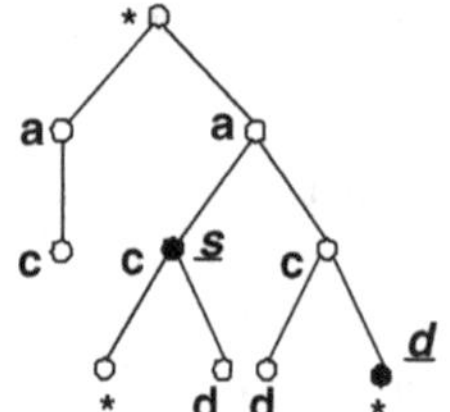

Fig. 2. An Example Tree Query

Definition 4. *Let* $P = (T, s, d)$ *be a tree query, and let* D *be a document. The semantics of* P *given* D*, denoted* $P(D)$*, is defined as the set*

$$\{(h(s), h(d)) \mid h \text{ is a containment mapping of } P \text{ in } D\}.$$

The semantics of $\emptyset$ *on* D*, i.e.,* $\emptyset(D)$*, is the empty set.*

Example 3. Figure 2 shows an example of a tree query. The semantics of this tree query given the document in Figure 1 is the set of pairs of that document exhibited in Example 2. We will show later in the paper that this tree query is actually equivalent with the Path$^+$ expression given in Example 2.

3 Equivalences of Query Languages

In this section, we show that Path$^+$, **T**, and Path$^+(\Pi_1, \Pi_2)$ are equivalent in expressive power by exhibiting a translation algorithm that translates an expression in one language to an equivalent expression in one of the other languages.

Proposition 1. *The query language* $\boldsymbol{T}$ *is at least as expressive as Path*$^+$*, and there exists an algorithm translating an arbitrary Path*$^+$ *expression into an equivalent expression of* $\boldsymbol{T}$ *(i.e., a tree query or* $\emptyset$*.)*

Proof. It is straightforward to translate the primitives to expressions of **T**.

Now, let E be a Path$^+$ expression for which P is the equivalent expression in **T**. If P equals $\emptyset$, then both $\Pi_1(E)$ and $\Pi_2(E)$ are translated into $\emptyset$. Otherwise,

```
Algorithm Merge1

Input: two disjoint labeled trees
    T1 = (V1, Ed1, λ1)
    and T2 = (V2, Ed2, λ2);
    nodes m1 ∈ V1 and m2 ∈ V2.

Output: a labeled tree or ∅.

Method:
    q = min(depth(m1, T1), depth(m2, T2))
    for k = 0, ..., q
        if the level-k ancestors of m1 and m2
          have incompatible labels, return ∅
    for k = 0, ..., q
        merge the level-k ancestors m1^k of m1
          and m2^k of m2 into a node labeled
          λ1(m1^k) + λ2(m2^k)
    return the resulting labeled tree.
```

Fig. 3. The Algorithm *Merge1*

```
Algorithm Merge2

Input: a labeled tree T = (V, Ed, λ) and
    nodes m1, m2 ∈ V;

Output: a labeled tree or ∅.

Method:
    let q1 = depth(m1, T);
    let q2 = depth(m2, T)
    if q1 ≠ q2 return ∅
    for k = 0, ..., q1 = q2
        if the level-k ancestors of m1 and m2
          have incompatible labels, return ∅
    for k = 0, ..., q1 = q2
        merge the level-k ancestors m1^k of m1
          and m2^k of m2 into a node labeled
          λ1(m1^k) + λ2(m2^k)
    return the resulting labeled tree
```

Fig. 4. The Algorithm *Merge2*

let $P = (T, s, d)$ be the tree query under consideration. Then $\Pi_1(E)$ is translated into $P_1 = (T, s, s)$, and $\Pi_2(E)$ is translated into $P_2 = (T, d, d)$.

Finally, let E_1 and E_2 be expressions for which P_1 and P_2 are the equivalent expressions in $\mathbf{T}$. If one of P_1 or P_2 equals $\emptyset$, then both $E_1; E_2$ and $E_1 \cap E_2$ are translated into $\emptyset$. Otherwise, let $P_1 = (T_1, s_1, d_1)$ and $P_2 = (T_2, s_2, d_2)$ the two tree queries under consideration.

(1) *Translation of composition.* First, apply the algorithm *Merge1* (Figure 3) to the labeled trees T_1 and T_2 and the nodes d_1 and s_2. If the result is $\emptyset$, so does the translation of $E_1; E_2$. Otherwise, let T be the returned labeled tree. Then $E_1; E_2$ is translated into the tree query $P = (T, s_1, d_2)$.

(2) *Translation of intersection.* First, apply the algorithm *Merge1* to the labeled trees T_1 and T_2 and the nodes s_1 and s_2. If the result is $\emptyset$, so does the translation of $E_1 \cap E_2$. Otherwise let T_{int} be the labeled tree returned by *Merge1*. Next, apply the algorithm *Merge2* (Figure 4) to the labeled tree T_{int} and the nodes d_1 and d_2. If the result is $\emptyset$, so does the translation of $E_1 \cap E_2$. Otherwise, let T be the labeled tree returned by *Merge2*. Then $E_1 \cap E_2$ is translated into the tree query $P = (T, s, d)$, where s is the node that resulted from merging s_1 and s_2, and d is the node that resulted from merging d_1 and d_2.

Example 4. Consider again the Path$^+$ expression given in Example 2:

$$\Pi_1(\downarrow); \Pi_2(\hat{d}; \uparrow; \hat{c}); \Pi_2(\hat{a}; \downarrow; \hat{c}); \uparrow; \Pi_2(\Pi_1((\downarrow; \hat{a}; \downarrow) \cap (\downarrow; \downarrow; \hat{c})); \downarrow); \downarrow; \Pi_1(\hat{c}; \downarrow; \hat{d}); \hat{c}; \downarrow$$

We will now translate this expression into a tree query. First write the expression as $E_1; E_2; E_3$, where

$$E_1 = \Pi_1(\downarrow); \Pi_2(\hat{d}; \uparrow; \hat{c}); \Pi_2(\hat{a}; \downarrow; \hat{c}); \uparrow$$
$$E_2 = \Pi_2(\Pi_1((\downarrow; \hat{a}; \downarrow) \cap (\downarrow; \downarrow; \hat{c})); \downarrow)$$
$$E_3 = \downarrow; \Pi_1(\hat{c}; \downarrow; \hat{d}); \hat{c}; \downarrow$$

The translation is illustrated in Figure 5. Figure B exhibits how expression E_2 is translated into a tree query. Figures B.1, B.2, B.3, B.4 and B.5 correspond to the translations of the subexpressions $\downarrow; \hat{a}; \downarrow$, $\downarrow; \downarrow; \hat{c}$, $(\downarrow; \hat{a}; \downarrow) \cap (\downarrow; \downarrow; \hat{c})$, $\Pi_1((\downarrow; \hat{a}; \downarrow) \cap (\downarrow; \downarrow; \hat{c}))$, and $\Pi_1((\downarrow; \hat{a}; \downarrow) \cap (\downarrow; \downarrow; \hat{c})); \downarrow$

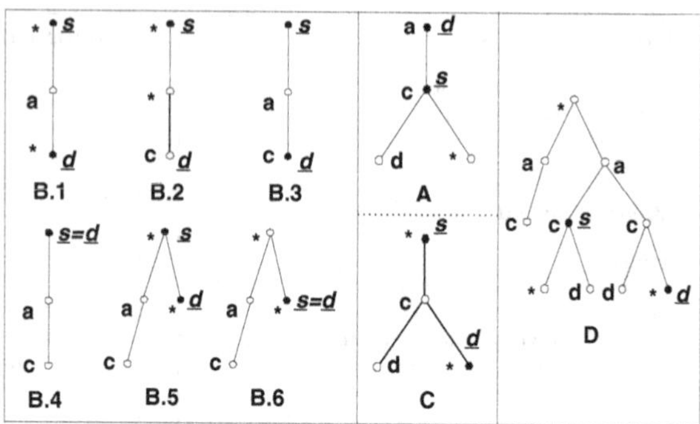

Fig. 5. Translation of the Tree Query in Exp. 4

Algorithm *Tree_to_Path*
Input: a tree query $P = (T, s, d)$;
Output: a Path$^+(\Pi_1, \Pi_2)$ expression E.
Method:
  if T is base case
    $E := basecase(T, s, d)$
  else if s is not an ancestor of d (**case 1**)
    $p :=$ the parent of s
    $T_1 :=$ the subtree of T rooted at s
    $T_2 :=$ the subtree resultant from removing all nodes in T_1 from T
    if s has no child and $\lambda(s)$ is wildcard
      $E := \uparrow; Tree\_to\_Path(T_2, p, d)$
    elseif d is the parent of s, d has no ancestor, no child other than s and $\lambda(d)$ is wildcard
      $E := Tree\_to\_Path(T_1, s, s); \uparrow$
    else $E := Tree\_to\_Path(T_1, s, s); \uparrow; Tree\_to\_Path(T_2, p, d)$
  else if s is not the root (**case 2**)
    $r :=$ the root of T
    $T_1 :=$ the subtree of T rooted at s
    $T_2 :=$ the subtree resultant from removing all strict descendants of s from T, with $\lambda(d)$ assigned to the wildcard $*$
    if s has no child and $\lambda(s)$ is wildcard
      $E := \Pi_2(Tree\_to\_Path(T_2, r, s))$
    else $E := \Pi_2(Tree\_to\_Path(T_2, r, s)); Tree\_to\_Path(T_1, s, d)$
  else if s is a strict ancestor of d (**case 3**)
    $p :=$ the parent of d
    $T_1 :=$ the subtree of T rooted at d
    $T_2 :=$ the subtree resultant from removing all nodes in T_1 from T
    if d has no child and $\lambda(d)$ is wildcard
      $E := Tree\_to\_Path(T_2, s, p); \downarrow$
    elseif s is the parent of d, s has no child other than d and $\lambda(d)$ is wildcard
      $E := \downarrow; Tree\_to\_Path(T_1, d, d)$
    else $E := Tree\_to\_Path(T_2, s, p); \downarrow; Tree\_to\_Path(T_1, d, d)$
  else if $s = d$ is the root(**case 4**)
    $c_1, \ldots, c_n :=$ all children of s
    for $i := 1$ to n **do**
      $T_i :=$ the subtree of T containing s, c_i, and all descendants of c_i, with $\lambda(s)$ assigned to the wildcard $*$
    if $\lambda(s)$ is wildcard $*$
      $E := \Pi_1(Tree\_to\_Path(T_1, s, c_1)); \ldots; \Pi_1(Tree\_to\_Path(T_n, s, c_n))$
    else $E := \Pi_1(Tree\_to\_Path(T_1, s, c_1)); \ldots; \Pi_1(Tree\_to\_Path(T_n, s, c_n)); \lambda(s)$
  return E

Fig. 6. The Algorithm *Tree_to_Path*

respectively. Figure B.6, finally, corresponds to the translation of E_2. Figures A and C exhibit the translation of the expressions E_1 and E_3 respectively (details omitted). Finally, Figure D exhibits the translation of the expression $E_1; E_2; E_3$.

Proposition 2. *The query language Path*$^+(\Pi_1, \Pi_2)$ *is at least as expressive as* $\boldsymbol{T}$*, and there exists an algorithm translating an arbitrary* $\boldsymbol{T}$ *expression into an equivalent expression of Path*$^+(\Pi_1, \Pi_2)$.

Proof. Clearly, $\emptyset$ is translated into $\emptyset$. We also have that (1) the tree query $((\{s\}, \emptyset), s, s)$ is translated to ϵ if $\lambda(s) = *$ and is translated to $\hat{\lambda(s)}$ otherwise; (2) the tree query $((\{s, d\}, \{(s, d)\}), s, d)$, where $\lambda(s) = \lambda(d) = *$, is translated to $\downarrow$; and (3) the tree query $((\{s, d\}, \{(d, s)\}), s, d)$, where $\lambda(s) = \lambda(d) = *$, is $\uparrow$. We collectively call (1), (2), and (3) the base cases, and in the algorithm exhibited in Figure 6 they are handled by the function *basecase*(T, s, d). For an arbitrary tree query $P = (T, s, d)$, a recursive translation algorithm is exhibited in Figure 6.

Example 5. Consider the tree query in Figure 5.D. Following the algorithms in Figure 6, this tree query can be translated into an equivalent Path$^+(\Pi_1, \Pi_2)$ expression: $\Pi_1(\downarrow); \Pi_1(\downarrow; \hat{d}); \hat{c}; \uparrow; \Pi_2(\Pi_1(\downarrow; \Pi_1(\downarrow; \hat{c}); \hat{a}); \downarrow); \hat{a}; \downarrow; \Pi_1(\downarrow; \hat{d}); \hat{c}; \downarrow$.

We can now summarize Propositions 1 and 2.

Theorem 1. *The query languages Path*$^+$*,* $\boldsymbol{T}$ *and Path*$^+(\Pi_1, \Pi_2)$ *are all equivalent in expressive power, and there exist translation algorithms between any two of them.*

4 Normal Form for Expressions in the Path$^+$ Algebra

Normalization is frequently a critical step in rule-based query optimization. It serves the purpose of unifying queries with the same semantics, detect containment among sub-queries, and establish the foundation for cost-based query optimization, in which evaluation plans are to be generated. As it will turn out, using this normal form, we can decompose a Path$^+$ query into sub-queries that can be expressed in DPath$^+(\Pi_1)$, a very small sub-fragment of Path$^+$ for which efficient evaluation strategies are available [7]. The full query can then be evaluated by joining the results of these DPath$^+(\Pi_1)$ expressions.

When we revisit Section 3, where the translation from queries in $\mathbf{T}$ to expressions in XPath$^+(\Pi_1, \Pi_2)$ is described, we observe that the result of the translation is either $\emptyset$, or ϵ, or has the following general form (composition symbols have been omitted for clarity):

$$C_{u_m} \uparrow \cdots \uparrow C_{u_1} \uparrow C_{top} \downarrow C_{d_1} \downarrow \cdots \downarrow C_{d_n},$$

where (1) $m \geq 0 \wedge n \geq 0$; (2) the C_i expressions, for $i \in u_1, \cdots, u_m, d_1, \cdots, d_n$, are of the form $[\Pi_1(D)]^*[\hat{l}]^?$, where the D expressions are DPath$^+(\Pi_1)$ expressions in the normal form; and (3) C_{top} is of the form $[\Pi_2(D)][\Pi_1(D)]^*[\hat{l}]^?$ where D is an expression in DPath$^+(\Pi_1)$ in the normal form. Observe that in particular, there are no $\cap$ operations present in the normal form, and that there appears at most one Π_2 operation. We say that a Path$^+(\Pi_1, \Pi_2)$ expression of this form is in *normal form*.

Example 6. Reconsider Example 2. Clearly, the expression

$$\Pi_1(\downarrow); \Pi_2(\hat{d}; \uparrow; \hat{c}); \Pi_2(\hat{a}; \downarrow; \hat{c}); \uparrow; \Pi_2(\Pi_1((\downarrow; \hat{a}; \downarrow) \cap (\downarrow; \downarrow; \hat{c})); \downarrow); \downarrow; \Pi_1(\hat{c}; \downarrow; \hat{d}); \hat{c}; \downarrow$$

is not in normal form (e.g., note that this expression contains an intersection operation and multiple occurrences of the Π_2 operation). In Example 4, we exhibited how this expression is translated in the tree query shown in Figure 2. In Example 5, we exhibited how this tree query is translated into the Path$^+(\Pi_1, \Pi_2)$ expression

$$\underbrace{\Pi_1(\downarrow); \Pi_1(\downarrow; \hat{d}); \hat{c}}_{C_{u_1}}; \uparrow; \underbrace{\Pi_2(\Pi_1(\downarrow; \Pi_1(\downarrow; \hat{c}), \hat{a}); \downarrow); \hat{a}}_{C_{top}}; \downarrow; \underbrace{\Pi_1(\downarrow; \hat{d}); \hat{c}}_{C_{d_1}}; \downarrow \underbrace{\quad}_{C_{d_2}}$$

This expression is in normal form, with the key ingredients identified below the expression.

We have the following theorem.

Theorem 2. *The* tree-to-path *algorithm of Figure 6 translates each tree query into an equivalent Path$^+(\Pi_1, \Pi_2)$ expression which is in normal form.*

Proof. $\emptyset$ is translated into the expression $\emptyset$. The tree query with a single node labeled with a wildcard is translated into the expression ϵ.

Case 1 of the translation algorithm deals with the generation of the upward fragment $C_{u_m} \uparrow \cdots \uparrow C_{u_1} \uparrow$ in the normal form expression; Case 2 deals with generation of the optional $\Pi_2()$ expression C_{top}; Case 3 deals with the generation of the downward fragment $\downarrow C_{d_1} \downarrow \cdots \downarrow C_{d_n}$ in the normal form expression; and Case 4 deals with the generation of the expression $[\Pi_1(D)]^*[\hat{l}]^?$ that is associated with a node in the tree query.

5 Resolution Expressiveness

So far, we have viewed Path$^+$ as a query language in which an expression associates to every document a binary relation on its nodes representing all paths in the document defined by that expression. We have referred to this view as the *query-expressiveness* of Path$^+$. Alternatively, it can be viewed as language in which, given a document and a pair of its nodes, one wants to navigate from one node to the other. From this perspective, it is more meaningful to characterize the language's ability to distinguish a pair of nodes or a pair of paths in the document, which we will refer to as the *resolution-expressiveness* of Path$^+$.

In this section, we first establish that two nodes in a document cannot be resolved by a Path$^+$ expression if and only if the paths from the root of that document to these nodes have equal length, and corresponding nodes on these paths are bisimilar, a property that has been called *1-equivalence* in [10]. The proof has the same structure as the proofs of similar properties for other fragments of Path in [10]. Next, we extend this result to the resolving power of Path$^+$ to pair of paths in a document.

We first make precise what we mean by indistinguishability of nodes.

Definition 5. *Let m_1 and m_2 be nodes of a document D.*

- *m_1 and m_2 are* expression-related *(denoted $m_1 \geq_{exp} m_2$) if, for each Path$^+$ expression E, $E(D)(m_1) \neq \emptyset$ implies $E(D)(m_2) \neq \emptyset$, where $E(D)(m_1)$ and $E(D)(m_2)$ refer to the sets $\{n \mid (m_1, n) \in E(D)\}$ and $\{n \mid (m_2, n) \in E(D)\}$, respectively.*
- *m_1 and m_2 are* expression-equivalent *(denoted $m_1 \equiv_{exp} m_2$) if $m_1 \geq_{exp} m_2$ and $m_2 \geq_{exp} m_1$.*

As already announced, we intend to show that the semantic notion of expression-equivalence coincides with the syntactic notion of 1-equivalence. Before we can give the formal definition of 1-equivalence of nodes, we need a few intermediate definitions. First, we define downward 1-relatedness of nodes recursively on the height of these nodes:

Definition 6. *Let m_1 and m_2 be nodes of a document D. Then m_1 and m_2 are* downward 1-related *(denoted $m_1 \geq^1_\downarrow m_2$) if and only if*

1. *$\lambda(m_1) = \lambda(m_2)$; and*
2. *for each child n_1 of m_1, there exists a child n_2 of m_2 such that $n_1 \geq^1_\downarrow n_2$.*

We now bootstrap Definition 6 to 1-relatedness of nodes, which is defined recursively on the depth of these nodes:

Definition 7. *Let m_1 and m_2 be nodes of a document D. Then m_1 and m_2 are* 1-related *(denoted $m_1 \geq^1 m_2$) if*

1. *$m_1 \geq^1_\downarrow m_2$; and*
2. *if m_1 is not the root, and p_1 is the parent of m_1, then m_2 is not the root and $p_1 \geq^1 p_2$, with p_2 the parent of m_2.*

Finally, we are ready to define 1-equivalence of nodes:

Definition 8. *Let m_1 and m_2 be nodes of a document D. Then m_1 and m_2 are* 1-equivalent *(denoted $m_1 \equiv^1 m_2$) if and only if $m_1 \geq^1 m_2$ and $m_2 \geq^1 m_1$.*

We can now establish that two nodes in a document cannot be resolved by a Path$^+$ expression if and only if the paths from the root of that document to these nodes have equal length, and corresponding nodes on these paths are bisimilar, a property that has been called *1-equivalence* in [10].

Theorem 3. *Let m_1 and m_2 be nodes of a document D. Then, $m_1 \equiv_{exp} m_2$ if and only if $m_1 \equiv^1 m_2$.*

Obviously, two nodes are bisimilar (called *downward 1-equivalent* in [10]) if they are downward 1-related in both directions. For the purpose of abbreviation, we extend 1-relatedness to pairs of nodes, as follows. Let m_1, m_2, n_1, and n_2 be nodes of a document D. We say that $(m_1, n_1) \geq^1 (m_2, n_2)$ whenever $m_1 \geq^1 m_2$, $n_1 \geq^1 n_2$, and $\text{sig}(m_1, n_1) \geq \text{sig}(m_2, n_2)$. Next, we extend this result to the resolving power of Path$^+$ to pair of paths in a document. The following theorem now states the the main result about the resolution expressiveness of Path$^+$.

Theorem 4. *Let m_1, m_2, n_1, n_2 be nodes of a document D. Then, the property that, for each Path$^+$ expression E, $(m_1, n_1) \in E(D)$ implies $(m_2, n_2) \in E(D)$ is equivalent to the property $(m_1, n_1) \geq^1 (m_2, n_2)$.*

The theorem states that when we find a pair (m_1, n_1) in the result of a query in Path$^+$, then we are guaranteed that any pair (m_2, m_2) such that $(m_1, n_1) \geq^1 (m_2, n_2)$, will also be in the result of the query, and vice versa.

6 Efficient Query Evaluation

Minimizing algebraic expressions and rewriting queries into sub-queries for which efficient evaluation plans are available is a practice used extensively in relational database systems. An example of this is selection push down for the selection conditions on which indices are available. It is natural that the same principle and procedure is followed in XML query processing and optimization.

6.1 Minimization of Tree Queries

The results on containment and minimization of tree queries can easily be derived using the theory developed in Section 5. In particular, each tree query can be translated to a unique, equivalent, and minimal tree query.

First, we extend the notion of containment mapping (Section 2.2). Thereto, let $P_1 = ((V_1, Ed_1, \lambda_1), s_1, d_1)$ and $P_2 = ((V_2, Ed_2, \lambda_2), s_2, d_2)$ be tree queries. A *query containment mapping* of P_1 in P_2 is a mapping $h : V_1 \to V_2$ such that

1. for all $m_1, n_1 \in V_1$, $(m_1, n_1) \in Ed_1$ implies that $(h(m_1), h(n_1)) \in Ed_2$;
2. for all $m_1 \in V_1$, $\lambda_1(m_1) \in \mathcal{L}$ implies that $\lambda_2(h(m_1)) = \lambda_1(m_1)$; and
3. $h(s_1) = s_2$ and $h(d_1) = d_2$.

Proposition 3. *Let P_1 and P_2 be tree queries. Then P_2 is contained in P_1 if and only if there is a query containment mapping of P_1 in P_2. [9].*

Now, let $P = ((V, Ed, \lambda), s, d)$ be a tree query. We define the *first reduction* of P, denoted $\text{red}_1(P)$, to be the tree query obtained from P by merging 1-equivalent nodes. For this purpose, we interpret P as a document $D = (V, Ed, \lambda')$, wherein the nodes are relabeled as follows: (1) $\lambda'(s) = \lambda(s)_s$; (2) $\lambda'(d) = \lambda(d)_d$; and (3) for all other nodes m in V, $\lambda'(m) = \lambda(m)$.

Lemma 1. *A tree query and its first reduction are equivalent.*

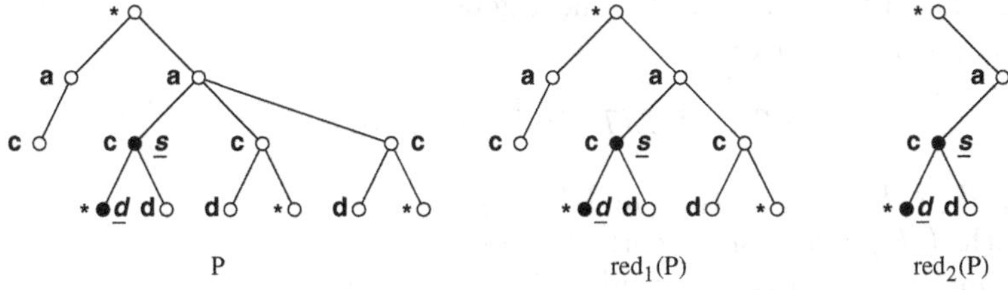

Fig. 7. A tree query and its first and second reductions

On the extended labels introduced above, we define an order which is the reflexive-transitive closure of the following: (1) for all $\ell \in \mathcal{L} \cup \{*\}$, $\ell_s \geq \ell$; (2) for all $\ell \in \mathcal{L} \cup \{*\}$, $\ell_d \geq \ell$; and (3) for all $\ell \in \mathcal{L}$, $\ell \geq *$.

We say that two extended labels ℓ_1 and ℓ_2 are compatible if either $\ell_1 \geq \ell_2$ or $\ell_2 \geq \ell_1$. For compatible extended labels ℓ_1 and ℓ_2, we define $\ell_1 + \ell_2 = \max(\ell_1, \ell_2)$.

Finally, we extend the notion of 1-relatedness from nodes of a document to nodes of a tree query with extended labels by replacing the condition $\lambda(m_1) = \lambda(m_2)$ in Definition 6 by $\lambda(m_1) + \lambda(m_2) = \lambda(m_2)$. We shall then say that m_1 and m_2 are 1-$*$-related and denote this by $m_1 \geq^1_* m_2$.

For P a tree query, we define the *second reduction* of P, denoted $\mathrm{red}_2(P)$, to be the tree query by deleting from $\mathrm{red}_1(P)$ in a top-down fashion every node m_1 for which there exists *another* node m_2 such that $m_1 \geq^1_* m_2$. Notice that the purpose of doing the reduction in two steps is to ensure that that the graph of the relation "$\geq^1_*$" is acyclic.

Lemma 2. *A tree query and its second reduction are equivalent.*

We can now show the following.

Theorem 5. *Let P be a tree query. Every (with respect to number of nodes) minimal tree query equivalent to P is isomorphic to $\mathrm{red}_2(P)$.*

Example 7. Figure 7 exhibits a tree query P and its first and second reductions, $\mathrm{red}_1(P)$ and $\mathrm{red}_2(P)$, respectively. The latter is the (up to isomorphism) unique minimal tree query equivalent to P.

6.2 Query Decomposition and Evaluation

In [7], the authors established the equivalence between the partitions on nodes (node pairs) of an XML document induced by its own structural features and the corresponding partitions induced by the DPath$^+(\Pi_1)$ algebra. Based on these findings, they showed that, with a $P(k)$-index [4] of $k > 1$, an index-only plan is available for answering any query in the DPath$^+(\Pi_1)$ algebra.

We now discuss how to take advantage of this result and the normal form we discovered for the Path$^+$ algebra to come up with an efficient query evaluation plan for queries in Path$^+$. Consider a Path$^+$expression *Exp* in its normal form, represented as a tree query, in the most generic case as in Figure 8. The normal form of *Exp* can be written as

$$E(T_{t,s})^{-1}; E(T_{t,t}); \Pi_2(E(T_{r,t})); E(T_{t,d}),$$

where $E(T_{t,s})$, $E(T_{t,t})$, $E(T_{r,t})$, and $E(T_{t,d})$ are the DPath$^+(\Pi_1)$ expressions corresponding to the tree queries $T_{t,s}$, $T_{r,t}$, $T_{t,t}$, and $T_{t,d}$, respectively. Since each of the four sub-queries is in DPath$^+(\Pi_1)$, efficient query evaluation with an index-only plan is available [7].

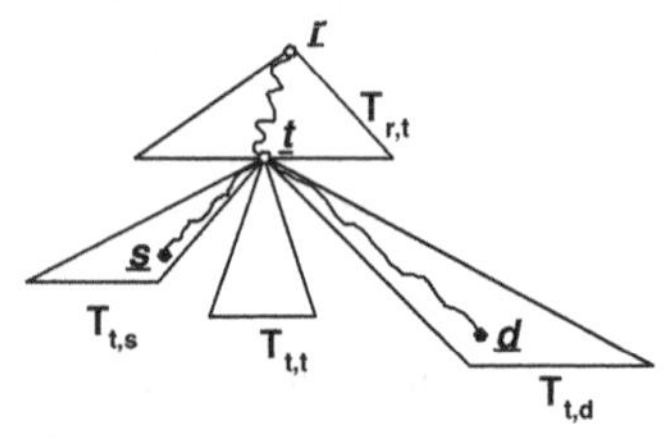

Fig. 8. General Structure of a Tree Query T

In conclusion, every Path$^+$ query can be evaluated efficiently with an index-only plan provided a $P(k)$-index [4] with $k > 1$ is available, and this with no more than three natural join operations, as guaranteed by the normal form. Indeed, for every document D, we have that

$$Exp(D) = E(T_{t,s})(D)^{-1} \bowtie E(T_{t,t})(D) \bowtie \Pi_2(E(T_{r,t}))(D) \bowtie E(T_{t,d})(D).$$

7 Discussion

This paper has been concerned with the translation of Path$^+$ expressions into equivalent Path$^+(\Pi_1, \Pi_2)$ expressions via a tree query minimization algorithm, followed by a translation algorithm from these queries into expressions in normal form. Furthermore, it was argued that such normal form expressions have sub-expressions that can be evaluated efficiently with proper index structures.

We now generalize the Path$^+$ algebra by adding set union and difference operations. Given a document specification $D = (V, Ed, \lambda)$, the *semantics* of a Path expression is defined by extending the definition of the semantics of a Path$^+$ expression with

$$E_1 \cup E_2(D) = E_1(D) \cup E_2(D) \qquad \text{and} \qquad E_1 - E_2(D) = E_1(D) - E_2(D).$$

Both in the operations present in the Path$^+$ algebra as in the set union and set difference, E_1, and E_2 now represent arbitrary Path expressions.

The set union operation alone does not alter the resolution expressiveness results presented in this paper, since set union operations can be *pushed out* through algebraic transformation, resulting into the union of expressions which no longer contains the set union operations. The set difference operation, however, significantly increases the resolution expressiveness of the language.

Consequently, Path expressions containing set union and set difference can in general no longer be expressed as tree queries, whence our minimization and normalization algorithms are no longer applicable. Expressions containing set union, but not set difference, however, can still be normalized using the algorithms discussed in this paper, after the set union operations are moved to the top. The resulting expression will then be a union of Path$^+(\Pi_1, \Pi_2)$ expressions.

Ancestor/descendant relationships can be expressed in most semi-structured query languages, and have been included as $\downarrow^*$ and $\uparrow^*$, in the XPath languages in some studies. However, we regard $\downarrow^*$ and $\uparrow^*$ merely as the transitive closure operation of the primitive operations $\downarrow$ and $\uparrow$, whose characteristics have been studied in the relational context. Furthermore, with proper encoding of the data—which represent the structural relationship of a semi-structured document—the ancestor/descentant relationship can be resolved via structural join [1],which is a value join on the structural encoding.

In conclusion, the results developed for Path$^+$ can be used to process more general path queries. In this regard, one can view the Path$^+$ algebra to the Path algebra as one can view the project-select-join algebra to the full relational algebra.

References

1. Al-Khalifa, S., Jagadish, H.V., Patel, J.M., Wu, Y., Koudas, N., Srivastava, D.: Structural joins: A primitive for efficient XML query pattern matching. In: ICDE (2002)
2. Amer-Yahia, S., Cho, S., Lakshmanan, L.V.S., Srivastava, D.: Tree pattern query minimization. VLDB J. 11(4), 315–331 (2002)
3. Benedikt, M., Fan, W., Kuper, G.M.: Structural properties of XPath fragments. Theor. Comput. Sci. 336(1), 3–31 (2005)
4. Brenes, S., Wu, Y., Gucht, D.V., Cruz, P.S.: Trie indexes for efficient XML query evaluation. In: WebDB (2008)
5. Chamberlin, D., et al.: XQuery 1.0: An XML query language, W3C (2003)
6. Clark, J., DeRose, S.: XML path language (XPath) version 1.0, `http://www.w3.org/TR/XPATH`
7. Fletcher, G.H.L., Van Gucht, D., Wu, Y., Gyssens, M., Brenes, S., Paredaens, J.: A methodology for coupling fragments of XPath with structural indexes for XML documents. In: Arenas, M., Schwartzbach, M.I. (eds.) DBPL 2007. LNCS, vol. 4797, pp. 48–65. Springer, Heidelberg (2007)
8. Gottlob, G., Koch, C., Pichler, R.: Efficient Algorithms for Processing XPath Queries. ACM Trans. Database Syst. 30(2), 444–491 (2005)
9. Götz, M., Koch, C., Martens, W.: Efficient algorithms for the tree homeomorphism problem. In: Arenas, M., Schwartzbach, M.I. (eds.) DBPL 2007. LNCS, vol. 4797, pp. 17–31. Springer, Heidelberg (2007)
10. Gyssens, M., Paredaens, J., Gucht, D.V., Fletcher, G.H.L.: Structural characterizations of the semantics of XPath as navigation tool on a document. In: PODS (2006)
11. Kaushik, R., Shenoy, P., Bohannon, P., Gudes, E.: Exploiting local similarity for indexing paths in graph-structured data. In: ICDE (2002)
12. Koch, C.: Processing queries on tree-structured data efficiently. In: PODS (2006)
13. Marx, M., de Rijke, M.: Semantic characterizations of navigational XPath. SIGMOD Record 34(2), 41–46 (2005)
14. Miklau, G., Suciu, D.: Containment and equivalence for a fragment of XPath. J. ACM 51(1), 2–45 (2004)
15. Paparizos, S., Patel, J.M., Jagadish, H.V.: SIGOPT: Using schema to optimize XML query processing. In: ICDE (2007)

Metamodel-Based Optimisation of XPath Queries*

Gerard Marks and Mark Roantree

Interoperable Systems Group, Dublin City University, Ireland
GMarks@computing.dcu.ie, Mark.Roantree@computing.dcu.ie

Abstract. To date, query performance in XML databases remains a difficult problem. XML documents are often very large making fast access to nodes within document trees cumbersome for query processors. Many research teams have addressed this issue with efficient algorithms and powerful indexes, but XML systems still cannot perform at the same level as relational databases. In this paper, we present a metamodel, which enables us to efficiently solve relationships between nodes in an XML database using standard SQL. By implementing the metamodel presented here, one can turn any *off-the-shelf* relational database into a high performance XPath processor. We will demonstrate the significant improvement achieved over three leading databases, and identify where each database is strongest in relation to XPath query performance.

1 Introduction

DataSpace applications permit the inclusion of multiple heterogeneous applications and storage systems to exist side-by-side in the same operational space. It is very useful for large applications where a single storage architecture is insufficient or where different user groups have differing requirements on the same datasets. It also permits different models and views of the same data. In our usage of a DataSpace architecture for sensor networks, we require that data is maintained in both relational and XML databases. Transaction management and updates are controlled using powerful relational technology while the XML store is used to share data and provide communication across applications. However, while XML databases offer high interoperability and sharing across user groups in the DataSpace, they suffer from poor query performance.

At the heart of the query optimisation problem in XML databases are *containment* queries. Containment queries are those that are based on containment and proximity relationships among elements, attributes, and their contents [20]. In XPath, *containment* queries return nodes based on *ancestor-descendant*, and/or *parent-child* relationships. Many research teams have addressed this issue with efficient algorithms and powerful indexes [1, 20, 13, 6, 17], but performance continues to be a problem.

* Funded by Enterprise Ireland Grant No. CFTD/07/201.

A.P. Sexton (Ed.): BNCOD 2009, LNCS 5588, pp. 146–157, 2009.

The limitation of XML-enabled relational databases to address this issue was noted in [13], and it is widely felt that the relational database kernel needs to be augmented with a new type of *join* operation [20, 13, 12] to compete with their native counterparts. [12] describes how RDBMSs (Relational Database Management Systems) execute XPath queries efficiently, by pushing the processing power of relational database systems to their limit, using *aggregation functions*, and *partitioned B-trees*.

As an alternative approach, we present an XPath processor using a *metamodel* for (*schema-independent*) indexing of XML documents or document collections, in any *off-the-shelf* relational database. In addition, we used only standard *B-tree* indexes and SQL (Structured Query Language). We exploit the multi-model database architecture of the DataSpace system to incorporate both relational and XML database technology in different applications in the database system.

Contribution. We provide a unique and efficient evaluation strategy for *containment* queries using the XPath [19] query language. Our purely relational approach can outperform industry leaders in this area using only standard *B-tree* indexes. By implementing the relevant features of the *metamodel* presented here, one can turn an *off-the-shelf* relational database into an XPath processor that can compete with any leading *native* or *XML-enabled* database. We validate our approach by comparing our XPath processor to three leading XML databases [5, 15, 9] in our experiments section.

2 Query Processing

The basis of all indexes for XML databases is a tree traversal technique that assigns a unique identifier to each node in the target dataset. In our metamodel, all nodes are assigned unique {*PreOrder*, *level*} pairs. In [16], it was shown that indexes based on PreOrder and level encoding of XML trees can support the evaluation of steps on all thirteen XPath axes. The *metamodel* is illustrated in Fig. 1 but for reasons of space, we cannot describe how all of the constructs are used. For a detailed description of the *metamodel* elements, please refer to the extended version of this paper [14]. In the remainder of this section, we will describe how XPath queries are mapped to SQL equivalents to provide significant performance gains. Those metadata constructs shown in Fig. 1 that are used in creating the performance gains are illustrated through a range of sample queries.

2.1 Query Mapping

A fundamental aspect of XPath processing is the relationship between *ancestor* and *descendant* nodes. Using a *pre/level/size* index, this relationship may be solved as follows: given nodes c, and v in an XML document, where c is the context node and v is an arbitrary node. Node v is a *descendant* of node c if [12]: $pre(c) < pre(v) \leq pre(c) + size(c)$.

However, in XPath we are usually working with sets of context nodes rather than a singleton. Thus, if there were 1000 context nodes in the set, the above

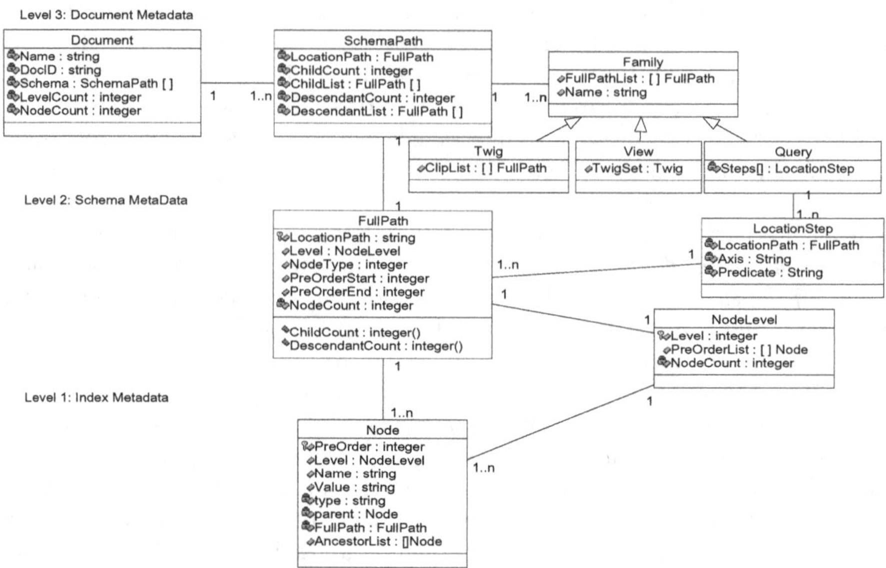

Fig. 1. XML Database Metamodel

```
select [distinct] Ancestor|Pre|Parent|Size|PreStart,PreEnd
from ANCESTORS|BIT|FULLPATH
where <condition_anc> | <condition_dec> | <condition_foll> |
<condition_prec> | <condition_child> | <condition_parent> |
<condition_prune>

<condition_anc>        := Type = 'type-value' [and Name = 'name-value']
                          and Pre in (condition-root)
<condition_dec>        := [DNAME = 'dname-value' and] Ancestor in (condition-root)
                          [and TYPE = 'type-value']
<condition_dec_root>   := Pre >= PreStart and pre <= PreEnd
                          [and name = 'name-value']
<condition_foll>       := [Name = 'name-value' and] Pre > MIN(condition-root)
                          + (select size from BIT where pre = MIN(condition-root))
<condition_prec>       := [Name = 'name-value' and] Pre < MAX(condition-root)
                          and Pre not in (Ancestors of MAX(condition-root))
<condition_parent>     := Pre in (condition-root)
<condition_child>      := [Name = 'name-value' and] Parent in (condition-root)
<condition_child_fast> := FullPath = 'fullpath-value'
<condition_prune>      := Name = 'name-value'
```

Fig. 2. XPath-SQL Template

processing logic would need to be repeated 1000 times to identify all possible *descendants* or *ancestors*. For this purpose, a number of algorithms were proposed (e.g. *structural join*, *staircase join*, *TwigStack*) to perform the *join* operation between a set of context nodes and target context nodes within the XML document. Unfortunately, these *join* algorithms are not found in *off-the-shelf* RDBMSs [12]. We are therefore limited to the relational *set* logic provided by RDBMS vendors.

In Fig. 2, we provide an informal BNF for the SQL expression that is derived from the XPath query. The XPath-SQL mapping is completely automated. There is a fixed *select-from-where* expression that employs one of nine internal conditions depending on the query classification. `ANCESTORS`, `BIT` (Base Index Table) and `FULLPATH` (Fig. 2) refer to the relational database tables in Fig. 3. Our classification list currently covers ancestor, descendant, preceding, following, child and parent axes with an extra condition to allow pruning of the search space. In a number of these expressions, a root processor is required to navigate to context nodes. The *root-condition* is the SQL representation of the *context node set* input to the current step in the path expression. For further details (due to space constraints), see Technical Report [14].

2.2 Ancestor Axis

One method of finding the ancestors of a context node is to store *pointer(s)* to its *parent* or *child* nodes(s). To find the ancestors of a context node, the *ancestor* algorithm from our earlier work [7] simply loops, selecting the node's parents recursively, until the *document node* is reached. The problem with this method is that the number of joins required is equal the node's level (i.e. distance from *document node*) within the document, and intermediate node evaluation is cumbersome. Intermediate nodes are the nodes evaluated along the path from parent to parent that only serve the purpose of identifying subsequent parents (i.e. nodes that do not have the target *NodeTest*). Fig. 3a, and 3b illustrate the relational implementation of the `NODE` structure using our *metamodel*. Fig. 3c is the relational implementation of the `FullPath` structure (*see* metamodel *Fig. 1*).

To reduce the number of joins required and intermediate node evaluation, we extend the *parent pointer* approach by storing a pointer from each node directly to its ancestors. The relational mapping of each node to its ancestors is shown in Fig. 3b. For example, taking the context node set {3, 4}. Noting that

PRE	LEVEL	SIZE	PARENT	NAME	PATH	TYPE
0	0	16	-1	R	/R	0
1	1	12	0	S	/R/S	1
2	2	2	1	T	/R/S/T	1
3	3	1	2	S	/R/S/T/S	1
4	4	0	3	S	/R/S/T/S/S	1
5	2	4	4	T	/R/S/T	1
6	3	1	5	A	/R/S/T/A	1
7	4	0	6	N	/R/S/T/A/N	1
8	3	1	7	S	/R/S/T/S	1
9	4	0	8	N	/R/S/T/S/N	1

(a) Base Index Table (BIT)

PRE	ANCESTOR	NAME	TYPE	DNAME
4	0	R	Anc	S
4	1	S	Anc	S
4	2	T	Anc	S
4	3	S	Anc	S
4	4	S	Self	S
3	0	R	Anc	S
3	1	S	Anc	S
3	2	T	Anc	S
3	3	S	Self	S
2	0	T	Anc	T

(b) Ancestors

PATH	NAME	LEVEL	PRESTART	PREEND	CARDINALITY
/R	R	0	0	0	1
/R/S	S	1	1	14	2
/R/S/T	T	2	2	15	3

(c) FullPath

Fig. 3. Relational mapping of an XML document

document order (as required by XPath [19]) is achieved at the application layer[1]. To find all their ancestors the following SQL is used: `SELECT DISTINCT ANCESTOR FROM ANCESTORS WHERE TYPE = 'A' AND PRE IN (3, 4)`. The resulting context nodes (Fig. 3b) are: {0, 1, 2, 3}. Some duplication arises due to the fact that nodes 0, 1, and 2 are *ancestors* of both nodes 3 and 4. These duplicate nodes are removed using the `DISTINCT` keyword.

Pruning. Storing *ancestors* in this way (Fig. 3b) leads to inevitable duplication. Considerable performance gains can be achieved by removing these duplicates, as the number of nodes that must be processed is reduced. The *document node* is an *ancestor* of all nodes and thus, we remove the duplicate entries from the *ancestor* table (*shaded rows* Fig. 3b). For example, if there were 1,000,000 context nodes in a set, and we must return all of their *ancestors*, then the *document node* would be returned 1 *million* times. This would slow the query optimiser down considerably. Instead, we store the *document node* once with: `TYPE` = 'D', and append it to all queries (see Technical Report [14]). In the presence of a *NodeTest* that is not a *wildcard*, we also filter ancestors by `NAME`. Using this mapping logic, we convert the XPath expression `//T/ancestor::S` (*find the S elements that are ancestors of T*) to:

```
SELECT DISTINCT ANCESTOR FROM ANCESTORS WHERE TYPE = 'A' AND NAME
= 'S' AND PRE IN (SELECT PRE FROM BIT WHERE PRE >= 2 AND <= 15
NAME = 'T')
```

The sub expression (`SELECT PRE FROM BIT WHERE PRE >= 2 AND <= 15 AND NAME = 'T'`) that solves the first step (`//T`) of the path expression is explained in §2.3.

2.3 Descendant Axis

The ancestor data structure maps every context node to its set of ancestors. Therefore, by default (Fig. 3b) every ancestor is mapped to its descendants. Consider, the SQL statement: `SELECT PRE FROM ANCESTORS WHERE ANCESTOR IN (8, 14, 12)`. This returns all descendants of nodes 8, 14, and 12 including itself (i.e. *descendant-or-self* axis). In our XML tree (Fig. 4) these are the nodes {8, 9, 12, 13, 14, 15, 16}. To find all the descendants of the context nodes excluding itself (i.e. *descendant* axis) we simply append: `AND TYPE != 'Self'` to the SQL statement. Thus, nodes {9, 13, 15, 16} are returned.

So far, we have described the evaluation of the *descendant* axis in the presence of *wildcards*, i.e. `/descendant::*`, `/descendant-or-self::*`. A little more work is required when the *step* in the XPath *path* expression contains the *descendant* axis and a *NodeTest* that is not a *wildcard*. Note that every *non-initial*

[1] Our experimental evaluation (§3.) showed us that the performance overhead of using *order by* in SQL queries is much greater than adding the result nodes to a *SortedSet* at the application layer. Therefore, in contrast to similar approaches [10, 12] we do not add the expensive *order by* clause to our SQL queries.

occurrence of the abbreviated *syntax* '//' is replaced by the *descendant-or-self* axis in an XPath processor [19]. The abbreviation for the *child* axis is '/'. Considering step two (//S) of the *path expression* //T//S.

The method to compute this is:

1. Select all the descendants of T (i.e. //T//*). Nodes {2, 3, 4, 5, 6, 7, 8, 9, 15, 16}, identified by outer circle in Fig. 4.
2. Select all the nodes in the BIT (Base Index Table) that have the name S. Nodes {1, 3, 4, 8, 12, 13, 14, 16}, identified by the thick inner black circle in Fig. 4.
3. Returning their common nodes. Nodes {3, 4, 8, 16}, identified by the thick inner black circle and the outer circle in Fig. 4.

The SQL statement generated for //T//S is:

```
SELECT PRE FROM BIT WHERE NAME = 'S'
INTERSECT
SELECT PRE FROM ANCESTORS WHERE ANCESTOR IN (SELECT PRE FROM BIT
WHERE NAME = 'T')
```

However, by adding the DNAME (Descendant Name) column to our ANCESTORS schema (Fig. 3b), we can remove the *hash join* required to INTERSECT the two sets. In addition, we can prune the set to only those elements with the name 'S'. The SQL now looks like this:

```
SELECT PRE FROM ANCESTORS WHERE DNAME = 'S' AND ANCESTOR IN
(SELECT PRE FROM BIT WHERE NAME = 'T')
```

Pruning. We can prune the search space for all queries that are *descendants* of the *document node*. Using our *FullPath* schema (Fig. 3c), we can immediately identify the first and last (PreOrder) locations of a given element or path within the document. Taking the XPath expression: //T, we can quickly find the first and last location of this node within the document tree:

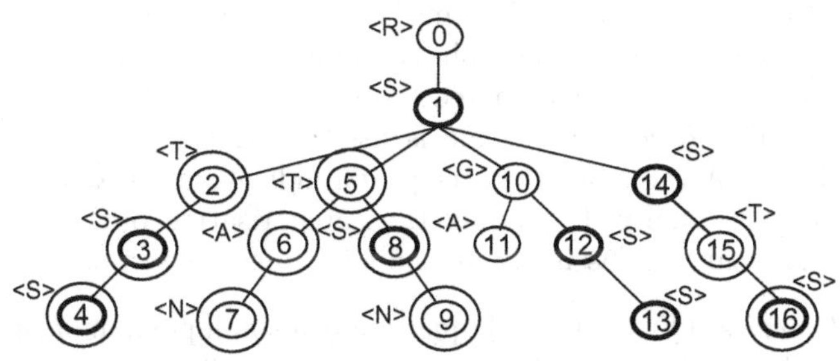

Fig. 4. PreOrder Encoded XML Tree

`SELECT PRESTART, PREEND FROM FULLPATH WHERE NAME = 'T'`. Nodes (2, 15).

Then, using our previous example (above) the SQL becomes:

```
SELECT PRE FROM ANCESTORS WHERE DNAME = 'S' AND ANCESTOR IN
(SELECT PRE FROM BIT WHERE PRE >= 2 AND <= 15 AND NAME = 'T')
```

This allows us to quickly identify the region of the document that contains the target nodes. This technique is particularly effective for XML documents that contain a small number of instances of the *NodeTest* (i.e. node NAME).

To evaluate the *parent* axis we exploit the *parent* pointer in the BIT (Fig. 3a). For the *child* axis we reverse the pointer from the parent nodes to their children. For further details of this process, and the remaining axes (e.g. *preceding*), we refer you (due to space constraints) to Technical Report [14].

3 Experiments

Experiments ran on identical servers with a 2.66GHz Intel(R) Core(TM)2 Duo CPU with 3.24GB of RAM. *MonetDB4-XQuery 4.26.4*, *eXist 1.0rc2* and *Sedna 3.1* ran on *Windows XP Professional* operating system, with *Java Virtual Machine* version 1.6. Our index (FASTX) was deployed using an *Oracle 10g* database, running on a *Fedora 7 Linux* platform. We used *eXist version 1.0* instead of *version 1.2.4* as earlier tests showed invalid results for a large number of queries across DBLP and XMark, making an accurate comparison impossible. However, eXist version 1.0 has the same kernel as version 1.2.4 [15], and is closest to the current version.

XMark [18] and DBLP [8], the most frequently used benchmarks for evaluating XPath query performance, were employed in our experiments. The DBLP benchmark contains over 12 million elements, 2 million attributes, 6 levels and has a size of 494Mb. XMark consists of over 3.5 million elements, 0.8 million attributes, 12 levels and has a size of 238MB.

3.1 XML Query Set

The query set is presented in Table 1. An 'Error' indicates that the query could not be processed by a database. The times are shown in Fig. 5, and 6, with the time (ms) displayed beneath each graph. Each query was executed six times and averaged, (with the first run eliminated) to ensure that all results were warm cache numbers. The *Logarithmic* scale is used along the y axis to make the comparison clearer. A value of '-1' in the graph indicates that an error occurred while processing the query. Extending the categorisation outlined in [3], the queries in Table 1 can be classified into the following categories:

- *Punctual queries* (Q9, Q10, Q11), query only a small portion of the database and have a high selectivity, thus they return a small number of matches.
- *Low selectivity queries*, queries that may return a large number of matches.

Table 1. Queries

	Query	FASTX	MonetDB	eXist	Sedna	Matches
Q1	//article//sub	1,515ms	187ms	344ms	1647ms	3,163
Q2	/dblp/inproceedings/booktitle	1,625ms	8,719ms	21,047ms	492,515ms	690,686
Q3	//inproceedings//title[.//i]//sub	52ms	312ms	1,157ms	58,000ms	151
Q4	/dblp/inproceedings[title]/author	22,063ms	23,453ms	**Error**	**+60min**	1,844,671
Q5	//inproceedings[title = 'Semantic Analysis Patterns.']/following::inproceedings	9,063ms	**Error**	20,171ms	1,549,329ms	19,984
Q6	//inproceedings[title = 'Semantic Analysis Patterns.']/preceding::book	360ms	2,046ms	4,171ms	1,424,078ms	1,262
Q7	//inproceedings[author = 'Jim Gray'][year = '1980']/following::article	7,469ms	**Error**	20,500ms	1,079,094	427,186
Q8	//sub/ancestor::inproceedings	79ms	234ms	422ms	28,156ms	619
Q9	//article[author='Serge Abiteboul']	2,328ms	1,891ms	4,546ms	114,187ms	61
Q10	//article[author='Serge Abiteboul'][year='1990']	78ms	1,878	4,906ms	204,190ms	4
Q11	//book[booktitle='Modern Database Systems']	78ms	94ms	141ms	1,290,063ms	1
Q12	/site/regions//item/location	1,297ms	703ms	1,422ms	17,095ms	45,675
Q13	//open_auction//description//listitem	2,875ms	4,828ms	1,254ms	27,734ms	35,500
Q14	//item[location][quantity][.//keyword]/name	7,312ms	594ms	1,266ms	42,198ms	31,575
Q15	/site//open_auction[.//bidder/personref]//reserve	5,578ms	297ms	516ms	33,573ms	11,275
Q16	//people//person[.//address/zipcode]/profile/education	62ms	219ms	281ms	6,026ms	6,635
Q17	//keyword/ancestor::description	2,656ms	15,781ms	3,532ms	266313ms	50,361
Q18	//description[.//text]//keyword/ancestor::parlist	8,984ms	15,734ms	**+60min**	1325703ms	35,506
Q19	//item[location]/following::keyword	3,016ms	3,046ms	**Error**	**+60min**	148,007
Q20	//item[location][quantity][.//keyword]/preceding::listitem	4,469ms	7,891ms	**Error**	**+60min**	61,260

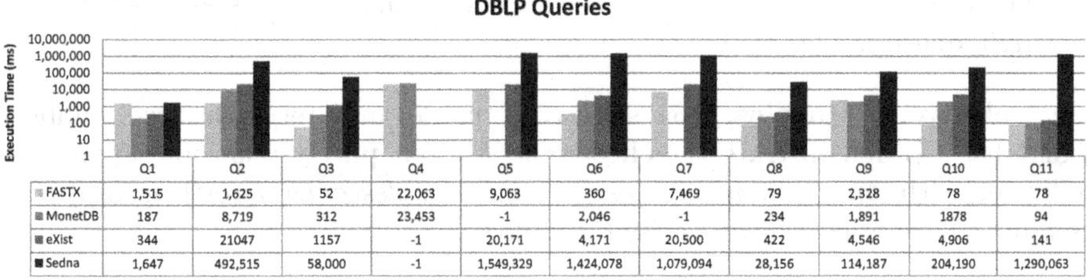

Fig. 5. Queries Over DBLP dataset

XMark Queries

Execution Time (ms): 10,000,000 / 1,000,000 / 100,000 / 10,000 / 1,000 / 100 / 10 / 1

	Q12	Q13	Q14	Q15	Q16	Q17	Q18	Q19	Q20
FASTX	1,297	2,875	7,312	5,578	62	2,656	8,984	3,016	4,469
MonetDB	703	4,828	594	297	219	15,781	15,734	3,046	7,891
eXist	1,422	1,254	1,266	516	281	3,532	5,258,500	-1	-1
Sedna	17,095	27,734	42,198	33,573	6,026	266,313	1,325,703	-1	-1

Fig. 6. Queries Over XMark dataset

Punctual Queries. Punctual queries have high selectivity, and return a small number of matches. This is the case for Q9, Q10, and Q11 as the predicate values limit the search space (e.g. to a particular year: *1990*, or book title: *Modern Database Systems*). FASTX, MonetDB and eXist perform very well in this category. The *bottom-up* approach used by FASTX enables the processor to jump directly to the target predicate values using a standard *B-tree* index, then work backwards. For example, to retrieve (Q9) *the books that have the*

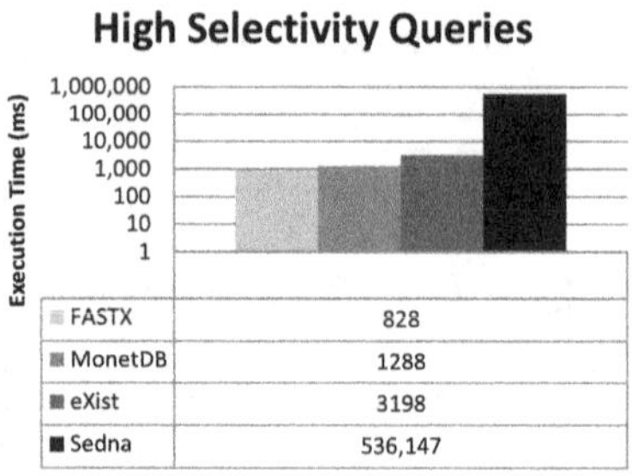

High Selectivity Queries

Execution Time (ms)

FASTX	828
MonetDB	1288
eXist	3198
Sedna	536,147

Fig. 7. Average for queries Q9, Q10, Q11

title 'Modern Database Systems', FASTX can jump straight to the nodes with the value *'Modern Database Systems'* (1 node), then find its parent (book). From queries 9-11, we can conclude that FASTX and MonetDB perform best for this type of query, eXist is slightly slower, and Sedna struggles even with *high selectivity* queries. This is shown in Fig. 7.

Low Selectivity Queries. Low selectivity queries may return a large number of matches. To demonstrate the effectiveness of our purely relational approach to dealing with *containment* queries (i.e. *ancestor-descendant* relationship) the remaining 17 queries are categorised as follows:

1. *Twig queries* (Q1-Q4, and Q12-Q16) that is, queries that only contain *descendant* and/or *child* axes.
2. Queries (Q8, Q18) containing the *ancestor* axis.
3. Queries (Q5, Q7, Q19) containing the *following* axis.
4. Queries (Q6, Q20) containing the *preceding* axis.

The *following* and *preceding* axes pose a major problem for *MonetDB*, *eXist* and *Sedna*. Even with the *Java Virtual Machine* set to the 1GB limit, *eXist* throws a *java.lang.OutOfMemoryError* for Q19 and Q20 even though the number of *matches* is only 148,007 and 61,260 respectively. One assumes that this is because a large number of intermediate nodes are stored in memory during processing. While *Sedna* does not encounter memory problems, it performs poorly when evaluating any query containing the *following* or *preceding* axes. Therefore, we conclude that *eXist* can only process queries containing the *following* or *preceding* axes when the number of matches returned is quite low (e.g. Q5, Q6). *Sedna* uses less memory than *eXist*, but takes a long time (15min+) to process queries containing the following and preceding axes, even when the selectivity is high (Q6).

MonetDB also finds the *following* and *preceding* axes difficult to process (e.g. Q5, Q7), and in some cases, the database server resets without warning (Q5, Q7). However, for queries that do not cause errors (Q6, Q19, Q20) *MonetDB* is quite efficient. Overall FASTX performs best for *following* and *preceding* using the method described in §2. This is shown in Fig. 5 and 6 (Q5, Q6, Q7, Q19, 20).

By modeling the *ancestor-descendant* relationship directly using a relational index, we (in many instances) outperform *MonetDB*, *eXist* and *Sedna* for *containment* queries (Q2, Q3, Q4, Q10, Q11). In Q2, for example, the boost in performance is achieved by the use of our FullPath construct, which allows us to process all three steps in the path expression simultaneously (see §2). The *staircase join* [13] algorithm (*MonetDB*), can process *containment* query relationships in a single document pass, and uses techniques such as *pruning*, and *skipping* to remove duplicate nodes within the hierarchy in order to reduce processing times. FASTX uses similar techniques (i.e. *pruning*, *skipping*, single document pass), but crucially, (unlike *MonetDB*) the relational database kernel does not need to be augmented. *eXist* and *Sedna* use the *Structural Join* [1] algorithm to solve this relationship [15,9]. Interestingly, *eXist* outperforms *Sedna* across all queries. This is interesting because, it shows that the Sedna's poor performance is related to the indexing of nodes, or encoding scheme (e.g. *encoding/decoding* node labels), as they both use the same algorithm to process the nodes once they have been retrieved.

Overall, FASTX performs better for *containment* queries when the number of duplicate paths to be processed is not very large. FASTX is slower for large volumes of duplicate paths due to the power of MonetDB's *staircase join* in pruning a larger number of duplicates, and skip a greater number of repeated paths (Q14, Q15). In some instances, eXist's *structural join* algorithm performs better than MonetDB's *staircase join* (Q17). Sedna performs worst than any of the others, across all queries. In addition, eXist's indexing structure, or DLN [4] encoding system outperforms Sedna's [9] index or SLS [2] encoding system, as they both use the *structural join* [1] algorithm to process *containment* queries. Finally, the XML-enabled relational databases (FASTX, MonetDB) perform better than the native databases (eXist, Sedna) in our overall experiments. This may indicate that native XML databases still cannot index nodes within XML documents as effectively as XML databases built on mature relational databases.

4 Related Research

In [12], they compare the performance of XPath queries across XML-enabled *off-the-shelf* RDBMSs, to those with a modified underlying kernel to suit XML tree structures (i.e. *TwigStack*, *structural join*, *staircase join*). This approach shows some similarities to ours: the ability to prune *context node* sets across the *following* axis using standard SQL and query rewrites. However, the core focus of this work relates to pushing the RDBMSs ability to store and index XML document nodes to the limit, with the use of *aggregation functions*, and *partitioned B-trees*. We differ greatly in our approach, as we use only standard *B-tree* indexes. We model hierarchical relationships within a relational database table (i.e. *ancestor-descendant* relationship), and our *metamodel* allows us to prune the search space, and remove duplicate nodes and paths to boost query performance.

The *join* algorithm outlined in [13] shows some similarities to ours, in that, as it has a pruning, and skipping stage. This allows the XPath processor to remove duplicates, and skip paths that are subsets of the paths returned by other nodes in the context node set. They however, make a local change to the relational database kernel, whereas we do not.

The *structural join* [1] is an extension of the *Multi-Predicate Merge Join* (MP-MGJN) [20]. The *structural join* is a *binary* join algorithm that treats each *ancestor-descendant* or *parent-child* pair separately. For example, for the query `//T[//S]/N`, this algorithm will solve `//T//S` then the output set of nodes is *joined* with `/N`. Similar to this, we solve these relationships in a *binary* (i.e. between two sets of nodes) fashion. However, we model the relationship in our *metamodel* and therefore, do not require any algorithms beyond those provided by the RDBMS.

[11] records schema information (PreOrder, PostOrder and level values) to provide a significant optimisation for XPath axes. However, they do not model the *ancestor-descendant* relationship for *containment* query evaluation. In addition, they do not employ the different levels of abstraction presented here and thus, attributes such as `GetPreStart` and `GetPreEnd` for the FullPath object allow us to limit the search space to achieve greater performance.

5 Conclusions

In this paper, we presented a method for optimising XML queries using a template system to map XPath constructs to SQL statements based on a metamodel for an XML repository. The metamodel describes how XML metadata (index data) is modelled and stored in relational tables.

Our experiments benchmarked our results against leading XML database implementations using a wide range of queries to categorise how different query types performed in different database systems. Our experiments show that while MonetDB outperforms other XML systems, it was possible for our system to improve on MonetDB for certain query types. Our current research is focused on addressing those query classes where MonetDB continues to outperform our existing implementation.

References

1. Al-Khalifa, S., Jagadish, H.V., Jignesh, M.P., Wu, Y., Koudas, N., Srivastava, D.: Structural Joins: A Primitive for Efficient XML Query Pattern Matching. In: ICDE, pp. 141–152 (2002)
2. Aznauryan, N.A., Kuznetsov, S.D., Novak, L.G., Grinev, M.N.: SLS: A Numbering Scheme for Large XML Documents. Programming and Computer Software 32(1), 8–18 (2006)
3. Barta, A., Consens, M.P., Mendelzon, A.O.: Benefits of Path Summaries in an XML Query Optimizer Supporting Multiple Access Methods. In: VLDB, pp. 133–144 (2005)
4. Bohmeand, T., Rahm, E.: Supporting Efficient Streaming and Insertion of XML Data in RDBMS. In: RDBMS. Proc. 3rd Int. Workshop Data Integration over the Web DIWeb 2004, pp. 70–81 (2004)

5. Boncz, P., Grust, T., van Keulen, M., Manegold, S., Rittinger, J., Teubner, J.: MonetDB/XQuery: a fast XQuery processor powered by a relational engine. In: SIGMOD 2006: Proceedings of the 2006 ACM SIGMOD international conference on Management of data, pp. 479–490. ACM, New York (2006)
6. Bruno, N., Koudas, N., Srivastava, D.: Holistic Twig Joins: Optimal XML Pattern Matching. In: SIGMOD 2002: Proceedings of the 2002 ACM SIGMOD international conference on Management of data, pp. 310–321. ACM, New York (2002)
7. Noonan, C.: Pruning XML Trees for XPath Query Optimisation. Master's thesis, School of Computing, Dublin City University (2007)
8. Computer Science Bibliography. Online Resource, `http://dblp.uni-trier.de`
9. Fomichev, A., Grinev, M., Kuznetsov, S.D.: Sedna: A Native XML DBMS. In: Wiedermann, J., Tel, G., Pokorný, J., Bieliková, M., Štuller, J. (eds.) SOFSEM 2006. LNCS, vol. 3831, pp. 272–281. Springer, Heidelberg (2006)
10. Grust, T.: Accelerating XPath Location Steps. In: SIGMOD 2002: Proceedings of the 2002 ACM SIGMOD international conference on Management of data, pp. 109–120. ACM, New York (2002)
11. Grust, T.: Accelerating XPath Location Steps. In: SIGMOD 2002: Proceedings of the 2002 ACM SIGMOD international conference on Management of data, pp. 109–120. ACM, New York (2002)
12. Grust, T., Rittinger, J., Teubner, J.: Why off-the-shelf RDBMSs are better at XPath than you might expect. In: SIGMOD 2007: Proceedings of the 2007 ACM SIGMOD international conference on Management of data, pp. 949–958. ACM, New York (2007)
13. Grust, T., van Keulen, M., Teubner, J.: Staircase Join: Teach a Relational DBMS to Watch Its (axis) Steps. In: VLDB 2003: Proceedings of the 29th international conference on Very large data bases, VLDB Endowment, pp. 524–535 (2003)
14. Marks, G., Roantree, M.: Metamodel-Based Optimisation of XPath Queries. Technical report, Dublin City University (2009), `http://www.computing.dcu.ie/~isg/publications/ISG-09-01.pdf`
15. Meier, W.: Index-driven XQuery Processing in the eXist XML Database. In: XML Prague (2006)
16. O'Connor, M.F., Bellahsène, Z., Roantree, M.: An extended preorder index for optimising xPath expressions. In: Bressan, S., Ceri, S., Hunt, E., Ives, Z.G., Bellahsène, Z., Rys, M., Unland, R. (eds.) XSym 2005. LNCS, vol. 3671, pp. 114–128. Springer, Heidelberg (2005)
17. Qin, L., Yu, J.X., Ding, B.: TwigList: Make twig pattern matching fast. In: Kotagiri, R., Radha Krishna, P., Mohania, M., Nantajeewarawat, E. (eds.) DASFAA 2007. LNCS, vol. 4443, pp. 850–862. Springer, Heidelberg (2007)
18. Schmidt, A., Waas, F., Kersten, M., Carey, M.J., Manolescu, I., Busse, R.: XMark: A Benchmark for XML Data Management. In: VLDB '02: Proceedings of the 28th international conference on Very Large Data Bases, VLDB Endowment, pp. 974–985 (2002)
19. XML Path Language 2.0. Online Resource, `http://www.w3.org/TR/xpath20`
20. Zhang, C., Naughton, J., DeWitt, D., Luo, Q., Lohman, G.: On Supporting Containment Queries in Relational Database Management Systems. In: SIGMOD 2001: Proceedings of the 2001 ACM SIGMOD international conference on Management of data, pp. 425–436. ACM, New York (2001)

Compacting XML Structures Using a Dynamic Labeling Scheme

Ramez Alkhatib and Marc H. Scholl

University of Konstanz, Box D 188 78457 Konstanz, Germany
{Ramez.Alkhatib,Marc.Scholl}@uni-konstanz.de

Abstract. Due to the growing popularity of XML as a data exchange and storage format, the need to develop efficient techniques for storing and querying XML documents has emerged. A common approach to achieve this is to use labeling techniques. However, their main problem is that they either do not support updating XML data dynamically or impose huge storage requirements. On the other hand, with the verbosity and redundancy problem of XML, which can lead to increased cost for processing XML documents, compaction of XML documents has become an increasingly important research issue. In this paper, we propose an approach called CXDLS combining the strengths of both, labeling and compaction techniques. Our approach exploits repetitive consecutive subtrees and tags for compacting the structure of XML documents by taking advantage of the ORDPATH labeling scheme. In addition it stores the compacted structure and the data values separately. Using our proposed approach, it is possible to support efficient query and update processing on compacted XML documents and to reduce storage space dramatically. Results of a comprehensive performance study are provided to show the advantages of CXDLS.

1 Introduction

XML [1] is becoming widely used for data exchange and manipulation. As a consequence, an increasing number of XML documents need to be managed. Therefore different languages have been proposed to query data from XML documents, among them XQuery and XPath [1] which are currently the popular XML query languages. They are both strongly typed, declarative, and use path expressions to traverse XML data. Therefore, efficiently processing path expressions plays an important role in XML query evaluation. There have been many proposals to manage XML documents. However, two common strategies are available to provide robust storage and efficient query processing. The first is based on labeling schemes that are widely used in XML query processing. These numbers represent the relationships between nodes, playing a crucial role in efficient query processing. However, some labeling schemes [2,3] have the problem that they either do not support updates to XML documents or need huge storage. The second strategy tries to reduce the size of XML documents through compaction techniques [4,5,6] that can improve the query speed by saving scan

A.P. Sexton (Ed.): BNCOD 2009, LNCS 5588, pp. 158–170, 2009.

time. Unfortunately, the problem in such methods is that they are not able to discover all the redundancy present in the structure of XML, and thus often do not yield the best compression result. Another significant drawback of such methods is that they do not support direct updates or direct querying, i.e. querying a compressed document without decompressing it. Our approach bridges the gaps between labeling schemes and compression technology to bring a solution for management of XML documents that yield performance better than using labeling and compression independently. In our work, we focus on the separation of content from structure of an XML document. It is coupled with an effective method for XML compacting based on the exploitation of the similarity of consecutive tags and subtrees in the structure of the XML documents. Also we use the ORDPATH labeling scheme [7] for gathering sufficient structural information from the compacted XML document. We then store the compacted XML in a way that allows fast access and efficient processing over secondary storage. The main contributions of this paper can be summarized as follows:

- CXDLS avoids unnecessary scans of structures or irrelevant data values by separating the XML structure from the content.
- CXDLS processes queries and updates directly over the compacted XML document by using the labeling technique.
- We evaluate the performance of CXDLS on a variety of XML documents. Our results show that the approach can outperform existing ones significantly.

This paper is organized as follows. Section 2 introduces related research on XML labeling and compacting techniques. In Section 3, our approach for XML compaction and management is described in detail. In Section 4, experimental results are presented with comparisons to other existing approaches in this field. Finally, Section 5 concludes the paper.

2 Related Work

In this section, we consider two main directions for efficient handling XML documents, namely compact representation of XML and labeling schemes.

XML compression. Recently, several algorithms have been proposed for XML compression. One of the first approaches to XML compression was the XMill compressor [8] that separates the content from the XML structure. This separation increases the data similarity in each of them and allows to achieve better data compression rates. The separation is reminiscent of earlier vertical partitioning techniques for relational data, which divide a table into multiple tables defined over subsets of the attributes [9]. This partitioning typically lets queries scan less data and thus improves query performance [10]. However, the separation in XMill is used for XML compression only and it does not allow direct querying of the compressed data. Buneman et al. [5] proposed an approach based on skeleton compression [6], which extends the separation idea of XMill for efficiently querying compressed data. The skeleton approach removes the redundancy of the document structure by using a technique based on the idea of sharing common subtrees and replacing identical and consecutive branches with one branch and a multiplicity annotation. In this approach, main memory data structures are used for the compressed skeleton, while external memory data structures hold textual contents. This approach still has some drawbacks: First, sometimes the compressed skeleton is still too large to fit into main memory. Second, compressed skeletons will always be scanned in their entirety to identify the relevant data vales.

XML labeling schemes. XML documents are usually modeled as labeled trees, where nodes represent tags or values and edges represent relationships between tags or tags and values. In order to facilitate query processing for XML data, several labeling schemes have been proposed. The basic idea of these schemes is to assign unique codes to the nodes in the tree in such a way that it takes constant time to determine the relationship between any two nodes from the codes. Therefore a good labeling scheme makes it possible to quickly determine the relationships between XML elements as well as to quickly access to the desired data. Existing labeling schemes can be classified into interval labeling and prefix labeling. In interval labeling [2] each node in the tree is labeled with a pair of numbers, with the first one being, e.g., the preorder number of this node and the second being the postorder number. While almost all region labeling schemes support XML query processing efficiently, their main drawback is the processing of dynamic updates of the tree structure. Insertion or deletion of new nodes would require relabeling existing nodes. In prefix labeling [7,3], each element is labeled by its path from the root, so that the label of a parent node is the prefix of the labels of all of its descendants. In schemes of this type, the size of labels is variable and depends on the tree depth. In contrast to interval labeling, the main advantage of prefix based schemes is their dynamics; labels of the existing nodes can remain stable in case of insertions and other updates. Therefore, we aim at preserving the benefits of the prefix labeling schemes and try to avoid their drawbacks. Several methods have been proposed to deal with this problem [11,12]. However the goal of those methods has been to improve the encoding to reduce label length, while our approach (CXDLS) exploits the properties of XML structures for compaction achieving minimal space consumption and it uses the ORDPATH labeling scheme to represent the XML document after the compaction retaining all the desirable features of ORDPATH. The goal of CXDLS is to combine the strengths of labeling and compression technologies. CXDLS bridges the gaps between them to get most benefits and avoid the drawbacks of those approaches yielding performance better than their independent use.

3 Compact Representation of XML

In this section, we introduce the labeling of nodes using the ORDPATH labeling scheme and we shall show their maintenance under insertions and deletions and we give some insight into their internal representation. After that we shall describe the main idea behind our approach for compacting the structure of XML documents and we illustrate how the ORDPATH labels are used for maintaining relationships between nodes after compaction.

3.1 The Labeling Scheme Used in CXDLS

The ORDPATH labeling scheme [7] is a particular variant of a hierarchical labeling scheme, it is used in Microsoft SQL Server's XML support. It is essentially an enhanced, insert-friendly version of Dewey tree labeling [3] that specifies a procedure for expressing a path as a binary bitstring. It aims to provide efficient insertion at any position of an XML tree, and also supports extremely high performance query plans for native XML queries. ORDPATH assigns node labels of variable length and only uses the positive, odd numbers in the initial labeling, even-numbered and negative integer component values are reserved for later insertions into an existing tree, the fragment root always receives label 1. The nth (n = 1, 2,) child of a parent node labeled p receives label p. (2* n - 1). When an insertion occurs, the negative ordinals support multiple inserts of nodes to the left of a set of existing siblings and the even number between two odd numbers is used as an intermediate node that does not count as a component that increases the depth of the nodes. The new node will be the child of the intermediate node, avoiding the need to relabel existing nodes at any time. An example of an XML document and its ORDPATH labels are given in Figure 1. Internally, ORDPATH labels are not stored as · -separated ordinals but using Unicode-like compact representations called prefix free encoding that is generated to maintain document order and allow cheap and easy node comparisons. This encoding consists of a set of L_i/O_i bitstring pairs: one for each component of the ORDPATH label. Each L_i bitstring specifies the length in bits of the succeeding O_i bitstring. Thus the L_i bitstrings are represented using the form of prefix-free encoding shown in

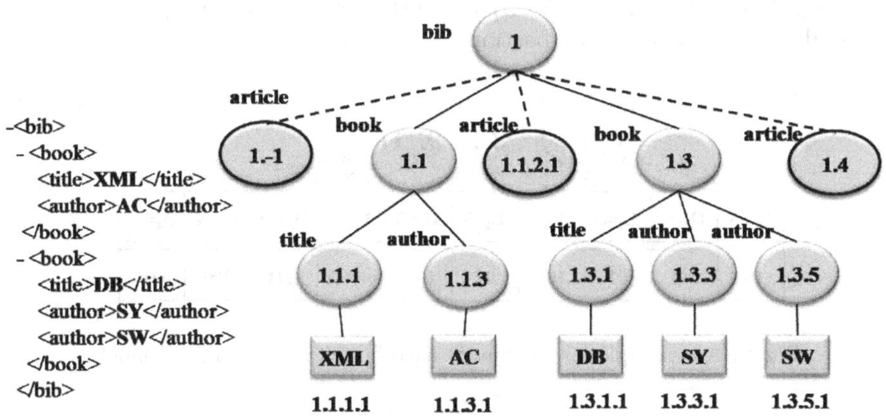

Fig. 1. A simple XML document with insertion of new nodes and corresponding ORDPATH labels

Table 1. L_i/O_i components can specify negative ordinals O_i as well as positive ones. They would be used to encode an ORDPATH label. The binary encoding is produced by locating each component value in the O_i value ranges and appending the corresponding L_i bitstring followed by the corresponding number of bits specifying the offset for the component value from the minimum O_i value for that range. Figure 2 displays the binary encoding of an ORDPATH label.

Table 1. Prefix-Free Encoding of the Bitstrings

Bitstring	L_i	O_i **Value range**
0000001	48	$[-2.8 \times 10^{14}, -4.3 \times 10^9]$
0000010	32	$[-4.3 \times 10^9, -69977]$
0000011	16	$[-69976, -4441]$
000010	12	$[-4440, -345]$
000011	8	$[-344, -89]$
00010	6	$[-88, -25]$
00011	4	$[-24, -9]$
001	3	$[-8, -1]$
01	3	$[0, 7]$
100	4	$[8, 23]$
101	6	$[24, 87]$
1100	8	$[88, 343]$
1101	12	$[344, 4439]$
11100	16	$[4440, 69975]$
11101	32	$[69976, 4.3 \times 10^9]$
11110	48	$[4.3 \times 10^9, 2.8 \times 10^{14}]$

In our approach we use the ORDPATH labeling scheme to label the XML documents, where for each element node of the structure is assigned *ORDPATHLB* that is a label obtained by the ORDPATH labeling scheme, while we assign to each text node[1] a text identifier *TID* that is the *ORDPATHLB* of its parent element. We use the ORDPATH labeling scheme because it has nice properties:

First, all relationships between nodes can be inferred from the labels alone. Second, it is easy to determine the order of the nodes. Third, ORDPATH allows the insertion of new nodes at arbitrary positions in the XML tree but nevertheless avoids relabeling existing nodes. Fourth, ORDPATH has an internal representation which is based on a compressed binary form. Finally, ORDPATH allows a faithful representation of the XML documents after compaction.

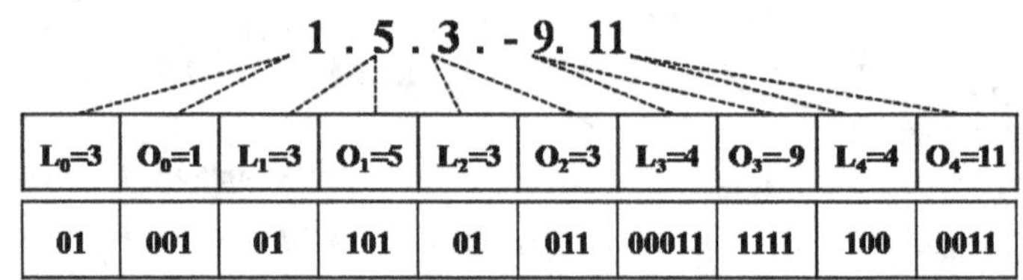

Fig. 2. An example for the binary encoding of an ORDPATH label

[1] For simplicity's sake, attribute nodes are disregarded in this paper,they can be treated like text nodes

3.2 XML Compaction

The basic idea of our approach combines the encoding scheme described in the previous section 3.1 and a new compaction technique to achieve a compact representation of XML documents for efficient management. The handling of navigational aspect of query evaluation, which only needs access to the structure, usually takes a considerable share of the query processing time, while character contents are needed for localized processing. Therefore separation the XML structure from its data values is very useful, because this separation typically lets queries scan less data and avoids unnecessary scanning of structures and thus improves query processing performance. Almost all previously proposed methods used the XML structure or a compressed version of the structure for navigational aspects of queries in main memory. However the XML structure can be very large in complex databases, thus it may not fit in main memory even in its compressed representation.

Our compaction Method helps to remove the redundant, duplicate subtrees and tags in an XML document. It separates the data values from the XML structure that is compacted based on the principle of exploiting the repetitions of similar sibling nodes in the XML structure, where "similar" means: elements with the same tag name. These similar nodes are replaced with one compacted node that is assigned a start label equal to the label of the first node of compacted nodes and an end label equal to the label of the last node of compacted nodes. Another principle is exploiting the repetitions of "identical" subtrees in the XML structure, where "identical" means: consecutive branches of trees which have exactly

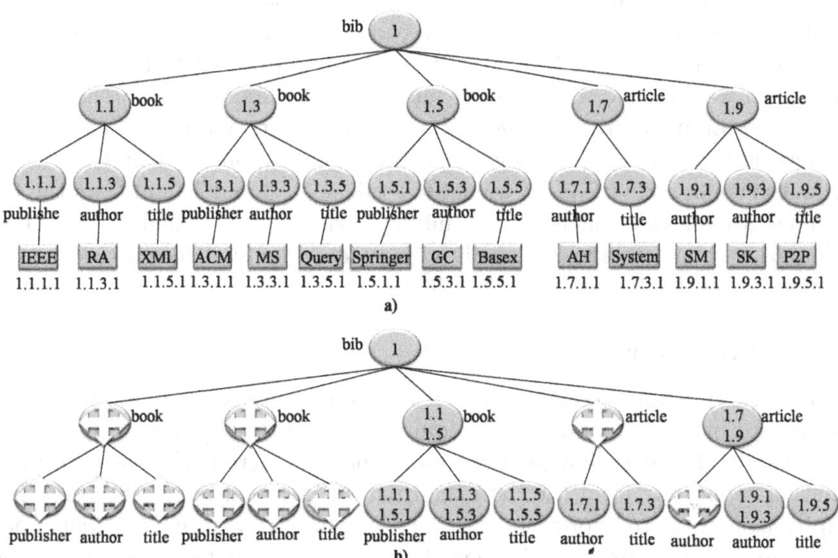

Fig. 3. The XML structure and its compacted form

the same substructure. These identical subtrees are also replaced with one compacted subtree, where each node in the compacted subtree is assigned a start label equal to the label of the node in the first subtree of compacted subtrees and an end label equal to the label of corresponding node in the last subtree of compacted subtrees.

Example 1. For the document of Figure 3(a) the compacted structure is shown in Figure 3(b); observe that the first, second and third subtrees which have the root "book" "1.1" , "1.3" and "1.5", are identical. Therefore, we can replace them by one subtree, where its root is assigned a start label equal to the label of the root in the first compacted subtree i.e "1.1" and an end label equal to the label of root in the third compacted subtree i.e "1.5" so that the new subtree has the root "book" and its label is "1.1, 1.5", and each child node of this subtree is assigned a start label equal to the label of the node in the first compacted subtree and an end label equal to the label of corresponding node in the third compacted subtree, so that the new subtree has child nodes which have the following labels respectively publisher "1.1.1, 1.5.1" author "1.1.3, 1.5.3" title "1.1.5, 1.5.5". Note that the nodes "article" with the labels "1.7" and "1.9" are similar, we also can replace them by one node "article" with label "1.7, 1.9" while the labels of their child node do not change. The process is recursively applied to all repetitive consecutive subtrees and tags in the XML structure. Figure 3(b) displays the compacted XML structure, where the crossed-out nodes will not be stored.

Using these labels we fully maintain all nesting information after the compaction, so the original XML document can be faithfully reconstructed. We consider for example the node "publisher" with the label "start = 1.1.1, end= 1.5.1". We can infer that this compacted node contains other nodes. To get the labels of decompacted nodes, we compare between each component of the start label with its corresponding in the end label to find all the odd numbers falling between them. Then we combine the resulting numbers from first two components with the resulting numbers from the second two components and so on. The process for inferring the labels is as follow: from the first components 1 and 1 it yields 1, from the components 1 and 5 we get the values (1, 3 and 5) and from the last components 1 and 1 it yields 1.We combine 1 with (1, 3, 5) yields "1.1", "1.3", "1.5" and we then combine this result with 1 yields the final decompacted nodes with the labels "1.1.1","1.3.1" and "1.5.1".

3.3 Data Structures

Based on our previous research [13,14] we strongly believe that improving the performance of XML management requires efficient storage structure and access methods. Towards that goal, we store the compacted XML structural information separately from the data information in the storage structure. This separation allows us to improve query processing performance by avoiding unnecessary scans of structures and irrelevant data values. But for query evaluation, we need to maintain the connection between structural information and value information. The ORDPATH labels can be used to reconnect the two parts of

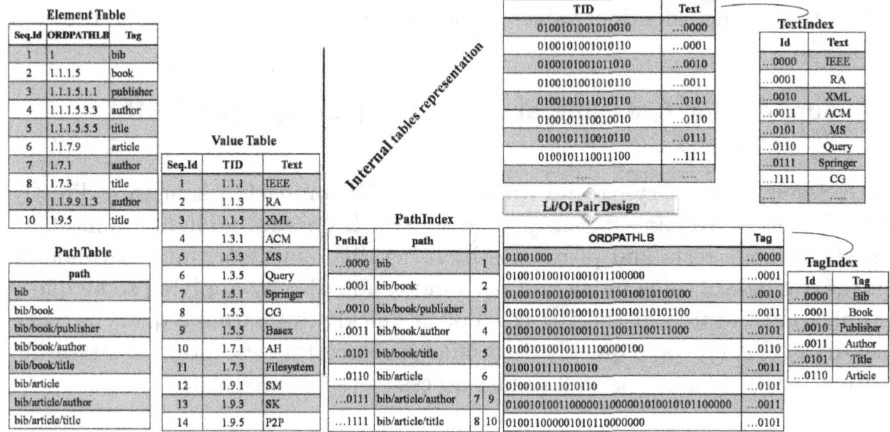

Fig. 4. Storage Structures for compacted XML document

information. Our storage model, shown in Figure 4, contains three tables: Element table, Value table and Path table.

The Element table stores the compacted structure of XML documents, where a sequence identifier, *ORDPATHLB* and tag name are stored for each element node of the compacted XML structure. We must note at this point that an *ORDPATHLB* of compacted node consists of start label and end label, which are stored as one label, where its first component is from the start label and its second component is from the end label and so on alternatingly. For example: the label "start=1.1.3,end= 1.5.3" of the compacted node "author" is stored as one label "1.1.1.5.3.3" in Element table. The tag names are indexed in a hash structures; the index entries are referenced by integer values. Thus each tag name is stored by its integer reference, the sequence identifier is implicitly given by the array position and each *ORDPATHLB* is stored in a byte array in binary form as described in Section 3.1. For example, the encoding of the *ORDPATHLB* "1.9" is the bitstring 010011000001. Because this bitstring is stored in a byte array, note that the last byte is incomplete. Therefore it is padded on the right with zeros to an 8bit boundary. Thus the stored bitstring is 0100110000010000. Note that all *ORDPATHLBs* start with "1.", therefore it is unnecessary to store this component explicitly, and thus we save 5 bits for each node of an XML document.

The Value table keeps all the data values of XML document, where sequence identifier, the own text identifier *TID* (see Section 3.1) and the text contents are stored for each leaf node. The sequence identifier is implicitly given by the array position and the *TID* is stored in a byte array in binary form and the text contents are indexed in hash structures and stored by their integer references.

The Path table stores all distinct paths in the structure of an XML document, where path means the sequence of elements from root node to any element node. The Path table is indexed in a hash structure; the index entries are referenced by integer values. To achieve better query performance, this index is extended by references to sequence identifier values of the Element table, resulting in an inverted index [15].

3.4 Support for Querying and Updating Compacted XML Structures

CXDLS supports querying and updating of the compacted structure directly and efficiently. We can efficiently process XPath queries, including almost all axis types, node tests and predicates by using different algorithms and indexes such as tag index, value index and path index and also by applying some query optimizations. Moreover the key concept in quickly evaluating a query is using the ORDPATH labels to quickly determine the ancestor-descendant relationship between XML elements as well as fast access to the desired data.

Example 2. For an XPath query containing slash "/" or double slash "//" can be easily answered by performing exact-match or prefix-match on the path index yield pathId(s) and *ORDPATHLB*(s) referenced by the resulting pathId(s). Let the XPath expression to be evaluated be Q1: /BIB/ARTICLE/TITLE. Thus, we only need to perform exact match on the path index yield the pathId = 8 and *ORDPATHLBs* "1.7.3", "1.9.5" referenced by this pathId. For a path expression containing the //-axis, such as Q2: //TITLE, it requires suffix match for TITLE in the path index yield the pathIds 5, 8 and the *ORDPATHLBs* "1.1.1.5.5.5" , "1.7.3" and "1.9.5" referenced by these pathIds. Note that *ORDPATHLB* equal to "1.1.1.5.5.5" is label of compacted node. To show the final results it requires decompaction using the process which we mentioned it in Section 3.2 yield three nodes having the *ORDPATHLBs* "1.1.5", "1.3.5" and "1.5.5". For path expressions containing a wildcard or //-axis in the middle of the path expression, such as Q3: /BIB/*/TITLE or Q4: /BIB//TITLE. In this case, exact-match for BIB and prefix-match for TITLE on the path index yield the same result of Q2.

Example 3. For an XPath query containing predicates, the path expression inside the predicate is rewritten. Child steps are converted to parent steps and descendant steps are converted to ancestor steps and vice versa [16]. This type of queries fits the pattern"path = value". For example, to answer the predicate query such as Q5: /BIB/ARTICLE[TITLE="P2P"], first the path expression /BIB/ARTICLE/TITLE is answered in a way similar to answering Q1 yield the *ORDPATHLBs* "1.7.3 " and "1.9.5 ". Next we check which one has a text value equal to"P2P" using exact match on the Text table. In this case only the *ORDPATHLB* "1.9.5" has data value equal to"P2P". To obtain the ARTICLE in the final result, we find the parent's *ORDPATHLB* of "1.9.5", that is, "1.9", which is easily inferred from the label only.

Example 4. For twig queries, such as Q6:/BIB/ARTICLE[TITLE="P2P"]//XX, we first answer the path expressions /BIB/ARTICLE[TITLE="P2P"] like Q5 then the ancestor-descendant relationship between the resulting nodes and XX can easily be determined by their labels.

Note that CXDLS supports all XPath axes in the same way as in the prefix labeling scheme because it is based on ORDPATH labeling, in which we can easily determine the relative order of nodes, the child-parent and the ancestor-descendant relationships by a byte comparison of two ORDPATH labels. It is important to note that the compacted labels also retain all such features of ORDPATH labels. For example, in Figure 3 article "1.9" is an ancestor of title "1.9.5" since "1.9" is prefix of "1.9.5". Also the compacted book "1.1.1.5" is an ancestor of publisher, author and title having the compacted labels "1.1.1.5.1.1", "1.1.1.5.3.3" and "1.1.1.5.5.5" respectively because book "1.1.1.5" is a prefix. The update behaviors of CXDLS are exactly as in ORDPATH, which guarantees the complete avoidance of relabeling for new insertions in any positions (see Section 3.1). In addition, the updated nodes are compacted, if they fulfill the conditions of compaction mentioned in Section 3.2. For deletions we can just mark as deleted the corresponding nodes in the structure without any relabeling. However since compacted nodes represent set of nodes in the uncompacted structure, we must consider some deletion cases which might require partial decompaction. For example, if we want to delete the node element specified by the XPath expression BIB/BOOK[2] which selects the node BOOK "1.3", the compacted node BOOK "1.1.1.5" needs decompaction to mark the selected node and any descendant nodes below it as deleted.

4 Performance Evaluation

In order to assess CXDLS, we implemented it in Java with JDK 1.6 and ran three sets of experiments (storage, query and update) on a 2.00GHz CPU, 2GB RAM, and 80 GB hard disk running Windows Vista. We carried out the experiments using real and synthetic XML data sets: 11MB, 111MB version of the XMark benchmark [17]. TreeBank [18] and Factbook. Shakespeare is a set of the plays of William Shakespeare marked up in XML for electronic publication. SWISS-PROT is a curated protein sequence database. PART and CUSTOMER are from the TPC-H Relational Database Benchmark [18].TOL is the tree structure of the Tree of Life web project [19] which has high depth XML tree more than 240 levels. We choose these datasets because they have different characteristics. Attributes are omitted for simplicity.

4.1 Consumption of Storage Requirements

We measure storage requirements of compacted structures to determine the influence of our compaction method on reducing the storage requirements. We also compared these results with the storage requirements of the original ORDPATH labeling scheme in [7] and our recent work called CXQU in [13] and different labeling schemes such as Dewey labeling scheme and pre/post [2] used in MonetDB/XQuery [20].In the implementation, we store for each node of the XML structure its label and tag name pointer and we use the same prefix-free encoding (see Table 1) for all mentioned prefix labeling schemes to do a fair

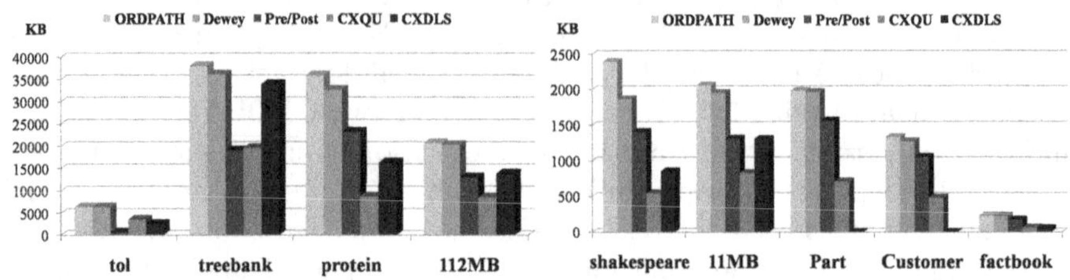

Fig. 5. Comparison of Storage Requirements

comparison between them. Also for pre/post labeling, we do not store the pre labels because they are implicitly given by the array position. The results shown in Figure 5 demonstrate that the success rate of the use of our compaction is very high. An interesting observation was made from the results. The storage requirements are determined by the degree of regularity of the structures of the XML documents. For this reason, it can be observed that the storage requirements are very small for the documents PART and CUSTOMER because they have regular structures. At the same time the storage requirements are still relatively small for other documents that have irregular structure or less regular structure when compared with other approaches,while our recent work CXQU [13] needs less storage requirements for this type of documents. The results from the last experiment show that our compaction method has efficient capabilities to reduce the storage requirements for both regular and irregular structures.

4.2 Query Performance

We investigate the query performance of CXDLS and compare it with the 32-bit version of MonetDB/XQuery 4.26.4 [20] and CXQU [13]. All approaches were experimentally evaluated using a set of queries which is compatible with XPath 1.0 currently supported by our system and comprises different kinds of XPath queries. Table 2 shows these queries for various XML datasets and presents the pure query evaluation times, excluding the times for parsing, compiling and optimizing the queries as well as serialization times. To gain a better insight into the query performance of all approaches, each query was repeated 10 times and the average of them was recorded as final result. These results confirm that CXDLS provides remarkably good query performance.

4.3 Update Performance

To evaluate the update performance, we measured the time for single node insertions at different positions of Hamlet XML document which is a Shakespeare's play. We chosen these documents due to their high compression ratio, thus the update time will be in a worst case. Hamlet has 5 ACTs, we add a new ACT before ACT[1], between ACT[4] and ACT[5], and after ACT[5] using "//ACT" as XPath expression with a predicate that selects the target node. We also measured the time for subtree insertions into different

Table 2. Elapsed time and XPath queries used in the comparison

Queries	MonetDB	CXQU	OUR	Hits
XMARK				
/site/regions/africa/item/description/parlist/listitem	9,30	0,54	0,30	57
/site/closed auctions/closed auction/price	7,60	2,39	3,34	975
/site/people/person/name	7,50	5,98	8,05	2550
/site/closed auctions//emph	5,50	6,38	5,45	1154
/site/regions//item/description	7,30	7,54	5,56	2175
//category/description/parlist/listitem	3,30	1,67	4,58	104
//people//person	4,50	4,24	3,63	2550
//africa//item	4,20	1,08	0,95	55
//africa//item[location = "United States"]	8,50	2,22	3,08	47
/site/regions/asia/item[shipping]/description	11,30	5,65	6,52	200
SHAKESPEARE				
/SHAKESPEARE/A_AND_C/PLAY/ACT/SCENE/SPEECH/SPEAKER	10,60	5,03	3,81	1179
/SHAKESPEARE/TAMING/PLAY//SCENE//SPEAKER	4,20	2,59	3,39	895
//PLAY/ACT/SCENE/SPEECH/LINE/STAGEDIR	57,70	5,0	4,35	618
//PLAY//EPILOGUE/STAGEDIR	5,30	2,82	4,7	5
//PLAY/*/*	30,90	11,32	14,64	1938
//PLAY//SCENE//STAGEDIR	11,20	29,64	22,65	6224
//PLAY/ACT/SCENE/*[2]	64,20	52,61	43,05	748
//PLAY/ACT[2]	8,50	8,54	5,34	37
//LINE[STAGEDIR="Aside"]	77,20	88,52	84,37	208

positions to the SHAKESPEARE XML document. We add a new play in three insertion positions of SHAKESPEARE selected by the XPath expressions: XP1:/SHAKESPEARE/A_AND_C, XP2:/SHAKESPEARE/MERCHANT, XP3:/SHAKESPEARE/WIN_TALE, this new play is inserted as a subtree that consists of 97 element nodes, 73 text nodes and has 6 levels. Figure 6 compares the average insertion times in milliseconds of CXDLS with Monet/XQuery and CXQU and demonstrates that the elapsed time for these updates in CXDLS is similar to CXQU and particularly fast. This result is expected because both use labeling schemes that adapt to updates. The difference of their elapsed time comes from the difference in the time that for locating the update point and generating the new labels takes.

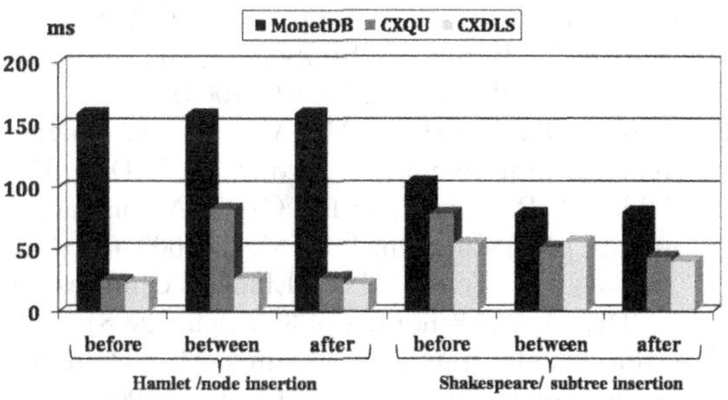

Fig. 6. The elapsed time for updates

5 Conclusions

In this paper, we proposed new approach called CXDLS that compacts the structures of XML documents by exploiting repetitive consecutive subtrees and tags in the structures and supports both update and query processing efficiently. The significant reduction in processing time and storage space is achieved by a combination of XML compacting and node labeling scheme. Our experiments verified that CXDLS improves performance significantly in terms of storage space consumption and query processing and updating execution time. As future work, we plan to extend CXDLS with more indexing structures and query processing algorithms and we will continue our investigations on variations of binary encoding forms that can further minimize the storage costs. We focused on XPath queries. For the future, we plan to extend our work to evaluating XQuery on compacted XML structures.

References

1. W3C, Extensible Markup Language (XML), XML Path Language (XPath), XQuery 1.0: An XML Query Language, `http://www.w3.org/TR/`
2. Grust, T.: Accelerating xpath location steps. In: SIGMOD Conference, pp. 109–120 (2002)
3. Tatarinov, I., Viglas, S., Beyer, K.S., Shanmugasundaram, J., Shekita, E.J., Zhang, C.: Storing and querying ordered xml using a relational database system. In: SIGMOD Conference, pp. 204–215 (2002)
4. Arion, A., Bonifati, A., Costa, G., D'Aguanno, S., Manolescu, I., Pugliese, A.: Efficient query evaluation over compressed xml data. In: Bertino, E., Christodoulakis, S., Plexousakis, D., Christophides, V., Koubarakis, M., Böhm, K., Ferrari, E. (eds.) EDBT 2004. LNCS, vol. 2992, pp. 200–218. Springer, Heidelberg (2004)
5. Buneman, P., Choi, B., Fan, W., Hutchison, R., Mann, R., Viglas, S.: Vectorizing and querying large xml repositories. In: ICDE, pp. 261–272 (2005)
6. Buneman, P., Grohe, M., Koch, C.: Path queries on compressed xml. In: VLDB, pp. 141–152 (2003)
7. O'Neil, P.E., O'Neil, E.J., Pal, S., Cseri, I., Schaller, G., Westbury, N.: Ordpaths: Insert-friendly xml node labels. In: SIGMOD Conference, pp. 903–908 (2004)
8. Liefke, H., Suciu, D.: Xmill: An efficient compressor for xml data. In: SIGMOD Conference, pp. 153–164 (2000)
9. Batory, D.S.: On searching transposed files. ACM Trans. Database Syst. 4(4), 531–544 (1979)
10. Ailamaki, A., DeWitt, D.J., Hill, M.D., Skounakis, M.: Weaving relations for cache performance. In: VLDB, pp. 169–180 (2001)
11. Böhme, T., Rahm, E.: Supporting efficient streaming and insertion of xml data in rdbms. In: DIWeb, pp. 70–81 (2004)
12. Härder, T., Haustein, M., Mathis, C., Wagner, M.: Node labeling schemes for dynamic xml documents reconsidered. Data Knowl. Eng. 60(1), 126–149 (2007)
13. Alkhatib, R., Scholl, M.H.: Cxqu: A compact xml storage for efficient query and update processing. In: ICDIM, pp. 605–612 (2008)
14. Alkhatib, R., Scholl, M.H.: Efficient compression and querying of xml repositories. In: DEXA Workshops, pp. 365–369 (2008)
15. Grün, C., Holupirek, A., Kramis, M., Scholl, M.H., Waldvogel, M.: Pushing xpath accelerator to its limits. In: ExpDB (2006)
16. Olteanu, D., Meuss, H., Furche, T., Bry, F.: XPath: Looking forward. In: Chaudhri, A.B., Unland, R., Djeraba, C., Lindner, W. (eds.) EDBT 2002. LNCS, vol. 2490, pp. 109–127. Springer, Heidelberg (2002)
17. Schmidt, A., Waas, F., Kersten, M.L., Carey, M.J., Manolescu, I., Busse, R.: Xmark: A benchmark for xml data management. In: VLDB, pp. 974–985 (2002)
18. University of Washington, XML Data Repository, `http://www.cs.washington.edu/research/xmldatasets`
19. Tree of Life web project, TOL, `http://tolweb.org/tree/home`
20. University of Amsterdam, Monetdb, `http://monetdb.cwi.nl/`

Evaluating a Peer-to-Peer Database Server Based on BitTorrent

John Colquhoun and Paul Watson

School of Computing Science, Newcastle University,
Newcastle-upon-Tyne, NE1 7RU, United Kingdom
{John.Colquhoun,Paul.Watson}@ncl.ac.uk

Abstract. Database systems have traditionally used a Client-Server architecture. As the server becomes overloaded, clients experience an increase in query response time, and in the worst case the server may be unable to provide any service at all.

In file-sharing, the problem of server overloading has been addressed by the use of Peer-to-Peer (P2P) techniques in which users (peers) supply files to each other, so sharing the load. This paper describes the Wigan P2P Database System, which was designed to investigate if P2P techniques for reducing server load, thus increasing system scalability, could be applied successfully in a database environment. It is based on the BitTorrent file-sharing approach.

This paper introduces the Wigan system architecture, explaining how the BitTorrent approach must be modified for a P2P database server. It presents and analyses experimental results, including the TPC-H benchmark, which show that the approach can succeed in delivering scalability in particular cases.

Keywords: P2P Computing, Database Systems.

1 Introduction

The scalability of applications that place a heavy load on database servers has again become the subject of intense commercial and research interest. Systems that allow thousands of simultaneous users to browse and purchase goods require highly-scalable, multi-tier systems, and so place great strain on the database tier. In another area, scientific researchers are now encouraged to provide open access to their databases so results can be widely shared, but this can cause performance problems if the data proves popular.

The limitations of server scalability as the number of simultaneous accesses increases used to be a problem in another area – file-sharing. However, that has been very successfully addressed in recent years by the introduction of Peer-to-Peer (P2P) techniques that harness the power of the clients in order to reduce the load on the server. This has lead to the design of extremely scalable, reliable and widely used applications.

In this paper we describe Wigan – a P2P database system designed to investigate whether the techniques used by file-sharing systems such as BitTorrent can be applied

A.P. Sexton (Ed.): BNCOD 2009, LNCS 5588, pp. 171–179, 2009.

to building highly scalable access to databases. We believe that this work is timely as almost all client computers, including desktop PCs, now have significant quantities of spare resources (CPU, memory, disk, network bandwidth) that could potentially be used to reduce the load on a DBMS. In Wigan, clients cache the results of their queries and these are then used to answer subsequent queries from themselves and other clients, so reducing the load on the server. This is not limited to exact query matches (as in Memcached [1]) – peers can answer queries that are a subset of the results they have cached. Designing Wigan has proved challenging due to the major inherent differences between accessing files and querying databases. The main differences are:

- Database queries include selects, projects and in some cases joins, whereas in file-sharing, files are accessed as a complete unit,
- Query results may contain, within them, the results of other queries, whereas in file-sharing this is not possible,
- Databases are updated whereas in file-sharing, files are considered immutable.

Clearly, there are many issues that must be resolved in designing a P2P database. This paper focuses on the key concepts of query matching at the Tracker and of query processing and is structured as follows. Section 2 provides an overview of BitTorrent, Section 3 introduces the Wigan architecture, Section 4 presents experimental results, Section 5 introduces related work while Section 6 concludes this paper.

2 BitTorrent Overview

BitTorrent [2] is a hybrid P2P file-sharing protocol [3]. The process of receiving a file in BitTorrent is called "downloading" and the corresponding process of providing a file to other peers is called "uploading." Similarly, peers engaged in these activities are known as "uploaders" and "downloaders." Uploaders advertise the file(s) they have copies of through a central component called a "Tracker." The Tracker acts as a directory, keeping track of which peers are downloading and uploading which files. Any peer that is advertising a complete file is known as a "seed", whilst any peer that is still in the process of downloading is known as a "leecher." There must be at least one seed present to introduce a file into the system and to place the first advertisement at the Tracker.

To start a download, a BitTorrent client will contact the Tracker and announce its interest in the file. Large files in BitTorrent are split into pieces, normally 256KB in size. The Tracker will provide a list of typically 50 random peers that already have some, or all, of the pieces. The downloader normally chooses the first piece at random and subsequent pieces in a rarest-first order. This allows rare pieces to spread further around the network. Once a downloader has received a complete piece, it is able to start uploading that piece to other downloaders. Thus, a BitTorrent leecher may be downloading and uploading different pieces of a particular file at the same time. A peer normally uploads to no more than five downloaders at any one time.

However, there are some peers that will operate according to a slightly amended lifecycle and will download but perform no uploading at all. These peers are called "Free Riders" and cause problems in BitTorrent and other file-sharing protocols because they consume resources but do not provide anything to other peers in return. BitTorrent's attempt to overcome this problem is to use a choking algorithm. "Choking" is the temporary refusal to upload a piece of a file to a particular downloader. The purpose of the choking algorithm is to ensure that those who provide little content into the system receive little in return.

3 Wigan Architecture

The Wigan system is derived from BitTorrent and hence the three major components in Wigan have the same names and basic roles as their counterparts in BitTorrent – the Seed, the Peers and the Tracker. Each is now discussed in turn.

3.1 The Seed

A Wigan seed possesses a complete copy of the database. Initially, the seed answers all queries (acting as if it were the server in a traditional client-server database). Once peers have begun to receive the results of queries then they advertise them at the Tracker and can then answer each other's queries where possible as described below. However, if at any time a downloading peer submits a query which cannot be answered by any of the other available peers, the query is answered by the seed.

3.2 The Peers

The peers are the equivalent of clients in traditional client-server systems – they send out queries and receive the results. However, they also cache the results of the queries in a local database server. This allows them to answer each other's queries, so taking the load away from the seed and providing greater scalability. The way in which they do this is governed by the Tracker (described below). As in BitTorrent, there is no assumption made about the amount of time the peers spend connected to the system – a peer may decide to disconnect at any time.

3.3 The Tracker

The central component in the Wigan system is the Tracker. There is one Tracker per database and this performs the same basic functionality as its namesake in BitTorrent in that it provides the downloading peers with a list of possible uploaders for the query they are requesting. However, due to the increased complexity of database queries when compared to file access, the Wigan Tracker has much more functionality and complexity. When a peer issues a query, it is sent first to the Tracker. This holds information on all the queries that have already been executed, along with the id of the peer that is caching the result. These "adverts" are stored in a canonical form representing the tables, columns and conditions on these columns for each query.

When a query arrives at the Tracker from a peer, it checks these adverts to see which other peers could answer the query. In Wigan, it is possible for a downloader's query to match exactly with an advertisement. In this, it is similar to Memcached [1]. However, a key difference is that Wigan goes beyond this and supports answering queries that are a proper subset of one or more advertisements. An example would be:

Query: `SELECT item FROM parts WHERE cost <= 10`
Advert1: `SELECT item FROM parts WHERE cost <= 10`
Advert2: `SELECT item FROM parts WHERE cost <= 15`

Both adverts can satisfy the query. We now describe in more detail the matching process. On arrival at the Tracker, the downloader's query is converted into the same canonical form as is used to store the adverts. The Tracker then retrieves all adverts which contain the tables and columns in the downloader's query. Note that if the downloader's query contains an aggregation, such as "MAX", the Tracker will retrieve both advertisements with the same aggregation and those which have the original column values. This is because an uploader with either the original column values or the same aggregation will be able to resolve the query, the only difference being that the latter will not have to perform the aggregation again when the query arrives because it already has the result.

This initial selection process removes the advertisements which do not have all of the required columns or contain none of the tables that appear in the downloader's query. The Tracker then examines all of the advertisements it has retrieved in the initial selection process to check that the conditions in the "WHERE" clause of the advertisement do not prevent the advertisement from resolving the downloader's query. If the query and an advertisement each contain a join, the Tracker must also check that the advertisement contains all of the tables in the downloader's query, not just some of them. The result of the final part of the selection process is a collection of adverts which can all resolve the downloader's query.

This collection of adverts may include a selection of different queries, given that we have already shown how one query may be resolved by an advert for a different query, providing that query is a subset of the advert. To enable a downloader to distinguish between adverts for different queries, the Tracker will group the adverts by query, stating for each query how many pieces the downloader should receive. This ensures the downloader is aware of when it can stop sending requests for data.

3.4 Downloading and Uploading

We now examine the process of downloading and uploading. A new downloader must contact the Tracker with the SQL query that it wishes to execute. The Tracker, using the processes described above, will return a list of suitable adverts grouped by query and the downloader must first select a group. For performance reasons, the downloader will choose those queries which exactly match the one it is searching for if this is possible or if it is not, start with the closest to an exact match.

The downloader contacts a randomly selected uploader peer from its chosen query group and submits a query for the first piece. If the uploader is able to accommodate a new downloader, it will perform the query and return all tuples from the first piece which matches the conditions of the query. A header with the query, piece number and a query ID is included so that if a downloader is receiving multiple queries simultaneously it can correlate responses to requests. Note that if there is no data in the first piece which matches the conditions of the query, the uploader will still send a response, containing just the header and no tuples. This prevents the downloader from assuming the response has gone missing because of a technical problem.

Once the first piece has arrived, the downloader stores the data in its local database and then makes a request for the next piece (potentially to a different peer). This process continues until the downloader has received all of the pieces. The downloader knows when this point occurs because the Tracker has informed it of the number of pieces. To improve performance, query requests for different pieces can be sent to a set of peers in parallel.

A BitTorrent peer can begin uploading as soon as it receives a complete piece of the file. However, a Wigan uploader cannot do this because it may be receiving data from an uploader advertising a different query. If the downloader has received its data from a peer advertising a different query, it will have to change the piece structure before it begins to upload.

For example, consider a university department's database, which has a Student table containing details of all students studying in the department. This table is split into 20 pieces. Initially, there is one seed containing the whole database and therefore a complete copy of the Student table. A new downloader requests the following query:

```
SELECT * FROM student WHERE tutor = 'Professor Lee'
```

During the download, the downloader will send 20 requests, one for each piece. For each request, the seed will send data from that piece containing details of all students whose tutor is Professor Lee. Let us assume that there are 20 such students. There is no guarantee of how these 20 students' details are distributed across the pieces. They may all be stored in one piece, in which case the downloader will receive 19 empty responses. This happens because the downloader has to request data from each piece; it does not know in advance which pieces will contain data matching the query. At this point there is no alternative option because the downloader has to query all 20 pieces. However, when the downloader makes this data available to others, it would not make sense from an efficiency point of view to have 20 pieces again. Instead, these resulting 20 tuples could all be grouped together in the minimum number of pieces.

Once a peer has made any required changes to the piece structure, it contacts the Tracker, stating it has received the query and informing it of how many tuples it received. The peer's advert is then stored at the Tracker and the peer becomes an uploader. Whilst uploading, it may receive requests periodically from downloading peers asking it to provide data. There is no specific amount of time that a peer has to upload for, as in BitTorrent, a peer can disconnect at any time.

4 Evaluation

Our initial implementation of the Wigan architecture was a simulator and is described in more detail in [4, 5, 6]. We then developed a "native" (non-simulated) version of Wigan using the algorithms implemented and tuned in the simulator. This implementation is also written in Java and uploading peers use OGSA-DAI [7] to expose their databases as a web service. Our implementation has been deployed on a small number of nodes and has been evaluated with both MySQL [8] and SQLServer [9] databases storing data from the TPC-H benchmark [10].

Our initial experiment was to test simple query execution. In this experiment, all peers were using MySQL databases. Eight peers submitted the same query in quick succession. This query ranged over a single table and had a small result set size. In addition to these eight peers, a seed and a Tracker were also required. In this experiment, both functionalities were performed by a single, ninth, peer. This was ready to provide data before any peers began submitting queries. The experiment was repeated five times and an average taken. The average response times are shown in Fig. 1.

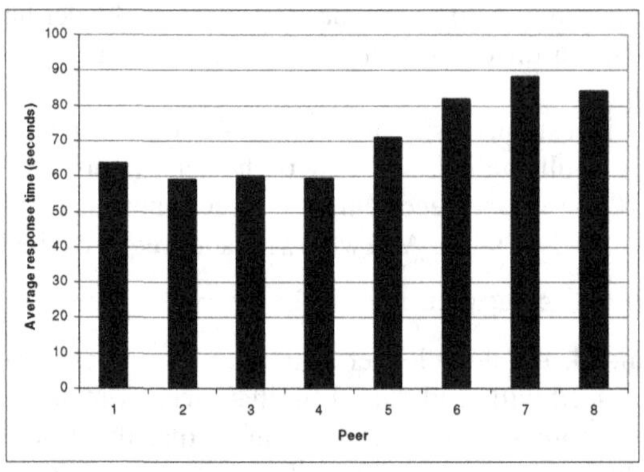

Fig. 1. Average Response Time for Eight Peers

The response times are quite high because, in this experiment, all peers have to download from the seed as there are no other peers available at the start of the download period. Although the query has a small result set size the seed has to examine every piece in the database table, as described in Section 3.4, which takes some time. Given that, like a BitTorrent peer, the seed is configured to only upload to a maximum of five downloaders simultaneously, a queue develops at the seed when the three final peers submit their queries. Therefore, the final three peers have to wait until those downloaders submitting queries earlier have received all of the query results. This is why these peers experienced longer response times than those starting before them.

In order to investigate how the system behaved when there were multiple peers available to answer queries, the experiment described above was altered so that a single peer submitted the query initially and then there was a delay before the remaining peers all submitted the same query in quick succession. This delay was set to one minute to allow the first peer more than enough time to complete its download. All other parameters, including the actual query, remained unchanged. Fig. 2 shows the corresponding results for this experiment.

It can be observed that from the second peer onwards, the response times drop considerably. This is because the first peer is available to upload the query to subsequent downloaders. The first peer has only the results of the query and not the entire table. These results can fit onto a single piece and hence downloaders must send only one request to the uploader in order to receive the full results.

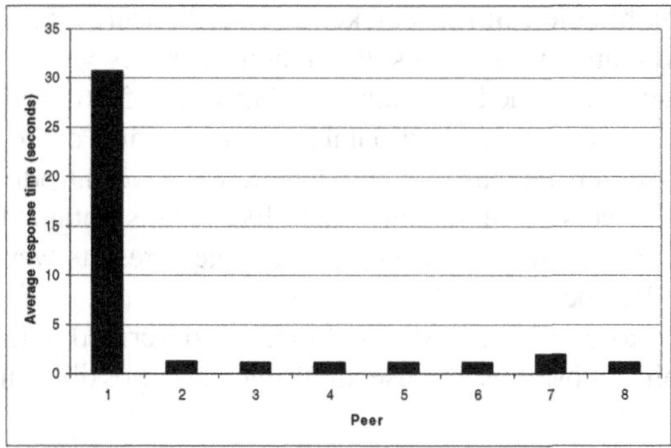

Fig. 2. Average Response Time for Eight Peers with One Starting Early

It can also be observed that the final three peers do not suffer the same increase in response time as those described in Fig. 1. This is because of the decrease in response times experienced by those peers starting earlier, which will have received their results and are advertising these at the Tracker by the time the final downloaders submit their queries. There is enough spare capacity in the system to accommodate all the downloaders, none of which have to queue.

5 Related Work

The potential of P2P computing has attracted some interest in the database community. Most of the existing P2P work [11, 12, 13, 14, 15, 16, 17, 18] views a P2P database as a collection of distributed databases and focuses on federating these databases, for example through schema integration. This is different to Wigan which focuses on single database server scalability.

The BioWired P2P database system [19] focuses on federating a collection of databases owned by different organisations, though they must have a global schema. However, if none of a peer's acquaintances are able to solve a query, this query is unanswerable in the current BioWired system. In contrast, by using a Tracker, Wigan ensures that clients will always receive a result if that is possible.

6 Conclusions

This paper has introduced the Wigan P2P Database System, a database architecture derived from the popular BitTorrent file-sharing protocol. This is, to our knowledge, the first P2P database system designed with a focus on scaling up the performance of a single database server, rather than on federating distributed databases. A central component known as a Tracker keeps a record of which peers have downloaded which queries and this information is used by query submitters to help them find peers which can resolve their queries. This query matching process is one of the key issues we had to address in this work. A special peer, known as the seed, possesses the complete database and can therefore answer any queries which are not held by the other peers. This combination of the Tracker and seed ensures that peers will always receive a correct and complete set of results to their queries.

Our previous results obtained through simulation [4, 5, 6] had shown that P2P techniques could be applied to scaling database servers, and could, in certain cases, outperform a client-server database. In this paper, we have introduced the initial results from the "native" version of Wigan which, like results obtained from the simulator in [4, 5], showed a drop in response time once query results were published at the Tracker by an uploading peer.

In addition to the basic query processing functions described in this paper, we have also investigated join algorithms for use in Wigan and possible means of handling data updates [4].

Acknowledgements

The work described in this paper was funded in part by the European Union through the IST 2002-004265 Network of Excellence, CoreGRID.

References

1. Memcached, `http://www.danga.com/memcached/`
2. BitTorrent, `http://www.bittorrent.com/index.html`
3. Schollmeier, R.: A Definition of Peer-to-Peer Networking for the Classification of Peer-to-Peer Architectures and Applications. In: First International Conference on Peer-to-Peer Computing (P2P 2001). IEEE Computer Society Press, Los Alamitos (2001)
4. Colquhoun, J.: A BitTorrent-Based Peer-to-Peer Database Server Newcastle University (2009), CS-TR No1135
5. Colquhoun, J., Watson, P.: A Peer-to-Peer Database Server based on BitTorrent. Newcastle University (2008), CS-TR No1089
6. Colquhoun, J., Watson, P.: A Peer-to-Peer Database Server. In: Gray, A., Jeffery, K., Shao, J. (eds.) BNCOD 2008. LNCS, vol. 5071, pp. 181–184. Springer, Heidelberg (2008)
7. OGSA-DAI, `http://www.ogsadai.org.uk/`
8. MySQL, `http://www.mysql.com`
9. SQL Server, `http://www.microsoft.com/sql/default.mspx`
10. Transaction Processing Council TPC-H Benchmark, `http://www.tpc.org/tpch/`
11. Bernstein, P.A., Giunchiglia, F., Kementsietsidis, A., Mylopoulis, J., Serafini, L., Zaihrayeu, I.: Data Management for Peer-to-Peer Computing: A Vision. In: Workshop on the Web and Databases, WebDB 2002, Madison (2002)
12. Serafini, L., Giunchiglia, F., Mylopoulos, J., Bernstein, P.A.: Local Relational Model: A Logical Formalisation of Database Co-ordination. University of Trento (2003), DIT-03-002
13. Franconi, E., Kuper, G.M., Lopatenko, A., Serafini, L.: A Robust Logical and Computational Characterisation of Peer-to-Peer Database Systems. In: Aberer, K., Koubarakis, M., Kalogeraki, V. (eds.) DBISP2P 2003. LNCS, vol. 2944, pp. 64–76. Springer, Heidelberg (2004)
14. Giunchiglia, F., Zaihrayeu, I.: Making Peer Databases Interact - A Vision for an Architecture Supporting Data Coordination. In: Klusch, M., Ossowski, S., Shehory, O. (eds.) CIA 2002. LNCS, vol. 2446, pp. 18–35. Springer, Heidelberg (2002)

15. Tatarinov, I., Ives, Z., Madhavan, J., Halevy, A., Suciu, D., Dalvi, N., Dong, X.L., Kadiyska, Y., Miklau, G., Mork, P.: The Piazza Peer Data Management Project. ACM Sigmod Record 32, 47–52 (2003)
16. Majkic, Z.: Weakly-coupled ontology integration of P2P database systems. In: The First International Workshop on Peer-to-Peer Knowledge Management. CEUR-WS.org, Boston (2004)
17. Rouse, C., Berman, S.: A Scalable P2P Database System with Semi-Automated Schema Matching. In: 26th IEEE International Conference on Distributed Computing Systems Workshops (ICDCSW 2006). IEEE Computer Society, Lisbon (2006)
18. Laborda, C.P.d., Popfinger, C., Conrad, S.: Digame: A Vision of an Active Multidatabase with Push–based Schema and Data Propagation. In: GI-/GMDS-Workshop on Enterprise Application Integration, EAI 2004 (2004)
19. Alvarez, D., Smukler, A., Vaisman, A.A.: Peer-To-Peer Databases for e-Science: a Biodiversity Case Study. In: 20th Brazilian Symposium on Databases, UFU, Federal University of Uberlândia (2005)

Towards Building a Knowledge Base for Research on Andean Weaving

Denise Y. Arnold[1], Sven Helmer[2], and Rodolfo Velásquez Arando[3]

[1] Instituto de Lengua y Cultura Aymara (ILCA), La Paz, Bolivia
[2] SCSIS, Birkbeck, University of London, London WC1E 7HX, United Kingdom
[3] Universidad de Aquino de Bolivia (UDABOL), La Paz, Bolivia

Abstract. We are working on a knowledge base to store 3D Andean textile patterns together with rich cultural and historic context information. This will allow ontological studies in museum collections as well as on ethnographic and archaeological fieldwork. We build on an existing ontology, extending it to incorporate more content and make it more accessible. This goes well beyond storing and retrieving textile patterns and enables us to capture the semantics and wider context of these patterns.

1 Introduction

Andean civilizations have used weaving for conveying information for a very long time. In contrast to alphabetic writing as developed in Europe and the Middle East, information is stored in patterns on materials such as cloth. Research has established that Andean people have employed this medium to document and communicate complex information. While some of these textile messages are understood by today's researchers, others are still waiting to be deciphered. One challenge in continuing this research is that the material for study is only available in a piecemeal fashion and without contextual information.

We plan to prepare information on Andean textiles in a way that puts it into a cultural, geographical, and historical context by providing a knowledge base that researchers can annotate and query. We draw on AI techniques for knowledge representation, reasoning, and sophisticated methods for searching. In addition to storing weaving patterns, our knowledge base will be fed with multimedia data such as digital photos, video, and text, providing context for the textile designs. This ranges from commentaries by living weavers to ethnographic and historical records. Also, previous approaches have been limited to surface features of textiles. We, on the other hand, plan to store 3D-data. This is important because it gives a deeper insight into weaving techniques (enabling us to identify historical and cultural associations) and particularly layered arrangements of the weft and warp also convey information (the surface patterns do not hold the whole information).

This project has a certain urgency to it, as the weaving practices, some of which have been handed down for millenia, are endangered. Younger generations no longer pick up weaving, meaning that this knowledge is being lost. We want

A.P. Sexton (Ed.): BNCOD 2009, LNCS 5588, pp. 180–188, 2009.

to contribute to preserving the rich cultural heritage of Andean weaving. In particular, we have the following aims:

- developing an ontological approach that permits a more logical systematization of the data and interpretations of former studies on Andean weaving
- surpassing the (software) limitations of the current 2D analysis techniques of Andean cloth through the development of 3D virtual reality fabric design and simulation tools

The remainder of this paper is organized as follows. The following section contains more background on the role of textiles in Andean culture. Section 3 briefly describes the approach we want to take in building the knowledge base. In Section 4 we discuss related work and Section 5 concludes the paper.

2 Background on Andean Weaving

In the Andes, the wider ramifications of textile design go beyond the immediate functional utility of cloth, having implications in cultural practice and learning methods, identity and territory, and importantly of cultural heritage. The international Coroma case, which sought the repatriation of 300 historical textiles taken from a Bolivian rural community without communal permissions, to be displayed in US Museums and put on sale in galleries, shows current limitations in textile documentation, classification methods, collecting methods, and legality, while revealing their value to local communities.

In cloth layout, the relative proportions of figurative and plain cloth reveal historical allegiances to the woven repertories of wider confederations, as well as indicating if textiles were woven in the highlands or valleys, and by older or younger generations. The colors of cloth indicate regional and historical identities, age and family groupings, and regional political identities. Cloth color also indicates the technological repertories of dying processes based on the natural resources available in plants, insects and minerals, and the time of year when these were collected. In the recent past, the detailed designs of flora, fauna, avifauna, and astral bodies, were directly inspired by the detailed observation of local resources, in such a way that textiles are often maps of regional territories, at particular times of the year.

In the language of local design patterns, textile features such as borders, figure orientation, style and color use, reveal changing social and cultural correlations. Design layouts also express the key symmetries used in particular cultures, in broader or minor patterns. The language of textile stripes frequently represents the flow of rivers, stone walls or boundary markers. In many weavings, band width (and higher warp count) indicates the quantity of the item represented, whereas band color indicates its content (water, maize, potato flowers). The relation of wider to narrower stripes also indicates generational hierarchy, commonly expressed as that between "mothers" and "offspring".

This language of width and color, found in both woven bags and the knotted *kipu* threads, is used to document the patrimony of raw food crops produced locally, or else preserved as toasted grains. The same language is also used to document the wider exchange patterns between highlands and lowlands, for example of highland salt for valley maize, when the colored stripes are ordered by the ecological zones where each product is found. "Raw" (natural) and "cooked" (dyed) colors have distinct usages.

Many different theoretical tendencies have sought to make sense of, and translate into broader terms, these matters of textile design. These have drawn on linguistics, semiotics, ethnography, archaeology, cultural geography, history of art and science, information studies, and math, among others.

Linguistic approaches tend to perceive Andean weavings as semiotic media, like language and written texts, hence artifacts through which cultural processes are negotiated and entextualized in wider semiotic systems. For instance, Halliday's social semiotics, applied to visual analysis, helped establish a "grammar of visual design" [1] that we intend to apply to textiles. Other linguistic studies argue that textiles embody not just grammars, but more general-purpose meaning systems [2] that work across cultural and linguistic boundaries in larger political formations in common. A current limitation of many linguistics-based studies is that they scrutinize individual acts of weaving as examples of sign-making or *parole*, while ignoring the *langue* of cloth as a more viable social medium for analyzing consistencies in regional repertoires of woven forms and their variations.

These perceived consistencies have also been approached from anthropology and the history of science. Recent studies of the *kipu* knotted threads [2,3] and cloth [4] view them as regional operative, information and documentary frameworks, that became standardized under Inca state control. For example, Urton's *kipu* studies identify both binary sequences (direction of spin, knotting) of individual artefacts, and segmentary hierarchies of information analysis in wider *kipu* repertories. Frank Salomón, drawing on the visual work of Edward Tufte, perceives *kipus* as information systems organized through cultural schema to express region-wide models that are homologues of decision-making and performance registries. However, both authors, by over-theorizing, ignore the regional ontologies we seek to explore, especially the regional metalanguage of primary fabric structures that appear to organize local cloth repertories [5].

3 Our Approach

In this section we present the general architecture of our system (see Figure 1 for an overview). Central to the design is the ontology, which provides the data model and terminology for the rest of the system. The content of the domain model will have to come from domain experts in the area of Andean weaving. Researchers will be able to load (annotated) data into the knowledge base, while users will be able to extract data via a portal, which hides the complexity of the underlying inference-based query engine.

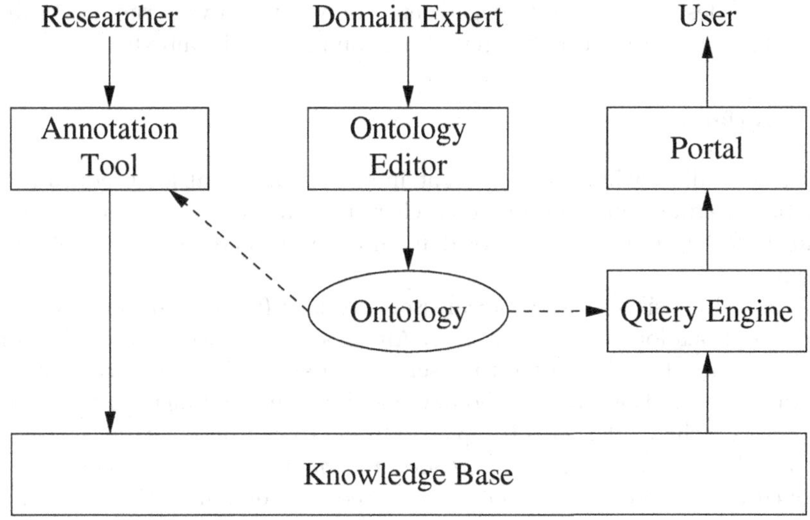

Fig. 1. Architecture of the system

3.1 Ontology

Attempts to standardize weaving terminology are currently mainly driven by (museum) collections, which tend to overlook the significances of weaving practices and cloth techniques and structures in their ethnographic contexts. Many collections are still cataloged in databases (such as Merlin) that consist of photographic records of samples, backed by registers of their probable precedence, date, size, material, collector and date of acquisition, with modest descriptions of any techniques used. In our ontology we want to describe the complex relationships between textiles and their ethnographic, geographical, historical, and archaeological contexts, enabling us to combine material from museum collections and from new fieldwork. While respecting infodiversity issues, our mediating systems must reinterpret the semantics of existing data structures in diverse collections, incorporating data regarding period, precedence, region and community, with weaving technique.

While there are projects which have developed ontologies from scratch, doing so is a very labor-intensive process. We will draw on existing developments in this area, using the reference model CIDOC-CRM as a basis [6]. Advantages of this model are that an RDF representation of the ontology already exists and that it was designed with the intention of being extensible. Our aim is to articulate an Andean textile heritage knowledge model with the recommendations in CIDOC-CRM, while modeling certain aspects in more detail. Textile patterns are much more than just images, We need to be able to describe weaving techniques and to express the productive basis (pastoral economy) as well as the multiple productive processes (rearing animals, shearing, classifying/cleaning wool, spinning, plying, washing, dying, looming up, weaving, and finishing cloth). In addition to this, we want to provide a platform for weavers to document and protect their repertory of designs. Rights management is only covered very rudimentarily in CIDOC-CRM, as are political, intercultural and regional contexts.

3.2 Interfaces

The system will provide three different interfaces: a portal for users to access the data, an annotation tool for researchers to add and update data, and last but not least an ontology editor for domain experts to create and maintain the ontology.

The portal consists of a graphical user interface (GUI) and more advanced (textual) options for experienced users. An important aspect of the GUI is not only ease-of-use, but the option to describe and search for 3D-textile patterns. This was motivated by Frame, who developed a structural method that relates early cloth's technical features (deep structures) to certain recurring surface designs [7]. The *Instituto de Lengua y Cultura Aymara*, ILCA, in Bolivia, recently developed a new tool, called *Sawu 3-D*, based on Computer Aided Geometric Design (CAGD) techniques, that can analyze textile structures in warp-faced cloth, and express them in 3D simulations with warp and weft counts, and relevant color coding (cf. [8]). For an example of the 3D-interface see Figure 2. We want to navigate seamlessly from these textile structures to specific samples, possible iconographic details, video illustrations of how these were woven, video

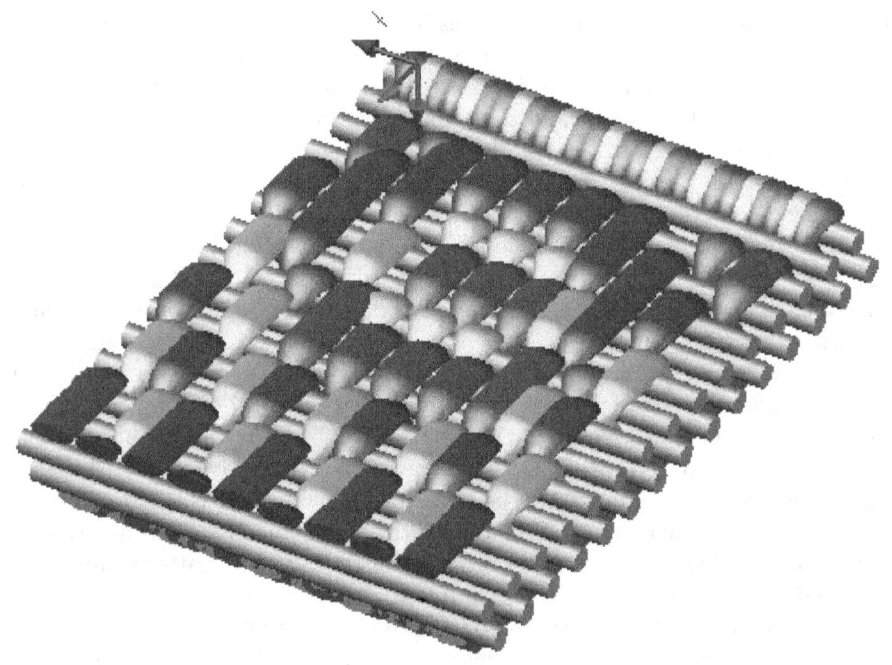

Fig. 2. Example of 3D interface

comments on their meaning, bibliographic references, the historical context of archaeological objects, and the time mapping of related pieces.

It is not enough to just load the raw data into the database. An annotation tool will provide functionality to preprocess data and establish connections to the existing data set with the help of the ontology. The 3D-interface will also help with inserting new patterns (from fieldwork) or establishing connections between existing patterns (from previously overlooked collection pieces), by allowing the comparison of textile samples with samples from other times and places. This helps in uncovering historical and regional connections that were not known before.

Creating and maintaining an ontology is a challenging task (even if an existing approach is used as a basis). We opt for Protégé as our ontology editor for the following reasons: it is a widely-used, open-source editor, it is extendible (a wide range of plugins are available), and it supports many different standards such as CLIPS, CORBA, and JDBC.

3.3 Query Engine

In addition to providing traditional database querying capabilities, we need to tackle three different tasks: first, the query engine has to be able to handle multimedia data such as audio and video files, second, we also need support for 3D-pattern-matching, and third, we need to be able to search on a semantic level, meaning we have to integrate an inference engined into the system.

Many vendors of (object-)relational database systems have integrated multimedia querying capabilities into their systems in the form of so-called Extenders or Cartridges (for an overview see [9]). For our purposes the functionality and performance of commercial systems in handling multimedia data is sufficient.

Integrating the 3D-textile pattern matching into the system is probably the most challenging task, as it is an inherently hard problem and there are no extensions for database systems that cover exactly these requirements. An approach based on voxels (three-dimensional pixels) seems very promising at the moment, as there exist querying techniques that are invariant with respect to certain geometric transformations such as translation and rotation [10]. An alternative to this solution could be a neural-network-based pattern matching approach. Neural networks have been used quite successfully for textile pattern-matching before, albeit for fault detection in fabrics [11]. Also most developments in weaving software and database systems to date have been done with reference to conventional looms (Jacquard), with a simple technical picking motion, and not to Andean looms, with complex 3D interlayering of cloth.

For the inference engine we need a solution that integrates it into the system efficiently (i.e. in terms of performance and scalability). Currently the most promising approaches are either using an inference engine that is already integrated into a relational system, such as Oracle 11g RDF/OWL [12], or employing an engine that can be implemented efficiently on top of a relational system, such as Minerva [13].

3.4 Storage Manager

For the storage manager of our system we face similar challenges as for the query engine. We need to be able to store, retrieve, and index multimedia data, 3D-textile patterns, and the ontological information in an efficient way. The central component of our storage manager will be an (object-)relational database system, as it offers scalability and transactional processing.

Querying in multimedia-capable databases is supported by indexing frameworks that offer advanced index types for multimedia types [9]. The Multimedia Data Cartridge of Oracle offers services for managing index structures based on Generalized Search Trees (GiST) [14], which also makes it possible to extend the indexing capabilities of the database system as needed.

Data objects in the voxel-approach described above are represented by so-called *shape descriptors*. Certain (numerical) features are extracted from an object and stored in a *feature vector*. Basically, this maps each objects to a point in a metric space. Assuming that this mapping is done correctly, the distance between two points in the metric space then measures the similarity of the two original data objects. In [10] such a technique based on partitioning the data objects (and extracting features from each partition) is described. The job of the database system, then, is to store and retrieve the feature vectors efficiently.

Most modern database systems offer bulk-loading functionality for efficiently inserting the initial data set into the database. Benchmarks run on standard Linux desktop PCs have shown that a Lehigh University Benchmark (LUBM) data set consisting of 25 million triples can be loaded and indexed in roughly 40 minutes [15].

4 Related Work

An *ontology* is an explicit specification of a conceptualization [16], which characterizes the objects and relationships between these objects for a given domain. For the purpose of describing ontologies many different languages such as Ontolingua [17], FLogic (Frame Logic) [18], Loom [19], RDF (Resource Description Framework) [20], PLIB (Parts Library) [21], and OMG's MetaObject Facility [22] have been developed (for an overview and a discussion on strengths and weaknesses, see [23]).

Many different ontologies have been built for specific domains ranging from medical applications [24], biodiversity analysis [25], biology [26], chemistry [26], all the way to woodworking and furniture [27]. Closest to our textile heritage domain comes the conceptual reference model CIDOC, which is an ontology for cultural heritage information [6].

5 Conclusion and Outlook

We are about to start a project to build a complete system as outlined in the previous section. Parts of the system, such as the Sawu 3-D tool for analyzing textile structures in three dimensions, have already been developed, while other parts are currently in the planning stage. The full-fledged project will also allow collaboration on a larger scale with museums in Peru, Chile, and Bolivia.

We are facing several challenges in implementing a knowledge-based system for research on Andean weaving. First we have to reorganize and consolidate existing textile collection data, which is quite diverse and spread out over many different locations. For a start, we will focus on three main historical axes: Inca (Cusco, La Paz, Killakas), Tiwanaku (Chilean and Peruvian coast), and Yampara (San Pedro de Atacama to the eastern valleys of Santa Cruz). Knowledge gained from this study will be used as a basis for the second step, developing an ontology that will serve as a common language to integrate data from many heterogenous sources. Finally, we plan to provide a visual interface to describe and search for complex three-dimensional textile patterns. This interdisciplinary project relies on the knowledge of ethnographer-linguists, archaeologists, museum curators, weavers, and computer scientists.

Acknowledgments. We thank the anonymous referees for their helpful suggestions, especially on the related work section.

References

1. Kress, G., van Leeuwen, T.: Reading Images: The Grammar of Visual Design. Routledge, London (1996)
2. Salomón, F.: The Cord Keepers: Khipus and Cultural Life in a Peruvian Village. Duke University Press, Durham (2004)
3. Urton, G.: Signs of the Inka kipu: Binary Coding in the Andean Knotted-String Records. University of Texas Press, Austin (1997)
4. Arnold, D., de Dios Yapita, J., Espejo, E.: Hilos sueltos: Los Andes desde el textil. Plural and ILCA, La Paz (2007)
5. Emery, I.: The Primary Structures of Fabric: An Illustrated Classification. Whitney Library of Design, New York (1986)
6. Doerr, M.: The CIDOC conceptual reference module: An ontological approach to semantic interoperability of metadata. AI Magazine 24(3), 75–92 (2003)
7. Frame, M.: The visual images of fabric structures in ancient Peruvian art. In: The Junius B. Bird Conference on Andean Textiles, The Textile Museum, Washington, D.C., pp. 47–80 (1986)
8. Sathendra, V., Adanur, S.: Fabric design and analysis system in 3d virtual reality. In: National Textile Center Annual Forum, Charlotte, North Carolina (2002)
9. Kosch, H., Döller, M.: Multimedia database systems: Where are we now? In: Int. Assoc. of Science and Technology for Development - Databases and Applications (IASTED-DBA), Innsbruck, Austria (2005)
10. Aßfalg, J., Kriegel, H.-P., Kröger, P., Pötke, M.: Accurate and efficient similarity search on 3D objects using point sampling, redundancy, and proportionality. In: Bauzer Medeiros, C., Egenhofer, M.J., Bertino, E. (eds.) SSTD 2005. LNCS, vol. 3633, pp. 200–217. Springer, Heidelberg (2005)
11. Ngan, H., Pang, G.: Novel method for patterned fabric inspection using Bollinger bands. Optical Engineering 45(8) (2006)

12. Chong, E., Das, S., Eadon, G., Srinivasan, J.: An efficient SQL-based RDF querying scheme. In: Proc. of the 31st Int. Conf. on Very Large Data Bases (VLDB), Trondheim, Norway, pp. 1216–1227 (2005)
13. Zhou, J., Ma, L., Liu, Q., Zhang, L., Yu, Y., Pan, Y.: Minerva: A scalable OWL ontology storage and inference system. In: Mizoguchi, R., Shi, Z.-Z., Giunchiglia, F. (eds.) ASWC 2006. LNCS, vol. 4185, pp. 429–443. Springer, Heidelberg (2006)
14. Hellerstein, J., Naughton, J., Pfeffer, A.: Generalized search trees for database systems. In: Proc. of the 21st Int. Conf. on Very Large Data Bases (VLDB), Zürich, Switzerland, pp. 562–573 (1995)
15. Das, S., Chong, E., Wu, Z., Annamalai, M., Srinivasan, J.: A scalable scheme for bulk loading large RDF graphs into Oracle. In: Proc. of the 24th International Conference on Data Engineering (ICDE), Cancún, Mexico, pp. 1297–1306 (2008)
16. Gruber, T.: Toward principles for the design of ontologies used for knowledge sharing? Int. J. Hum.-Comput. Stud. 43(5-6), 907–928 (1995)
17. Farquhar, A., Fikes, R., Rice, J.: The Ontolingua server: A tool for collaborative ontology construction. In: Proc. of 10th Knowledge Acquisition for Knowledge-Based Systems Workshop (KAW), Banff, Canada (1996)
18. Kifer, M., Lausen, G., Wu, J.: Logical foundations of object-oriented and frame-based languages. Journal of the ACM 42(4), 741–843 (1995)
19. MacGregor, R.: Inside the loom description classifier. SIGART Bulletin 2(3), 88–92 (1991)
20. World Wide Web Consortium: Resource Description Framework, RDF (2004), `http://www.w3.org/RDF/`
21. Pierra, G.: Context-explication in conceptual ontologies: the PLIB approach. In: Proc. of 10th ISPE Int. Conf. on Concurrent Engineering (ISPE CE), Madeira, Portugal, pp. 243–253 (2003)
22. Iyengar, S.: A universal repository architecture using the OMG UML and MO. In: Proc. of 2nd Enterprise Distributed Object Computing Conference (EDOC), San Diego, California (1998)
23. Corcho, Ó., Gómez Pérez, A.: A roadmap to ontology specification languages. In: Dieng, R., Corby, O. (eds.) EKAW 2000. LNCS, vol. 1937, pp. 80–96. Springer, Heidelberg (2000)
24. Rector, A., Rogers, J., Pole, P.: The GALEN high level ontology. In: 13th Int. Congress Medical Informatics Europe (MIE), Copenhagen, Denmark (1996)
25. Jones, A.C., Xu, X., Pittas, N., Gray, W.A., Fiddian, N.J., White, R.J., Robinson, J., Bisby, F.A., Brandt, S.M.: SPICE: A flexible architecture for integrating autonomous databases to comprise a distributed catalogue of life. In: Ibrahim, M., Küng, J., Revell, N. (eds.) DEXA 2000. LNCS, vol. 1873, pp. 981–992. Springer, Heidelberg (2000)
26. Stenzhorn, H., Schulz, S., Beißwanger, E., Hahn, U., van den Hoek, L., Mulligen, E.V.: BioTop and ChemTop - top-domain ontologies for biology and chemistry. In: Sheth, A.P., Staab, S., Dean, M., Paolucci, M., Maynard, D., Finin, T., Thirunarayan, K. (eds.) ISWC 2008. LNCS, vol. 5318, Springer, Heidelberg (2008)
27. Nicola, A.D., Missikoff, M., Navigli, R.: A software engineering approach to ontology building. Information Systems 34(2), 258–275 (2009)

The Adaptation Model of a Runtime Adaptable DBMS

Florian Irmert, Thomas Fischer, Frank Lauterwald, and Klaus Meyer-Wegener

Friedrich-Alexander University of Erlangen and Nuremberg,
Department of Computer Science,
Chair for Computer Science 6 (Data Management),
Martensstrasse 3,
91058 Erlangen, Germany
{florian.irmert,thomas.fischer,frank.lauterwald,kmw}@cs.fau.de

Abstract. Nowadays maintenance of database management systems (DBMSs) often requires offline operations for enhancement of functionality or security updates. This hampers the availability of the provided services and can cause undesirable implications. Therefore it is essential to minimize the downtime of DBMSs. We present the CoBRA DB (**Co**mponent **B**ased **R**untime **A**daptable **D**ata**B**ase) project that allows the adaptation and extension of a modular DBMS at runtime. In this paper we focus on the definition of an adaptation model describing the semantics of adaptation processes.

1 Introduction

In recent years the database community has realized that common database systems do not fit into every environment [6]. The obvious solution is the development of specialized DBMSs for each environment. However this approach is not suitable with respect to development cost, time to market and maintenance.

Tailor-made DBMSs [5] try to answer this challenge by adapting a DBMS towards a specific environment by providing a common code base from which customized DBMSs may be derived. Changing the functional range of such a DBMS however requires a shutdown and redeployment of a new version. Taking an application offline is often not feasible in some environments.

In the CoBRA DB (**Co**mponent **B**ased **R**untime **A**daptable **D**ata**B**ase) project [2] we propose an approach to tailor a DBMS at runtime. It uses a kind of DBMS "construction kit" and basic modules that are necessary in every DBMS. A DBMS can be assembled by choosing the appropriate modules for the intended functionality of the system. An important challenge is the modification of modular DBMSs at runtime. In our prototype it is possible to add, exchange, and remove modules while the database system is running. As a foundation we have developed an adaptation framework [3] that provides the exchange of components at runtime in a transparent and atomic operation. In this paper we present the adaptation model of CoBRA DB and its adaptation types.

A.P. Sexton (Ed.): BNCOD 2009, LNCS 5588, pp. 189–192, 2009.

2 CoBRA DB Runtime Environment

Runtime adaptation as a prerequisite to the proposed adaptable DBMS enables addition, removal and exchange of DBMS components at runtime without any downtime of the whole system.

To meet these requirements we have designed a runtime environment [3] based on a service-oriented component model [1,4]. The runtime environment is based on the OSGi Service Platform [4]. A component in the CoBRA runtime environment implements at least one service. The architecture of the CoBRA DB itself with its functional properties must therefore be sufficiently described by its services. To enable the adaptation to different environments, components implementing the same service can be exchanged at runtime. The resulting requirements for runtime adaptation are met by a transparent dynamic proxy concept which ensures the atomicity of the adaptation and the consistency of the state transfer from one component to the replacing one. Details can be found in [3].

3 Adaptation Model

Based on the prerequisites given by the runtime environment we developed an adaptation model in order to formalize the constraints for possible adaptations. This includes the structural model of the CoBRA DB architecture as well as the modeling of the different adaptation possibilities.

For the concrete specification of the adaptation model of the CoBRA DB we have chosen a service-oriented point of view. Figure 1 shows the structure of our model which consists of three design levels. The functionality of the DBMS is modeled by the specification of services (L3) like a *PageService* providing page-oriented access. The service descriptions of L3 must remain valid, even over adaptations. One or more services are embodied as components on L2, which are themselves implemented in one or more classes (L1). The components defined on L2 are the object of adaptation.

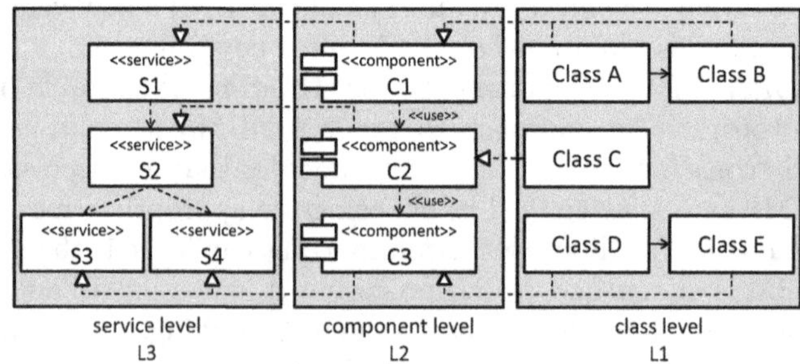

Fig. 1. Model hierarchy

To reflect the different changes possible in this model hierarchy we have identified different adaptation types.

3.1 Adaptation by Component Exchange

Components that implement the same services are interchangeable through *adaptation by component exchange* and may therefore only differ in their non functional properties. In figure 2 service P is realized by component C1. During the adaptation C1 is replaced by component C2 (which also implements service P).

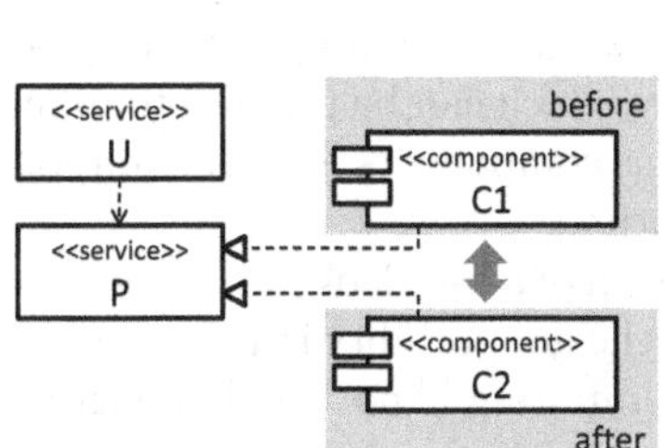

Fig. 2. Component exchange

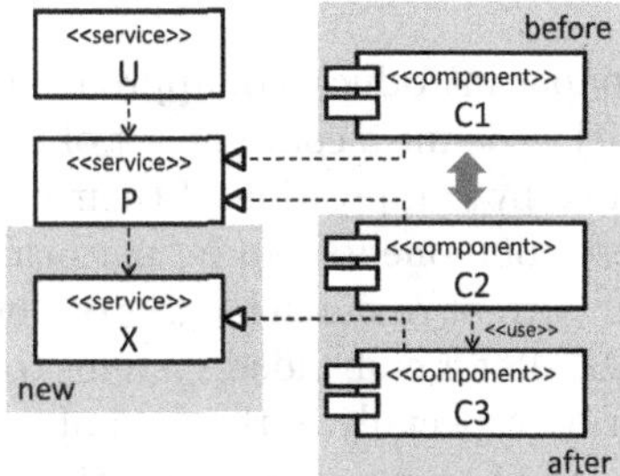

Fig. 3. Addition of a service

An example for this adaptation type is the exchange of the buffer replacement strategy, e.g. a FIFO strategy can be replaced by LRU (both implement the same service, but with a different algorithm).

Changes on L3 are also possible by component exchange on L2. Figure 3 shows the addition of a new service X implemented by component C3. Service P is realized by both component C1 and component C2 with the difference that C2 needs C3. Therefore C2 may only be deployed if a component that implements X is available. Obviously it is crucial to serialize the install operations in a suitable manner (C3 has to be installed before C1 can be replaced by C2). Removal of services on L3 is performed analogously.

3.2 Addition/Removal of Decorator Components

Another adaptation type is the addition or removal of decorator components. The scenario depicted in figure 4 shows the addition of a component C2 which acts as a decorator for C1. Both implement the same service P but the addition of C2 improves the non functional properties for the realization of P. An example is the addition of a buffer component to an I/O component in order to speed up the response time of P.

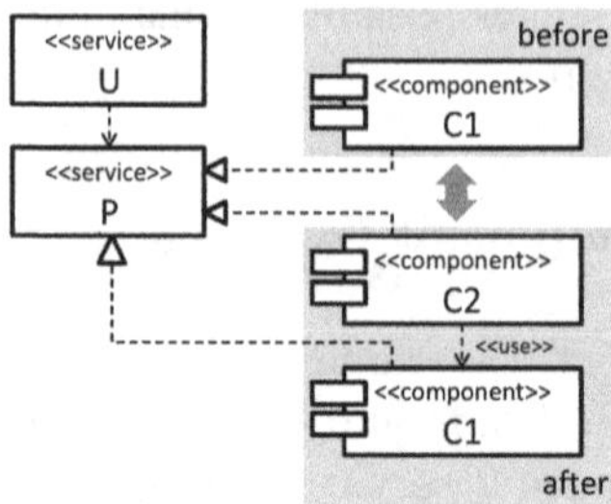

Fig. 4. Addition of a component

4 Conclusion and Future work

This paper introduces an approach to model adaptation in a DBMS. We specify a model to define a common set of functionality which can be tailored to different scenarios by composition of their implementing components and the adaptation of these components can be performed at runtime.

Our current research focuses on the finalization of a full-fledged runtime adaptable DBMS for embedded systems to evaluate the adaptability in more real world scenarios, especially in the field of pervasive computing. Another challenge is the definition of transaction semantics in adaptive systems and the enhancement of the framework to cover crosscutting concerns in order to build an adaptable transaction management.

References

1. Cervantes, H., Hall, R.S.: Autonomous adaptation to dynamic availability using a service-oriented component model. In: ICSE 2004: Proceedings of the 26th International Conference on Software Engineering, Washington, DC, USA, pp. 614–623. IEEE Computer Society, Los Alamitos (2004)
2. Irmert, F., Daum, M., Meyer-Wegener, K.: A new approach to modular database systems. In: Software Engineering for Tailor-made Data Management, pp. 41–45 (2008)
3. Irmert, F., Fischer, T., Meyer-Wegener, K.: Runtime adaptation in a service-oriented component model. In: SEAMS 2008: Proceedings of the 2008 International Workshop on Software Engineering for Adaptive and Self-Managing Systems (May 2008)
4. OSGi Alliance. OSGi Service Platform core specification, release 4 (August 2005)
5. Rosenmüller, M., Siegmund, N., Schirmeier, H., Sincero, J., Apel, S., Leich, T., Spinczyk, O., Saake, G.: FAME-DBMS: Tailor-made data management solutions for embedded systems. In: Proceedings of the EDBT 2008 Workshop on Software Engineering for Tailor-made Data Management, pp. 1–6. University of Magdeburg (2008)
6. Stonebraker, M., Cetintemel, U.: One Size Fits All: An idea whose time has come and gone. In: ICDE 2005: Proceedings of the 21st International Conference on Data Engineering, Washington, DC, USA, pp. 2–11. IEEE Computer Society, Los Alamitos (2005)

Schema Merging Based on Semantic Mappings

Nikos Rizopoulos and Peter McBrien

Dept. of Computing, Imperial College London, London SW7 2AZ
nr600@doc.ic.ac.uk, pjm@doc.ic.ac.uk

Abstract. In model management, the Merge operator takes as input a pair of schemas, together with a set of mappings between their objects, and returns an integrated schema. In this paper we present a new approach to implementing the Merge operator based on ***semantic mappings*** between objects. Our approach improves upon previous work by (1) using formal low-level transformation rules that can be translated into higher-level rules and (2) specifying precise BAV mappings, which merge schemas without any information loss or gain.

1 Introduction

Using the terminology of **model management** [2], the process of **data integration** requires the use of three operators. Firstly, ModelGen is used to translate schemas into some common data modelling language. Secondly, a Match operator is applied to two schemas and produces a set of mappings (sometimes called correspondences) between the objects found in the two schemas. Thirdly, the Merge operator takes this set of mappings and performs transformations on the schemas to produce a new integrated schema. In this paper, we propose an implementation of Merge based on **semantic mappings**, such as equivalence, subsumption, disjointness, *etc*, between objects.

In our approach, we have exhaustively identified all possible semantic mappings between pairs of objects, and defined formal Merge rules for all possible cases. Our rules merge schemas without any information loss or gain, and they form a framework for merging schemas represented in high level modelling languages, such as ER or relational [10]. Additionally, our fine-grained analysis identifies the cases where Merge can be fully automated and the cases which might require the user's intervention depending on the application setting.

2 The Merge Operator

Using a standard data integration approach [1], our Merge process is divided into three phases: (a) naming conforming, (b) unioning and (c) restructuring, with every phase having a set of formally defined rules. Here, we give an example of a rule for each phase.

Our approach uses a low-level graph language as the common modelling language, called the HDM [9], which represents data as *Nodes,Edges* and *Constraints*. For the purposes of this paper, we only use the subsumption constraint $\subseteq$, however our approach makes use of all the primitive constraint constructs in [4].

A.P. Sexton (Ed.): BNCOD 2009, LNCS 5588, pp. 193–198, 2009.

The transformation language we use in Merge is **both-as-view (BAV)** [9,8]. BAV is made up of primitive transformations, series of which form **pathways** which define mappings between schemas. There is the add transformation, together with its inverse delete transformation, which have the semantics that the object added/deleted is fully derivable from the other objects. In the case of nodes and edges, this means there must be a query q supplied, using a list comprehension [6] language called IQL [7], that gives the extent of the node or edge derived from the extent of nodes and edges already in the schema. Use of these transformations therefore has the **intentional domain preservation property (IDPP)** — there is no information gained or lost by their use. Another primitive transformation is rename, which also satisfies the IDPP. Finally, the extend transformation adds a schema object that is only partially derivable from the existing schema (i.e, query q returns a subset of the extent of the object). Use of this transformation does not satisfy the IDPP, and we wish to avoid its use in the rules defining our Merge operator.

2.1 Naming Conforming, Unioning and Restructuring

In the **naming conforming** phase, naming conflicts between the objects of the schemas are resolved. Given that there may be both homonym and synonym [1] conflicts between nodes and between edges, this leads us to have four rules: **Node Merge**, **Edge Merge**, **Node Distinction** and **Edge Distinction** rules. The Node Merge and Edge Merge rules resolve the synonym conflict, of having two equivalent ($\stackrel{S}{=}$) nodes or two equivalent edges, that do not have identical names. In such a case, the rules assign to the two equivalent objects a common name by generating two rename transformations. Formally, the Node Merge rule is the following:

$$\frac{\begin{array}{l}\mathsf{hdm{:}node{:}}\langle\langle \mathsf{N_1}\rangle\rangle \stackrel{S}{=} \mathsf{hdm{:}node{:}}\langle\langle \mathsf{N_2}\rangle\rangle\\ \neg\ \mathsf{identicalNames}(\mathsf{N_1},\mathsf{N_2})\\ \mathsf{commonName}(\mathsf{N_1},\mathsf{N_2},\mathsf{N'})\end{array}}{\begin{array}{l}\mathsf{rename}(\mathsf{hdm{:}node{:}}\langle\langle \mathsf{N_1}\rangle\rangle,\mathsf{hdm{:}node{:}}\langle\langle \mathsf{N'}\rangle\rangle)\\ \mathsf{rename}(\mathsf{hdm{:}node{:}}\langle\langle \mathsf{N_2}\rangle\rangle,\mathsf{hdm{:}node{:}}\langle\langle \mathsf{N'}\rangle\rangle)\end{array}}$$

All our rules take the form of conditions, defined above the horizontal line, which if satisfied cause the BAV transformations, below the line, to be generated.

The Node Distinction and Edge Distinction rules resolve the homonym conflict of having two non-equivalent objects with identical names. In this case, the objects have to be assigned distinct names to make them distinguishable.

The **unioning** phase is superimposing the two schemas; thus each pair of equivalent nodes, now with identical names, collapses into a single node. The superimposition of schemas in BAV is performed by a series of extend transformations. Note that this is the only place where our Merge operator uses extend, since it is stating here what *is not* directly available from one schema, but should only be sourced from the other.

In the rest of the unioning phase, subsumption, intersection and disjointness mappings between nodes are examined. The set of rules of this phase is the **Addition of Inclusion**, **Addition of Intersection** and the **Addition of Union** rules. The purpose of these rules is to identify any possible concepts that do not appear explicitly in the schemas but are implicitly defined. In addition, any appropriate constraints between the nodes are added. For example the Addition of Inclusion rule adds the $\subseteq$ constraint for each subsumption ($\stackrel{s}{\subset}$) mapping:

$$\frac{\text{hdm:node:}\langle\langle N_2\rangle\rangle \stackrel{s}{\subset} \text{hdm:node:}\langle\langle N_1\rangle\rangle}{\text{add(hdm:con:}\langle\langle \subseteq, \text{hdm:node:}\langle\langle N_2\rangle\rangle, \text{hdm:node:}\langle\langle N_1\rangle\rangle\rangle\rangle)}$$

The Addition of Intersection rule says that if an intersection mapping holds between two nodes, the common sub-domain is explicitly represented by an added *intersection node*. The Addition of Union rule adds the *union node* of two disjoint nodes.

In the final **restructuring** phase of Merge, the existence of equivalence, subsumption, intersection and disjointness mappings between edges are examined. The set of formally defined rules of this phase includes **Redundant Edge Removal** rules, **Optional Edge Removal, Specialisation of Edges, Addition of Intersection Edge** and **Addition of Union Edge** rules. The purpose of these rules is to minimize duplication and simplify the schema. For example, the Redundant Edge Removal can be applied when there are two equivalent edges and thus one is redundant:

$$\frac{\begin{array}{l}\text{hdm:edge:}\langle\langle e_1, N_1, N'_{1/2}\rangle\rangle \stackrel{s}{=} \text{hdm:edge:}\langle\langle e_2, N_2, N'_{1/2}\rangle\rangle\\ \text{hdm:node:}\langle\langle N_2\rangle\rangle \stackrel{s}{\subset} \text{hdm:node:}\langle\langle N_1\rangle\rangle\\ \text{constraints}(N_1, e_1, \text{Cons})\end{array}}{\begin{array}{l}\text{genDeleteCons(Cons)}\\ \text{moveDependents}(e_1, e_2)\\ \text{delete(hdm:edge:}\langle\langle e_1, N_1, N'_{1/2}\rangle\rangle, \text{hdm:edge:}\langle\langle e_2, N_2, N'_{1/2}\rangle\rangle)\end{array}}$$

The predicate constraints(N_1,e_1,Cons) instantiates the variable Cons with the list of constraint objects between N_1 and e_1. The rule calls the predicate genDeleteCons(Cons) which deletes Cons. moveDependents(e_1,e_2) replace the references to e_1 with references to e_2 by generating a series of add and delete transformations.

3 Merge properties

Our Merge process is based on formally defined rules. As seen in Table 1, we have exhaustively examined all possible cases of semantic mappings between objects. The first row of the table is concerned with the mapping x between nodes. The rest of the rows of the table are concerned with the mapping x between two edges.

The thorough examination of all possible mappings between two objects has identified cases where no simplification is possible, illustrated as NSP cells in the table, thus no rules have been defined for these cases. For example, edge disjointness cases can only be simplified if the nodes in at least one end of the edges are disjoint. This condition is necessary in order for the Addition of Union Edge rules to satisfy the IDPP. Additionally, we have identified cases where the semantic mappings specified between the objects are not possible, illustrated as SRNP cells in the table. For example, if the nodes at one end of the edges are disjoint then the edges cannot be equivalent based on the definition of equivalence [10].

The most important property of our rules is the IDPP. The fine-grained granularity of the BAV transformations and the low-level constraints used in the proposed approach allow us to reason about the correctness of the rules and demonstrate that the rules do

Table 1. Summary of merge rules possible for all combinations of the semantic mappings

case	y	z	x: $\stackrel{s}{\equiv}$	x: $\stackrel{s}{\subset}$	x: $\stackrel{s}{\cap}$	x: $\stackrel{s}{\not\cap}$
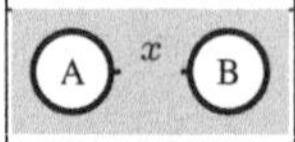			Node Merge/Distinction	Addition of Inclusion	Addition of Intersection	Addition of Union
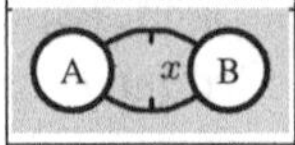			Edge Merge/Distinction	NSP	Addition of Edge Intersection	NSP
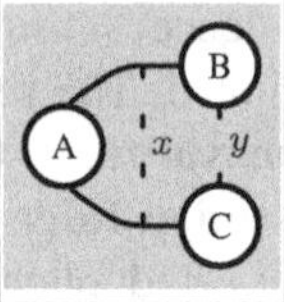	$\stackrel{s}{\subseteq}$		Redundant Edge Removal	Optional Edge Removal	Addition of Edge Intersection	NSP
	$\stackrel{s}{\cap}$		Specialisation of Edges	NSP	Addition of Edge Intersection	NSP
	$\stackrel{s}{\not\cap}$		SRNP	SRNP	SRNP	Addition of Union Edge
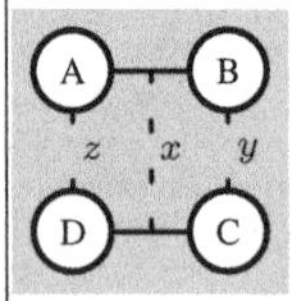	$\stackrel{s}{\subseteq}$	$\stackrel{s}{\subseteq}$	Redundant Edge Removal	Optional Edge Removal	Addition of Edge Intersection	NSP
		$\stackrel{s}{\cap}$	Specialisation of Edges	Optional Edge Removal	Addition of Edge Intersection	NSP
		$\stackrel{s}{\not\cap}$	SRNP	SRNP	SRNP	Addition of Union Edge
		$\stackrel{s}{\supset}$	Redundant Edge Removal	NSP	Addition of Edge Intersection	NSP
	$\stackrel{s}{\cap}$	$\stackrel{s}{\subseteq}$	Specialisation of Edges	Optional Edge Removal	Addition of Edge Intersection	NSP
		$\stackrel{s}{\cap}$	Specialisation of Edges	NSP	Addition of Edge Intersection	NSP
		$\stackrel{s}{\not\cap}$	SRNP	SRNP	SRNP	Addition of Union Edge
	$\stackrel{s}{\not\cap}$	$\stackrel{s}{\subseteq}$	SRNP	SRNP	SRNP	Addition of Union Edge
		$\stackrel{s}{\cap}$	SRNP	SRNP	SRNP	Addition of Union Edge
		$\stackrel{s}{\not\cap}$	SRNP	SRNP	SRNP	Addition of Union Edge

not violate the IDPP. For example, the Redundant Edge Removal rule does not violate the IDPP since the extent of the deleted edge e_1 can be retrieved from the equivalent edge e_2, as specified by the query attached to the delete transformation.

A by-product of the IDPP in the application of the rules, is that the order of their application is insignificant. Even if, for example, an object which is deleted by a rule R_1 is necessary for the application of another rule R_2, the IDPP ensures that the domain of the deleted object can still be retrieved from the current schema and therefore R_2 can still be applied. A recursive query re-writing process replaces each appearance of the deleted object in R_2 with the query specified in the object's delete transformation.

Finally, having exhaustively identified the possible rules, we can examine which rules and more specifically which parts of the rules can be fully automated and which parts need user intervention. The interesting predicates are the naming predicates used when a new object is added onto the schema. These are: commonName/3, distinctNames/4 and uniqueName/1. Possible automatic implementations of commonName/3 are: (a) identifying the common substring of the objects' names and (b) using the name of the object that belongs to the user's preferred schema. Predicate distinctNames/4 can also be automated quite simply: the distinct names can be produced by simply prefixing the name of the objects with the name of the schema they belong to. Similarly, the uniqueName/1 predicate can be fully automated by identifying a random name or by concatenating the names of the objects the rule is dealing with. In some cases, such implementations might be sufficient, *e.g.* in a meta-search engine where the merged schema is not presented to the user. On the other hand, if the schema is available to the user or it is used in some other data integration scenario, such a fully automated implementation might be problematic.

4 Summary and Conclusions

In this paper, we have presented an overview of our generic and formal framework to the Merge operator, which is used to produce a single integrated schema based on two existing schemas and supplied semantic mappings between the schema objects.

Our Merge process is based on formal rules, whose soundness and completeness can be proved, and whose fine-grained granularity enables the identification of the steps that can either be fully or semi-automated. Our approach is defined as rules over a low-level modelling language, the HDM, and it can form a framework for high-level model schema merging as illustrated in [10].

The work most related to ours is [3], where a generic Merge operator is proposed based on work in [5]. Two types of correspondences, equivalence and similarity, between nodes can be specified, but Merge cannot interpret the general similarity correspondence. Correspondences between edges cannot be specified. Instead, our approach examines both nodes and edges and in addition deals with a much wider set of mappings between them. However, in [3] multiple-type conflicts are resolved, which we do not consider. Cardinality conflicts though are implicitly resolved in our approach based on semantic mappings between edges. Additionally, [3] does not attempt to reduce structural redundancies and simplify the merge schema as we do in our approach.

References

1. Batini, C., Lenzerini, M., Navathe, S.: A comparative analysis of methodologies for database schema integration. ACM Computing Survers 18(4), 323–364 (1986)
2. Bernstein, P.: Applying model management to classical meta data problems. In: Proc. CIDR 2003 (2003)
3. Bernstein, P., Pottinger, R.: Merging models based on given correspondences. In: Proc. 29th VLDB (2003)
4. Boyd, M., Mçbrien, P.: Comparing and transforming between data models via an intermediate hypergraph data model. In: Spaccapietra, S. (ed.) Journal on Data Semantics IV. LNCS, vol. 3730, pp. 69–109. Springer, Heidelberg (2005)
5. Buneman, P., Davidson, S., Kosky, A.: Theoretical aspects of schema merging. In: Pirotte, A., Delobel, C., Gottlob, G. (eds.) EDBT 1992. LNCS, vol. 580, pp. 152–167. Springer, Heidelberg (1992)
6. Buneman, P., et al.: Comprehension syntax. SIGMOD Record 23(1), 87–96 (1994)

7. Jasper, E., Tong, N., McBrien, P., Poulovassilis, A.: View generation and optimisation in the AutoMed data integration framework. In: Proc. Baltic DB&IS 2004. Scientific Papers, vol. 672, pp. 13–30. Univ. Latvia (2004)
8. McBrien, P., Poulovassilis, A.: A uniform approach to inter-model transformations. In: Jarke, M., Oberweis, A. (eds.) CAiSE 1999. LNCS, vol. 1626, pp. 333–348. Springer, Heidelberg (1999)
9. Poulovassilis, A., McBrien, P.: A general formal framework for schema transformation. Data and Knowledge Engineering 28(1), 47–71 (1998)
10. Rizopoulos, N., McBrien, P.: A general approach to the generation of conceptual model transformations. In: Pastor, Ó., Falcão e Cunha, J. (eds.) CAiSE 2005. LNCS, vol. 3520, pp. 326–341. Springer, Heidelberg (2005)

A Database System for Absorbing Conflicting and Uncertain Information from Multiple Correspondents

Richard Cooper and Laura Devenny

Computing Science, University of Glasgow,
18 Lilybank Gardens,
Glasgow G12 8QQ
rich/devennyl@dcs.gla.ac.uk

Abstract. This paper discusses a database system which absorbs assertions about the data from a community of correspondents capturing also the uncertainty of the assertion and taking account of the potential unreliability of the correspondent. The paper describes a system compromising the capture of such assertions, the ability to impose an authorised version of a value, the maintenance of a reliability measure for each correspondent and a querying system which returns the most likely values.

Keywords: Collaboration, Uncertain Data, Reputation, Probabilistic Data.

1 Introduction

One of the most interesting uses of a database system is to hold a body of information brought together from a community of people with a shared interest. However, gathering information of this kind faces problems in that those supplying information may be more or less certain of it and may also be more or less reliable. Conflicts are likely in the information supplied and the best answers to queries are harder to judge.

We take as an example a group of aficionados wished to create a definitive web site about literature, each being to supply some information, but with more or less certainty. They must supply a degree of *confidence* in the information they contribute. Details from a book they own could be provided with 100% confidence while something they vaguely remember could be supplied with 10% confidence.

Humans are fallible and even information provided with 100% confidence can be totally wrong, so it is important to distinguish and record those more likely to be wrong - the *reliability* of a correspondent. This has much in common with the notion of reputation in commercial systems such as EBay [8].

Querying data is more complex. For each data item that might be returned there are likely to be several possible values each of which has been asserted with a different degree of confidence by correspondents with differing levels of reliability. Rather than just take an *ad hoc* decision on which version is correct, it is better to absorb all opinions and to make a probabilistic estimation of the likely correct version.

To do this requires a database capable of storing versions of data with an associated likelihood of correctness. The fundamental unit of a query result is the assertion

A.P. Sexton (Ed.): BNCOD 2009, LNCS 5588, pp. 199–202, 2009.

that a particular *property* of a particular *entity* has a particular *value* with a particular *confidence*, these being derived from all the assertions about this property. Handling this requires a probabilistic database [4] which can be queried in different ways: returning, for instance, the most likely guess as achievable in TriQL [7] does, the k most likely values [6] or only those values with a degree of confidence above a certain limit.

However, sometimes a value will become known with absolute confidence. We call this the "*authorised version*", coming about either through the an unimpeachable authority or by the evolution of a consensus among the community. Authorised versions must be stored separately, since they are different in kind to other values, even those with one correspondent asserting a value with 100% confidence. Recording *correspondent reliability* can now be based on agreement with the authority as well as agreement with other correspondents.

In this paper, we describe a database system designed to hold multiple versions of any data item, each associated with a likelihood of being true. The system is updated by assertions from correspondents and these are stored as the basic data, with the confidence values for each property being derived from these. Each is entered with a confidence rating and each correspondent has an associated reliability rating.

The rest of the paper describes how the database is built up and the information consolidated and then how querying works, before looking at related work and suggesting some extensions.

2 The Database

The work described here is part of an on-going project on Information Extraction from messages and uses an entity data model to hold data values [3]. A particular feature to manage data involved in human discourse is that it has two kinds of "key": a standard database key for entity identification and linkage in a database (*Dkey*) and the kind of term used for identification in human communication (*Hkey*).

The database is described in terms of a set of *entities* and their *properties*. Entities are grouped into entity type, each having DKeys and Hkeys. An *item* is one property of a particular entity and has a slot for the *authorised value*, if there is one. The fundamental storage structure in the database is not the table but the *assertion*, which consists of five aspects – the entity, E, and property, P, the correspondent, Cr, the value, V, asserted and the confidence of the correspondent, Cf. For any item, there will be several values stored, each asserted by different correspondents with different levels of confidence. Data is also stored about correspondents with a variety of identification fields, such as name, e-mail, etc. and a field for *reliability*.

To test the system, a web interface has been set up to capture assertions, record authorised versions and to respond to queries. To use the system, a schema must be entered by storing the meta-data. The correspondent selects a domain of interest, the entity and property involved and enters a value and a confidence level. The authorised versions are captured separately in a similar manner. The assertions gathered in this way are used to update correspondent reliability. This remains at 100% until a clash of opinion is encountered at which point the reliability is calculated as described in a later section of the paper.

3 Using the Information

The confidence estimation of each item value is calculated from the number of times that value is asserted and on the correspondents' reliability and their confidence in the assertions. To start with, we fully trust all correspondents - i.e. they all have 100% reliability. An overall confidence is calculated in a series of steps:

1. Any item with an authorised value has 100% confidence for that value and 0% confidence for any other value.
2. Items with only one assertion have the confidence that the correspondent has specified which may of course also be 100%.
3. An item with conflicting values all asserted with 100% confidence has its confidence calculated to be the percentage of correspondents asserting that value.
4. If conflicting assertions are made with varying degrees of certainty, then the confidence is calculated as the ratio of the sum of confidences of assertions suggesting that value to the sum of confidence values of all assertions involving that item.
5. Finally we introduce the notion of unreliability of correspondents and take the confidence we have in an assertion now to be the product of the confidence with which the assertion is made with the correspondent reliability:

Thus we end up with a range of values for each item, each with a probability of correctness which sums to unity and are in a position to return a range of different query results with a varying degree of detail.

Correspondent reliability is calculated from the assertions made by the correspondent. An assertion in which the correspondent was less sure should affect the reputation less than one with 100% confidence. The reputation score of an assertion is calculated as the sum of the product of the confidence percentages that the system and the correspondent have in the value and the product of the inverses of these. For instance, if a correspondent has 80% confidence in a value which the system thinks has a 60% chance of being right, the reputation is enhanced by the 80% times 60% indication of a correct value and also by the 20% times 40% indication of an incorrect value, leading to a value 56%. For any correspondent, their reputation is now simple to calculate as the average of the contributions of the individual assertions.

Querying the data can be achieved in a number of ways. The *best guess* result is simply the value for each item returned by the query with the highest degree of confidence. We can display the result either with or without the confidence we have in each value. The *best supportable guess* is the same, except that only values with confidence above a certain selected level are returned. If we are less sure about any of the values then we return a null value which represents "don't know" if there is conflict and "no information" if no-one has suggested a value at all. The returning interface also supports the display of *all values with confidence greater than a certain value* ordered by and annotated by their confidence and similarly, the *k most likely values* can be requested [6]. Finally *all of the values* can be requested again ordered by and annotated by their confidence, as above.

4 Summary and Conclusions

We have presented a form of probabilistic database [1] for capturing uncertain and conflicting data from a group of people who are attempting to build an information source for an area of shared interest, the most obviously related example being the more extensive TRIO system [7]. However, TRIO works at the implementation level of relations rather than at the conceptual level and has no mechanism for determining the source of confidence of an assertion (such our reliability measure), rather concentrating on how the confidence should be managed once it has been stored. MYSTIQ [2] is another related system which allows queries to be expressed with imprecision and ranks the query results by confidence and can query inconsistent data, produced, for instance by combining two data sources.

Reputation systems as used in on-line auctions [5] are the basis of our notion of correspondent reliability. Zachariah et al. report on two systems for managing reputation from the ratings of others - one for a loosely connected community and the other for a highly connected community [8]. Their algorithms are more sophisticated than ours and show one way in which we might evolve the system.

Other planned extensions are: the assertion of a *range of values* for a single valued item; the management of *multi-valued properties*; assertions made on *multiple items simultaneously*; varying the reliability score a correspondent according to *information domain*; providing more sophisticated algorithms for *combining data* for the various aspects of the system; and integrating it with our Information Extraction Engine [3]. More detail is available in a fuller version of this paper available on request.

References

1. Adar, E., Re, C.: Managing Uncertainty in Social Networks. IEEE Data Engineering Bulletin 30(20) (2007)
2. Boulos, J., Dalvi, N., Mandhani, B., Mathur, S., Re, C., Suciu, D.: MYSTIQ: A System for finding more answers by using probabilities. In: SIGMOD (2005)
3. Cooper, R., Ali, S., Bi, C.: Extracting Information from Short Messages. In: Montoyo, A., Muńoz, R., Métais, E. (eds.) NLDB 2005. LNCS, vol. 3513, pp. 388–391. Springer, Heidelberg (2005)
4. Dalvi, N., Suciu, D.: Management of Probabilistic Data: Foundations and Challenges. In: PODS 2007 (2007)
5. Houser, D., Wooders, J.: Reputation in Auctions: Theory, and Evidence from eBay. Journal of Economics & Management Strategy 15(2) (2006)
6. Re, C., Dalvi, N., Suciu, D.: Efficient Top-k Query Evaluation on Probabilistic Data. In: ICDE 2007 (2007)
7. Widom, J.: Trio: A System for Data, Uncertainty, and Lineage. In: Aggarwal, C. (ed.) Managing and Mining Uncertain Data. Springer, Heidelberg (2008)
8. Zacharia, G., Moukas, A., Maes, P.: Collaborative Reputation Mechanisms in Electronic Marketplaces. In: Proceedings of the 32nd Hawaii International Conference on System Sciences (1999)

Semantic Exploitation of Engineering Models: An Application to Oilfield Models

Laura Silveira Mastella[1], Yamine Aït-Ameur[2], Stéphane Jean[2], Michel Perrin[1], and Jean-François Rainaud[3]

[1] Ecole des Mines de Paris, Paris, France
[2] LISI/ENSMA and University of Poitiers, Futuroscope, France
[3] Institut Français du Pétrole, Rueil-Malmaison, France
{laura.mastella,michel.perrin}@ensmp.fr
{yamine,jean}@ensma.fr
j-francois.rainaud@ifp.fr

Abstract. Engineering development activities rely on computer-based models, which enclose technical data issued from different sources. In this heterogeneous context, retrieving, re-using and merging information is a challenge. We propose to annotate engineering models with concepts of domain ontologies, which provide data with explicit semantics. The semantic annotation makes it possible to formulate queries using the semantic concepts that are significant to the domain of the engineers. This work is inspired from a petroleum engineering case study and we validate our approach by presenting an implementation of this case study.

Keywords: Ontologies, Ontology-based databases, Semantic annotation, Oilfield engineering models.

1 Introduction

Engineering development activities produce a huge quantity of technical data that can be expressed in various types of models: database tables, programming modules, mathematical expressions, and so on. Retrieving and re-using information created in such heterogeneous models is a challenge. The engineering area, studied in this work, is the petroleum exploration, and, in particular, the activity of *oil & gas reservoir modelling*. Considering a typical reservoir modelling workflow, geoscientists rely on three-dimensional representations of the earth underground (called *reservoir models* or *oilfield models*) to take important decisions about oil-reservoir operations.

The proposal of this work is an approach based on *semantic annotation of engineering models*. We envisage the use of semantic annotation for: (*i*) making the expert knowledge explicit in the model and (*ii*) querying raw data using semantic concepts. To carry out this approach, we consider the use of an Ontology-Based Database (OBDB), that stores data and ontologies in a common and shared database. An implementation for the oil & gas reservoir modelling activity and some initial results are presented to illustrate how this approach enables emergence of the semantics of the concepts manipulated by engineering models.

A.P. Sexton (Ed.): BNCOD 2009, LNCS 5588, pp. 203–207, 2009.

2 Background

The last decade has seen the emergence of the use of *ontologies*, in order to provide explicit and formal semantics to specific domains [1]. Several tools support ontology-based annotations creation over resources (web pages, textual documents, multimedia files). From a comparative analysis of semantic annotation projects, available in [2], we understand that most of these tools still rely on knowledge stored in HTML pages, XML documents or in other textual resources. None of the annotation tools proposed so far, enable the annotation of *engineering models* (or, more generally, annotation of computer-based models). As a matter of fact, no technique allowing to complete computer-based models by formal comments or explanations, nor to attach more semantics to the technical data produced by modelling tools is available. Indeed, a big part of a company's knowledge can be found in text repositories, such as projects documentation and reports. Nevertheless, engineering models keep storing some strategic knowledge that cannot be lost. Next section proposes an approach for addressing these issues.

3 Proposed Approach

In order to make experts' knowledge explicit in engineering models, we propose to annotate these models with domain ontologies concepts and/or instances. The *engineering models annotation* process must consider the following elements: (*i*) ontologies and their instances; (*ii*) engineering models and their data and (*iii*) annotations of the engineering models, which establish links between the (*i*) and (*ii*).

(*i*) Knowledge related to the considered specialized fields has been designed and formalized as *domain ontologies* and stored in an ontology-based database.

(*ii*) We are interested in persisting engineering data in the same database where ontologies are stored. But it is not desirable to represent the engineering meta-data using constructs of ontologies, since we do not expect engineering models to have the same features as those that are currently proposed for ontologies (e.g., subsumption between concepts). Constructs of engineering meta-data should be different from those used to define ontologies (such as `owl:Class` in OWL language), because the two entities have different purposes. For these reasons, an *Engineering Meta-model* is defined. This Engineering Meta-model encodes the minimum necessary set of features that allows a uniform description of engineering models (file name, identificator, main composite objects, etc.). These constructs make it possible to represent the structure in which data are organized. The main constructs for building engineering meta-data are `#DataElement` and `#DataAttribute` (part (2) of Fig. 1).

(*iii*) Finally, we provide resources for linking engineering meta-models to the concepts of ontologies. In this context, each end-user may have a different interpretation of the model instances. For the same dataset, different annotations expressing each user's opinion probably exist. They must be uniquely identified.

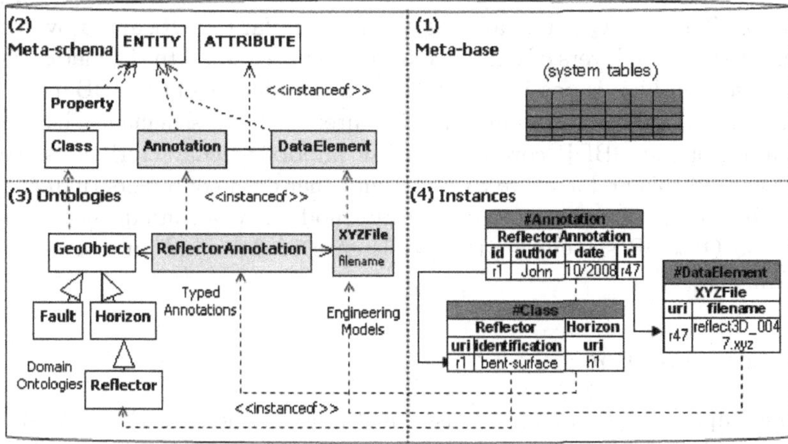

Fig. 1. Extension of OntoDB (2) and implementation of the case study (3 and 4)

One user should be able to annotate several data elements with one ontology concept, and vice-versa. As a consequence, a N-to-N relationship for the annotation elements is required. In this approach, annotation becomes a top-level entity, separated from the ontological concept and from the entity being annotated. The introduced annotation entity has also its own attributes, such as creation date, author name, version information, etc. Therefore, a *Meta-model for Annotation* is also required. The `#Annotation` construct creates a link between the construct of ontology concepts and the `#DataElement` construct through the relations `#annotates` and `#isAnnotatedBy`. The added meta-models are illustrated in part (2) of the UML diagrams of Fig. 1.

4 Case Study: Annotating Oilfield Models

In order to implement the case study, we store the whole data and knowledge manipulated by engineers in a persistent infrastructure. For this purpose, we use *ontology-based databases.*

4.1 Ontology-Based Databases (OBDBs)

Ontology-Based Databases (OBDB) address the persistence of ontologies while taking advantage of the characteristics of databases (scalability, safety, capability to manage a huge amount of data, etc.)[3]. The *OntoDB* system [4] makes a clear *separation of modelling layers*. The approach enables the extension of the core-model with constructors of other ontology models (e.g, RDF, OWL) and also the separation of the instances from their data structure and from their meta-model. The architecture of OntoDB is composed of four parts (see Fig. 1): system tables (1), meta-schema constructs (2), ontologies (3) and instances (1).

In order to exploit the OntoDB system, the *OntoQL* language has been proposed in [5]. The OntoQL language has a syntax similar to SQL, and provides operations at the three layers of OntoDB, from the logical level to the meta-schema level. Consequently, it is possible to extend the core-model of OntoDB using OntoQL Definition Language operators, which alter the meta-schema level. Support of evolution of the OBDB core-model is an important characteristic, since we need to extend this architecture to represent other data containers than the ontology meta-model (i.e., an *annotation* meta-model). As a consequence, we have chosen the OntoDB system for the persistence of data and ontologies in our approach.

4.2 Implementation

The first implementation step consists in extending the OntoDB's core-model (which already contains the ontology constructs `#Class` and `#Property`) to include the constructs of the **Engineering Meta-Model** and of the **Annotation Meta-Model** proposed in the previous section (see part (2) of Fig. 1). When the meta-model is set up, the oilfield meta-data have been defined using the new constructs for Engineering Meta-Models. For the case study, we chose a format known as *XYZ Format*, which represents raw data as 3D points. The OntoQL statement Q1 exemplifies the creation, by means of the added construct `#DataElement`, of an `XYZFile` element, with `filename` and `surfaceName` of type `String`, and multiplicity 1 as attributes (part (3) of Fig. 1). Fig. 1, part (4), shows an instance of such data element representing the file `reflect3D_0047.xyz`.

```
Q1: CREATE #DataElement XYZFile
     (PROPERTIES (filename String 1 1, surfaceName String 1 1))
```

The advantage of representing the technical data as instances within OntoDB is the capability to store both data and ontologies in the same repository, offering the possibility to create the link between the two.

Next, with the help of the end-user, the annotations that represent the experts' interpretation about field data are created. In the present case study, a well known annotation rule set up by experts is used: data contained in an `XYZ file` are interpreted by geologists as corresponding to some *Seismic Reflector*. *Seismic Reflector* is a term from the *GeoSeismics* domain, and it is represented as the ontology concept `Reflector`. Therefore, by means of the added construct `#Annotation`, the OntoQL statement Q2 creates an annotation-type that links elements of type `XYZFile` to concepts of type `Reflector` ((see part (3) of Fig. 1)).

```
Q2: CREATE #Annotation ReflectorAnnotation
     (XYZFileURI REF(XYZFile), ReflectorURI REF(Reflector))
```

Part (4) of Fig. 1 shows an instance of the typed-annotation *ReflectorAnnotation*, which refers to an instance of the meta-data `XYZFile` and an instance of the ontology concept `Reflector`.

4.3 Exploitation of the Extended OntoDB Architecture

At this point of the work, it is possible to query field data using concepts from the domain ontologies. To illustrate this querying capability, query Q3 retrieves the filename `reflect3D_0047.xyz`, which is interpreted as the *Seismic Reflector* identified by URI `r1`:

```
Q3: SELECT filename from XYZFile JOIN ReflectorAnnotation
      ON XYZFile.oid = ReflectorAnnotation.annotates.oid
     WHERE ReflectorAnnotation.isAnnotatedBy.oid =
      (select Reflector.oid from Reflector where Reflector.URI = 'r1')
```

Thanks to the new proposed constructs, the semantic concerning the engineering models, which is usually implicit within data, can be added in the database and retrieved by means of semantic queries.

5 Conclusions and Future Work

This paper has presented an extension of ontology-based databases that handles the semantic annotation of data elements issued from engineering models. As a consequence, we have obtained a homogeneous representation of the whole data and knowledge manipulated by engineers. This approach makes it possible to formulate queries that use domain specific semantic concepts instead of enforcing users to understand how data are stored within the database.

As future work, we intend to explore the multidisciplinary aspect of this domain. We aim at correlating data issued from various fields of expertise, by means of ontology mappings and subsumption relations.

Acknowledgments. This work is sponsored by The CAPES Foundation, Ministry of Education of Brazil (process no. 4232/05-4).

References

1. Gruber, T.: Toward principles for the design of ontologies used for knowledge sharing. Int. Journal of Human and Computer Studies 43(5/6), 907–928 (1995)
2. Uren, V., Cimiano, P., Iria, J., Handschuh, S., Vargas-Vera, M., Motta, E., Ciravegna, F.: Semantic annotation for knowledge management: Requirements and a survey of the state of the art. In: Web Semantics: Science, Services and Agents on the World Wide Web, vol. 4 (2006)
3. Broekstra, J., Kampman, A., van Harmelen, F.: Sesame: A Generic Architecture for Storing and Querying RDF and RDF Schema. In: Horrocks, I., Hendler, J. (eds.) ISWC 2002. LNCS, vol. 2342, pp. 54–68. Springer, Heidelberg (2002)
4. Dehainsala, H., Pierra, G., Bellatreche, L.: OntoDB: An ontology-based database for data intensive applications. In: Kotagiri, R., Radha Krishna, P., Mohania, M., Nantajeewarawat, E. (eds.) DASFAA 2007. LNCS, vol. 4443, pp. 497–508. Springer, Heidelberg (2007)
5. Jean, S., Aït-Ameur, Y., Pierra, G.: Querying ontology based database using ontoql (an ontology query language). In: Meersman, R., Tari, Z. (eds.) OTM 2006. LNCS, vol. 4275, pp. 704–721. Springer, Heidelberg (2006)

Ontology-Based Method for Schema Matching in a Peer-to-Peer Database System

Raddad Al King, Abdelkader Hameurlain, and Franck Morvan

IRIT Laboratory, Paul Sabatier University
118, route de Narbonne, F-31062 Toulouse Cedex 9, France
{alking,hameur,morvan}@irit.fr

Abstract. In a P2P DBS, the databases are often developed independently so their schemas are highly heterogeneous. Creating matching rules (henceforth MR) between a given mediated schema and each peer schema[1] at the design-time is not suitable for a volatile P2P environment; in which, a peer may participate in the system only once. For this reason, the MR must be done at the run-time. Schema designers are often the only persons knowing about the semantics of their schemas. At the run-time, one (or both) schema designer(s) could not be available; hence the user must be able to create the MR to support his/her changing requirements. Given that the semantics of a domain ontology is explicitly explained and in order to help the user to create the MR, we propose a schema matching method based on a domain ontology which plays a similar role as that played by a given mediated schema.

Keywords: P2P Databases, Schema Matching, Domain Ontology.

1 Introduction

A peer-to-peer database system (hereafter P2P DBS) could be seen as a large-scale distributed system, in which, the nodes (called peers) are autonomous and can join and leave the system in a completely decentralized way. Each node has its own database system (DBS) consisting of a database management system (DBMS), and one or more databases that it manages. Given that the relational model is the most used by DBMS today, we focus our study on relational databases.

1.1 Problem Definition

In P2P DBS, peer autonomy means that each peer has an autonomous administration on its resources. It develops and maintains its databases independently, thus leading to the schema heterogeneity problem which is still an open research area in spite of the big efforts [9] that made to solve it. The schema heterogeneity could appear at semantic and structural levels. While the first one involves: synonymy, polysemy and data value precision, the second occurs when two relations belonging to two schemas are

[1] We use "peer schema" and "local schema" interchangeably throughout the paper.

A.P. Sexton (Ed.): BNCOD 2009, LNCS 5588, pp. 208–212, 2009.

semantically similar but they are represented by different attributes (e.g., Doctor (Name, Salary) against Doctor (Name, Address)). Schema matching is the process of creating MR for reconciling the differences between two given schemas. Generally, determining the correspondences between two heterogeneous schemas in a fully automatic way is not possible [3, 6]; thus making schema matching a heuristic process that requires imperatively a human intervention. Any schema matching solution must reduce the time taken by a human to create the MR. The question here is when a human must intervene and who must create the MR: the designer or the user?

1.2 Related Work

In some existing projects of P2P DBS [1, 7], schema heterogeneity problem is solved due a mediated schema. Creating MR between local/mediated schemas must occur at the design time. The designers of both schemas must negotiate with each other to create the MR which are used at the run-time to translate automatically a user query. This solution is suitable for applications requiring sharing data for a particularly long duration. In such applications, it is worthy to negotiate between schema designers to create MR which are often hidden to users. In a volatile P2P environment, an unpredicted peer may participate in the system only once. There is no guarantee that this peer can join the system ever again. For this reason, the MR must be done at the run-time. The semantics of a given schema is not explicitly documented, it is found in the mind of the schema designer. At the run-time, schema designers could not be available so the user must be able to create the MR.

In the opposite of the solutions proposed in [1, 7] and in [4, 5] which cannot be easily extended to support user changing requirements; the authors of [8] suppose that the user must participate in the schema matching process. On each peer, the user creates for each relation name (and for each attribute name) of his/her peer schema a set of synonymous words. With the absence of a mediated schema, a synonym-based matching strategy has been proposed to match heterogeneous schemas at the run-time during the localization phase. The matching process occurs on the remote peers that may not have any matched schema so this solution could consume remote peer resources for no benefit.

Given that a domain ontology is explicitly explicated and considered as a user-friendly data model, section 2 presents an approach for query processing in an ontology-based P2P DBS. While section 3 explains our method for matching heterogeneous schemas, section 4 summarizes the paper content and presents our future works.

2 Query Processing in the OP2DB System

OP2DB is an ontology-based P2P DBS using a domain ontology (noted DO) to reconcile the semantic differences between the peer schemas. DO forms the only interface to interact between different peers in the system. OP2DB uses Chord protocol [10] to maintain the interconnection between all peers in the system. The user of OP2DB creates locally ontology/schema MR in order to be used for rewriting the terms of a query, sub-query or result written on local schemas into corresponding terms of the DO. When a user participates in the system, he/she must advertise the

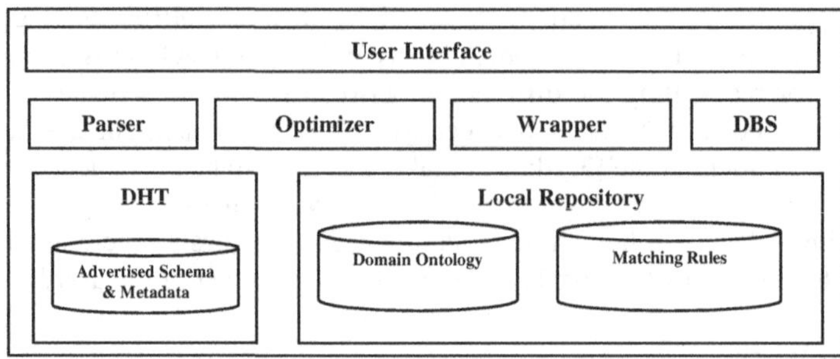

Fig. 1. Node software architecture

shared part of his/her peer schema. For this reason, Chord protocol generates for each relation a key. For each key, there is a responsible having as role to locate the sources storing the relation represented by the key. Each peer is responsible for injecting its keys at the moment of participation in the system and for deleting these keys at the moment of its departure. Based on figure 1, we summary the query processing approach in the OP2DB system. A user submits his/her query within a *User Interface*. After *Parsing* the query, *Query Reformulation* phase is carried out by a *Wrapper* which uses the MR created by the user. The objective of this phase is to make query understood by the remote peers. That is why the semantic and structural conflicts between the peer schema and the DO must be solved in this phase. After that, *Query Localization & Metadata Obtaining* phase locates relevant data sources. This phase is carried out by an extended version of chord protocol. The chord protocol is enriched by structure indexes [2] allowing each responsible for a given key to select relevant sources storing relation having the same structure of the relation represented by the key. Consequently, in this phase, the structural heterogeneity between peer schemas is solved and the relevant peers storing the required relations are selected. Furthermore, in this phase, the metadata required for the next phase are obtained. Next, a *Global Optimization* phase is carried out on the peer initiating the query. It is similar to global optimization in the data integration systems. The only different is that, in P2P environments, with the absence of a global catalog, global optimization is based on information obtained during the localization phase. The output of the previous phases is sub-queries written on the terms of the DO and sent to remote peers in order to enter in the *Local Optimization* phase and then in the *Execution* phase. Finally the results must be assembled and delivered to the user.

3 Schema Matching in the OP2DB System

Given that OWL (Web Ontology Language) is widely used today for modeling the Web resources, the OP2DB system uses OWL-based DO to make the user able to create MR between his/her peer schema and the DO at the run-time. The user must be able also to modify the MR in order to change the advertised part of his/her peer schema. To address this issue, the user must create for each relation name (and for each attribute name) a set of synonyms (e.g., for a relation named as “Doctor”, the user creates “Physician” and “Consultant”). Based on these synonyms, we propose similarity formulas helping to match a relation (respectively attribute) of a peer schema with its corresponding class (respectively datatype property) in the DO. We explain the similarity formulas as follows. Let R(a_1, a_2,... a_I) be an advertised relation where $a_{1\le i\le I}$, is the set of its attributes. Let C(d_1, d_2, ...d_J) be a class belonging to the DO where $d_{1\le j\le J}$ is the set of its datatype properties. The total similarity $0\le Sim_{Total}(R,C)\le 1$ between R and C is given by the next equation:

$$Sim_{Total}(R,C) = \alpha * SimRC + (1-\alpha) * SimAD \quad \text{with} \quad 0 \le \alpha \le 1 \tag{1}$$

While SimRC represents the similarity at the relation/class level, SimAD represents the similarity at the attribute/ datatype property level. α is a value indicating the importance of the relation/class level compared with the attribute/ datatype property level. If the both levels have the same importance then α = 0.5.

SimRC Term: This term is calculated by using the synonyms created by the user and the Tversky's measure [11]. The Tversky's measure assumes that the more common characteristics two objects (a and b) have and the less non common characteristics they have, the more similar the objects are. The Tversky's measure is done by equation 2 where A is a set of characteristics of a, B is a set of characteristics of b.

$$Sim(a,b) = \frac{|A \cap B|}{|A \cap B| + \beta * |A \setminus B| + (1-\beta) * |B \setminus A|} \quad \text{with} \quad 0 \le \beta \le 1 \tag{2}$$

β is a value indicating the importance of common characteristics compared with non common characteristics. By adapting Tversky's measure with our context, we obtain the equation 3, in which, Syn(R) is the set of R's synonyms and Syn (C) is the set of subclasses and equivalent classes of C.

$$SimRC = \frac{|Syn(R) \cap Syn(C)|}{|Syn(R) \cap Syn(C)| + \beta * |Syn(R) \setminus Syn(C)| + (1-\beta) * |Syn(C) \setminus Syn(R)|} \tag{3}$$

We consider the equivalent classes of C as C's synonyms. The C's subclasses are considered also as C's synonyms. For instance, in a medical ontology, "Dermatologist" could be used as a synonymous word of "Doctor". But, we cannot do the inverse thing. In other words, "Doctor" cannot be used as a synonymous word of "Dermatologist" because a "Doctor" is not necessarily a "Dermatologist".

SimAD Term: SimAD calculates the similarity at the attribute/ datatype property level. It is done by equation 4 where Sim(a_i, d_j) is the Tversky similarity between the both sets Syn(a_i) and Syn(d_j). Sim(a_i, d_j) is done by the equation 3 by replacing the Syn (R) by Syn(a_i) and Syn(C) by Syn(d_j).

$$SimAD = \frac{1}{I} * \sum_{i=1}^{I} \underset{1\le j\le J}{Max}(\, Sim(a_i, d_j)) \tag{4}$$

For a given relation R of the peer schema, we calculate Sim_{Total}(R, C) value for each class C of the DO. The classes having Sim_{Total}(R, C) value bigger than a given threshold are candidate to match the relation R. These candidate classes with their datatype properties, datatype values and their definition will be delivered to the user in order to create the MR concerning the relation R.

4 Summary and Ongoing Work

In this paper, we have presented an approach for query processing in an ontology-based P2P DBS and a method for schema matching in this system. Based on synonymous words added by the user, similarity formulas have been proposed to match a given relation belonging to a peer schema with other corresponding classes in the domain ontology on which the P2P DBS is based. As a future work, we will carry out experimental studies showing the benefits of our proposed method which will be extend to be adapted with P2P DBS based on multi-ontologies.

References

[1] Akbarinia, R., Martins, V.: Data Management in the APPA P2P System. In: Proceedings of the HPDGrid 2006, Rio de Janeiro, Brazil (2006)
[2] Al King, R., Hameurlain, A., Morvan, F.: Ontology-Based Data Source Localization in a Structured Peer-to-Peer Environment. In: Proceedings of the IDEAS 2008, vol. 299, pp. 9–18. ACM, New York (2008)
[3] Aumueller, D., et al.: Schema and Ontology Matching with COMA++. In: Proceedings of the ACM SIGMOD 2005, pp. 906–908 (2005)
[4] Gianolli, P.R., et al.: Data Sharing in the Hyperion Peer Database System. In: Proceedings of the VLDB 2005, pp. 1291–1294 (2005)
[5] Halevy, A., et al.: The Piazza Peer Data Management System. In: Proceedings of the TKDE 2004, vol. 16(7), pp. 787–798 (2004)
[6] Halevy, A.: Why Your Data Won't Mix: Semantic Heterogeneity. ACM Queue 3(8), 50–58 (2005)
[7] Huebsch, R., et al.: The Architecture of PIER: an Internet-Scale Query Processor. In: Proceedings of the CIDR 2005, pp. 28–43 (2005)
[8] Ng, W.S., et al.: PeerDB: A P2P-based System for Distributed Data Sharing. In: Proceedings of the ICDE 2003, pp. 633–644. IEEE, Los Alamitos (2003)
[9] Shvaiko, P., Euzenat, J.: A survey of schema-based matching approaches. In: Spaccapietra, S. (ed.) Journal on Data Semantics IV. LNCS, vol. 3730, pp. 146–171. Springer, Heidelberg (2005)
[10] Stoica, I., et al.: Chord: A Scalable Peer-to-Peer Lookup Service for Internet Applications. In: Proceedings of the ACM SIGCOMM 2001, pp. 149–160 (2001)
[11] Tversky, A.: Features of similarity. Psychological Review 84(2), 327–352 (1977)

Ideas in Teaching, Learning and Assessment of Databases: A Communication of the 7th International Workshop on Teaching, Learning and Assessment of Databases (TLAD 2009)

Anne James[1], David Nelson[2], Karen Davis[3], Richard Cooper[4], and Alastair Monger[5]

[1] Department of Computing and the Digital Environment, Coventry University, UK
a.james@coventry.ac.uk
[2] Department of Computing, Engineering and Technology, University of Sunderland, UK
david.nelson@sunderland.ac.uk
[3] Electrical and Computer Engineering Department, University of Cincinnatti, USA
karen.davis@uc.edu
[4] Department of Computing Science, University of Glasgow, UK
rich@dcs.gla.ac.uk
[5] School of Computing and Communications, Southampton Solent University, UK
al.monger@solent.ac.uk

Abstract. This paper is a record of the Seventh International Workshop on the Teaching, Learning and Assessment of Databases (TLAD 2009). Based on the contributions received, the paper describes the efforts that academics based in the UK and elsewhere around the world have made towards finding and disseminating new and interesting ways to enhance the teaching and learning of the database subject area. This year most of the submissions centred around the following areas: issues in teaching for databases and how to resolve these; methods for teaching enterprise systems development and the relevance of databases; interactive and collaborative learning support environments; and tools to aid learning in the database area.

Keywords: Databases, Teaching Methods, Learning Support, Tools.

1 Introduction

The teaching of databases is central to all Computing Science, Software Engineering, Information Systems and Information Technology courses. For the previous six years, Teaching, Learning and Assessment of Databases (TLAD) workshops have brought together those teaching database modules in order to share good learning, teaching and assessment practice and to explore which methods are particularly appropriate to teaching and learning in database and related modules.

The seventh workshop was held in conjunction with the 26th British National Conference on Databases (BNCOD 2009) in Birmingham. The workshop concentrated on novel teaching and assessment approaches for database and database related

A.P. Sexton (Ed.): BNCOD 2009, LNCS 5588, pp. 213–224, 2009.

modules, as well as covering other issues related to the teaching, learning and assessment of databases. Contributions were sought in the following areas: novel teaching areas and their relationship to databases; database and enterprise level computing; creating suitable curricula for the rapidly evolving ways in which database technology is used; use of appropriate and novel assessment techniques; use of appropriate and novel database technologies to enrich the student experience; ensuring that theory and practice are balanced in database teaching; exploiting innovative teaching methods in order to enhance the student learning experience; exploiting effective learning and teaching mechanisms for distance learning; pedagogical and organisational frameworks and paradigms; adaptivity in learning systems; integrating the teaching of databases from schools, through universities to the CPD of those in employment; teaching of internet programming and their use of databases; teaching of ubiquitous and pervasive application programming and their use of databases; relationship between XML and databases; and teaching methods for data mining, data warehousing and OLAP.

A number of very relevant and interesting contributions were received this year which generally fell into the following four main categories: issues in teaching databases and methods of resolution; database methods for enterprise systems development; interactive collaborative learning; and visualisation and tools to support database learning. As might be expected, some submissions had aspects that were relevant to more than one category.

The remainder of this paper is organized as follows. Section 2 covers issues in teaching databases and methods of resolution. Section 3 describes work in teaching enterprise systems development and its relevance for databases. Section 4 covers ongoing activities in interactive collaborative learning and support environments, particularly as these relate to database teaching. Section 5 discusses tools that have been developed to aid database learning. Finally, section 6 offers a conclusion on current activities in database teaching, learning and assessment.

2 Issues in Teaching Databases and Methods of Resolution

When teaching databases, many of us have found that students have problems understanding some important concepts. For instance in entity-relationship modelling students sometimes are unable to identify weak entities, they may not identify key attributes correctly, they may miss relationships or get cardinalities wrong. When writing SQL they may fail to appreciate the need for precision in naming, they may not understand the domain rule when specifying values, they find the "group by" clause particularly hard and they may not understand how to link between tables for a complex query.

With suitable practice and assimilation we imagine these skills will be acquired, but interestingly, a recent European survey found that one of the skills companies consider to be most lacking in new IT graduate recruits was database design [1]. Connolly [2] believes this is due to the abstract nature of analysis and design and also the lack of skills in areas such as team and project work. Technical understanding may also have some bearing.

Karen Renaud, Huda Al Suaily and Richard Cooper of the University of Glasgow, UK, have looked into this area in detail [3]. They have studied particular common errors that students make when doing practical work in databases and have also carried out a survey of database lecturers to try to ascertain the areas in which students have most difficulty and why. An important aspect they consider is the tension between the need to teach core underlying topics of database technology and the pressure to produce students with the marketable skills necessary to meet business database requirements such as knowing how to use a particular product. Having noted areas of difficulty, Renaud et al looked at theories of learning such as those of Bloom [4] and Gorman [5] and noted that previous studies have shown that what one learns early on tends to stick more [6]. The contributors believe that the reason students sometimes have difficulty applying database skills, such as entity-relationship modelling and SQL, is that they have often not yet assimilated core database concepts such as functional dependencies, keys and relations. The authors argue that it is thus very important to impart core knowledge first and later students can build skills up on top of that core knowledge.

The authors provide us with a phased pattern of database teaching designed to be spread over a number of years which is based on Gorman's simpler taxonomy [5] and informed by work on applying Bloom's taxonomy to database design [7]. Their recommendation is that in the first year one should teach the underlying concepts like functional dependencies, relations, transactions, primary keys, indexing, storage structures, recovery and concurrency. In the following year, practical database skills like E-R Modelling and SQL can be taught. Finally, in the following year the skills can be contextualised and grounded in real-life settings. This pattern should produce confident database practitioners.

Sue Barnes and Joanne Kuzma of the University of Worcester, UK, have made a study of two occurrences of the same database module [8]. The student intake differed in terms of previous experience, attendance patterns, grades, participation and attitude. Barnes and Kuzma attempt to identify the reasons for this difference. They also evaluated some of the teaching activities in the module in terms of their success and the perception and feedback from the students, and recommended alterations to the module based on their experiences.

Pirjo Moen of the University of Helsinki, Finland, provided an interesting discussion of the use of concept maps to support database concept learning [9]. Concept maps are diagrams showing relationships between different concepts within a certain theme. In a concept map, each concept is connected to other concepts and linking words are used to connect the concepts. Concept maps are known to develop students' logical thinking and study skills and have been widely used in many sectors.

Moen points out that at university level, concept maps have been used successfully in many disciplines, such as chemistry, biology, medicine, psychology, history and computer science. Concept maps have even been applied in teaching databases, e.g., in teaching E-R modelling.

As a trial of their efficacy in aiding the learning of database concepts, Moen used the technique as a learning and assessment method in a distributed database course. Students were required to build a concept map for the concepts of each weekly topic.

Groups of students then compared their maps so that individual students were able to see different ways to structure the concepts. Finally individual students revised their concept maps according to what they had learnt through the group comparison. CMapTools software was used for building the concept maps [10].

The final outcome was that students felt they had learned the fundamental issues and concepts more thoroughly through the use of concept maps than they would have done with more traditional learning methods. They found that the database concepts were easy to represent in a concept map, but that the use of maps for describing small details and different processes were more difficult. Moen concluded that concept maps are a good tool to enhance learning of database concepts.

3 Teaching Enterprise Systems Development and Databases

Anne James of Coventry University, with industry collaborators, Steve Bushell of Mercato Solutions, Birmingham and Peter Robbins of Probrand, Birmingham, discussed how to enhance database teaching and learning using real-world applications [11]. The context of this discussion was a final year module on enterprise systems development run at Coventry University. The module builds on a second year database module which provides an in-depth coverage of entity-relationship modelling, normalisation and relational databases, with some introduction to recovery, security and concurrency. The idea in this final year module is to put the theory learnt earlier to good practice by including activities based on real-world innovative applications.

To help with the supply of innovative applications to support the teaching of data management in an enterprise context, Coventry University is collaborating with two small, very successful, and rapidly growing local companies. One of the local companies has developed an e-procurement procedure that can capture a supply chain and allow purchasers to make best purchase decisions based on real-time pricing. The system encompasses a patented process. The second company markets and develops the software system to support the process. It is planned that students taking the module will develop a close relationship with the companies during their study.

In terms of applied database learning, the e-procurement process will be explored in detail by the students and in particular the importance of data capture will be emphasised. This study will enhance the students' understanding of workflow, a very important part of enterprise systems development. Students will also be given the opportunity to use and examine the e-procurement software. A major part of the process is the cleansing and warehousing of data. As a supply chain encompasses many different suppliers each with their own database, getting the data into a normalised form which all parties can understand and work from is a daunting and far from straightforward process.

Integration is an issue for many companies in many areas of application and it becomes more pertinent as our global databanks grow and as we wish to increase electronic intercommunication. Experience of a real problem of this kind will help the students understand the issues involved in data warehousing. The area of scalability will be covered by looking at sizing of the data sets in systems and how events such as new suppliers and purchasers joining the supply chain are handled. This area will lead to consideration of distribution of data sets across servers and how this may be best achieved in different contexts. In turn this will lead to understanding how decisions may be made on ICT infrastructure for an enterprise.

Linking databases to software applications will be covered by examining components of the e-procurement system and gaining understanding of how data may be linked to a variety of third party software products such as SAP and SAGE. The concept of rules and triggers in data management will also be covered as the E-procurement system contains business logic and alerts based on price changes or new agreements which result in identification of best purchase decisions. An innovative feature of the E-procurement system is a daily online auction in which suppliers may offer their best price that day. This real-time on-line interaction raises interesting issues for concurrency management and transaction handling in databases.

The idea of tying some aspects of this module so closely with an innovative product of a local commercial enterprise is to give students a feel for the real issues they will face in developing databases and commercial systems. Direct experience of the e-procurement and supply chain system will enhance the students' knowledge and serve them well in their career search upon graduation. The approach is in line with a move towards activity led learning where students learn actively rather than passively [12, 13]

Alastair Monger, Sheila Baron and Jing Lu of Southampton Solent University, UK, discussed Oracle Application Express (APEX) [14]. They expanded on work presented at TLAD'08 [15] by Tomlinson and Gardner on teaching enterprise database application development using APEX [16]. APEX [17] is essentially a web-based development tool for building web-based database applications. Form, Report, Home and Login pages with default functionality can be developed quickly using wizards, and these pages can then be developed into usable and accessible applications using PL/SQL, trigger and other established Oracle technology.

Monger et al focused on the following areas: administration, scalability and reliability for teaching; teaching and learning of introductory database application development; promotion and monitoring of engagement and feedback of learning; and teaching and learning of more advanced database application development. They described an evaluation that was carried out based primarily on student performance and APEX data derived from the delivery of a level 1 "Introduction to Databases" unit. The APEX development environment was structured around development and production workspaces for each class. The Oracle-supplied OEHR (OE – Order Entry, HR – Human Resource) sample objects schema was used. The administration of the student roles such as developers, applications and users was an interesting aspect.

The contributors showed how APEX features such as accessibility, application windows and monitoring statistics can be used to promote and monitor engagement and feedback of learning. The monitoring statistics included, for example, the number of days and hours a student is using APEX and the number of page views. Monger et al also evaluated how effective APEX was for learning other more advanced database application development concepts such as concurrency, database access control and database auditing.

A positive view was provided overall in relation to the four criteria and the view of Tomlinson and Gardner [15] that APEX is an appropriate tool for the teaching and learning of databases was supported.

4 Interactive and Collaborative Learning

Building on previous work [18, 19], Karen Davis and Angela Arnt of the University of Cincinnati, USA, presented novel methods for teaching database design concepts through the use of interactive collaborative experiences in the classroom. The activities and their assessment were described in terms of a blueprint for learning [20] that aligns course content with cognitive processes and knowledge domains based on Bloom's revised taxonomy [21]. The course blueprint design has four knowledge dimensions: factual; conceptual; procedural; and meta-cognition. In the knowledge dimension teaching goals, learning objectives and learning experiences are important as well as evaluation plans that link assessment to achievement of the learning outcomes.

The novel methods include in-class activities to motivate each topic. For example the course begins with writing basic SQL queries on the first day and this is used as a foundation throughout the course for discussing topics such as data models, data architectures, query processing and optimisation, normalisation, physical storage and transaction processing. Other examples include teaching aggregate SQL queries using a paper manipulative technique and teaching normalisation through flawed relational designs. Student learning is assessed via project and examination questions as well as through student reflection on their learning experience.

Tugrul Esendal and Matthew Dean of De Monfort University presented a model for teaching internet programming with databases [22]. They advocated the teaching and learning of databases within a software development model using a software engineering approach, and moving from introductory programming and database fundamentals in one year to 3-tier, object-oriented solutions in the next year. The 3-tier solution includes internet programming in the top tier, business intelligence in the middle tier, and database interfacing in the bottom tier. The environment used was Visual Studio.NET, with ASP.NET, VB.NET, and ADO.NET for the three tiers, respectively.

An innovation of Esendal and Dean's approach is peer learning, supported by a web-based, in-house tool that gives individuals feedback on the quality of their code. "Quality" means adherence to various guidelines and targets. The complete model operates in a "figure-of-eight" framework, in an upward spiral. The importance of feedback in formative assessment is noted [23].

The approach operates over two years with fundamental programming and database processing via highly abstracted, pre-set interfaces in the first year and user-created SQL, database objects, and custom-made own classes in the second year. The second year is also the year when the learning of databases culminates in the development and delivery of database applications drawn from real life, e-auctions being a typical example. There is less peer learning in the second year than in the first year as students become more independent, although some still takes place, via wikis, laboratory assistantships and group contributions to the module library of database classes. The difficulty many students have with programming is recognized and thus considerable time is devoted to this in year 1 [24].

Richard Cooper [25] described an approach that involved the use of coursework that draws together many diverse strands of database programming in the area of data intensive internet application design. The problem with teaching this area is that there are so many technologies contributing to the final solution: languages for client side and server side scripting, database access, protocols for message passing, web services and relational and XML data models. Students must be introduced to many types of programming task. Each of these is individually a straightforward and familiar task, i.e. they are simple SQL queries, small JavaScript functions and regularly structured server side programs. However when brought together the development task may seem overwhelming.

The coursework was designed to allow the students to exercise the various technologies in a coherently structured way. There are so many different tasks to complete that it would be beyond the range of an average student to produce more than a minimal exercise. Consequently, the work is carried out in groups of three. Although it is expected that each group member may well concentrate on only one part of the design, they are encouraged to study the whole of the submission so that they understand both the whole structure and details of parts that they may not have been much involved in producing. In order to maintain the interest of the whole class, an application is chosen where the structure is distributed and for which each group can act as a node in the network.

The exercise breaks down to completing the following steps: (i) creating a database; (ii) designing a consistent page structure; (iii) creating a cascading style sheet; (iv) writing pages for a user management system: (v) writing further pages to deliver the functionality for customers; (vi) adding some privileged pages for staff; (vii) producing a standalone content management system.; (viii) creating an XML file of the core data; (ix) storing the XML file in an XML repository; and (x) absorbing the XML files from other groups. The students then have to come up with two reports: a group report describing the application including E-R diagrams and the WebML design diagrams for site views, style guidelines and navigation diagrams; and a solo report discussing the appropriateness of the technology and how the site could be evolved for different client devices. The authors concluded by suggesting a number of improvements for the next iteration of the module which include: evolving the assessment to include active learning so that the students engage more deeply in the learning process; further consideration of accessibility issues; consideration of security issues such as SQL Injection and cross-side scripting; and switching to utilize peer-to-peer architectures so that students further appreciate the challenges of distributed architectures.

5 Visualisation and Tools to Support Learning

John Wilson and Fraser Hall from the University of Strathclyde described a learning tool that enables users to develop a correct mental model of the XPath language which can then be used to generate queries that can then be incorporated into other systems [26]. The effort is informed by previous related work [27, 28].

The interface provides a tabbed page with segments for tutorial, visualisation and test activities. The tutorial page supports step-through of a set of examples. At each stage in this step-through process, the user is presented with a data structure and conducted through the construction of an XPath query. On the practice page, the user can create their own query and view the results of that query in the visualisation section. On the testing page, the user is provided with a series of questions and asked to construct XPath expressions against a specified data structure. The query is evaluated against the data structure at the request of the user. For incorrect solutions, the correct parts of the query are highlighted to the user, who is provided with another opportunity to formulate the correct query.

The tutoring system has been evaluated with both XML-experienced and inexperienced students and found to be effective in developing an understanding of XPath. It provides the opportunity for students to review material and was generally well received by users. Although the system presents a successful framework for conceptualising XPath queries, it also highlights the need for more complex interactions that could identify consistent student errors [29].

Motivated by ideas for automated tutoring in database distance learning [30], Carsten Kleiner of the University of Applied Sciences and Arts in Hannover, Germany, introduced the concept for automated grading of exercises in introductory database system courses [31]. The contributor referred to a tool called SQLify [32] for partial automated grading of SQL queries. SQLify focuses heavily on peer-reviews and interaction in order to improve student learning. Kleiner suggests that a web-based tool like WebCAT [33] would be more suitable for large classes where a less interactive approach would be necessary. WebCAT includes facilities for both automated and manual grading. Some ideas were presented of the requirements of a web-based automatic grading tool for SQL. Criteria such as syntax, style and efficiency were discussed as possible axes for automated grading.

Christian Goldberg of the Martin-Luther University Halle-Wittenberg in Germany discussed semantic errors in database queries [34]. Errors in SQL queries can be classified into syntactic errors and semantic errors. Semantic errors can be further classified into those where we must understand the domain and task to determine whether an error has occurred or not and those where we can detect an error without knowing the domain or goal of the query (e.g. WHERE grade = "A" AND grade = "F") [35]. Consideration was given to techniques and tools that could be built to catch both syntactic and semantic errors of the sort where one needs not know the domain and goal [36].

Josep Soler, Imma Boada, Ferran Prados, Jordi Poch and Ramon Fabregat of the University of Girona, Spain described a web-based e-learning tool for the continuous assessment and skills improvement of main database course topics [37]. The ACME-BD is a web-based e-learning-tool for skills training and automatic assessment of main database course topics. It integrates five different modules that automatically correct exercises of: entity-relationship diagrams [38], relational database schemas [39], database normalisation, relational algebra and SQL queries [40]. A teacher enters database problems into the system and this generates personalised workbooks with different exercises for each student. Students propose a solution and enter it into the system using the interface specifically designed for this purpose. The solution is automatically corrected and immediate feedback is provided. In case of errors advice about how to correct them is given. All the information about student solutions is recorded by the system and it can be used for student assessment. The platform has been used since 2005 and student results have improved. The authors discussed their experiences with the tool.

6 Conclusion

The 7th Workshop on Teaching, Learning and Assessment in Databases (TLAD 2009) bore witness to a healthy interest in the subject area with interesting contributions from many countries. Current interest lies particularly in pedagogic models to inform the teaching and learning of database systems in an effective way, ideas for bringing in real-world applications to the database teaching arena, methods for teaching the wider applications of database such as internet systems development, and computerized tools for supporting learning and assessment. The 7th workshop has provided many interesting ideas to inform our teaching and assessment of databases.

Acknowledgement

The TLAD 2009 Workshop Co-chairs, Anne James and David Nelson would like to thank the TLAD 2009 Programme Committee for their considerate work in reviewing abstracts and full submissions to tight deadlines. Membership of the TLAD 2009 Programme Committee is as shown below:

Anne James (Coventry University)
David Nelson (University of Sunderland)
Fang Fang Cai (London Metropolitan University)
Jackie Campbell (Leeds Metropolitan University)
Richard Cooper (University of Glasgow)
Karen Davis (University of Cincinnati)
Barry Eaglestone (University of Sheffield)
Mary Garvey (University of Wolverhampton)
Caron Green (University of Sunderland)
Petra Leimich (University of Abertay)
Nigel Martin (Birkbeck University of London)
Pirjo Moen (University of Helsinki)
Alastair Monger (Southampton Solent University)
Uday Reddy (University of Birmingham)
Mick Ridley (University of Bradford)
Nick Rossiter (University of Northumbria)
John Wilson (University of Strathclyde)

References

1. Connolly, T.M., Stansfield, M.H., McLellan, E.: Using an Online Games-Based Learning Approach to Teach Database Design Concepts. Electronic Journal of e-Learning 4(1), 103–110 (2005)
2. Connolly, T.M., Begg, C.E.: A Constructivist-based Approach to Teaching Database Analysis and Design. Journal of Information Systems Education 17(1), 43–53 (2006)
3. Renaud, K., Al Shuailu, H., Cooper, R.: Facilitating Efficacious Transfer of Database Knowledge and Skills. In: 7th Workshop on Teaching, Learning and Assessment in Databases, Birmingham, UK, HEA (2009) (to be published)
4. Bloom, B. (ed.): Taxonomy of Educational Objectives, the Classification of Educational Goals - Handbook I: Cognitive Domain. McKay, NewYork (1956)
5. Gorman, M.E.: Types of knowledge and their Roles in Technology Transfer. Journal of Technology Transfer 27(3), 219–231 (2002)
6. Little, A.C., Penton-Voak, S., Burt, M., Perrett, D.I.: Investigating an Imprinting-like Phenomenon in Humans: Partners and Opposite-sex Parents have similar Hair and Eye Colour. Evolution and Human Behavior 24(1), 43–51 (2003)
7. Mohtashami, M., Scher, J.M.: Application of Bloom's Cognitive Domain Taxonomy to Database Design. In: ISECON (Information Systems Educators Conference), Philadephia, USA (2000)
8. Barnes, S., Kuzma, J.: Empirical Case Study in Teaching First-Year Database Students. In: 7th Workshop on Teaching, Learning and Assessment in Databases, Birmingham, UK, HEA (2009) (to be published)
9. Moen, P.: Concept Maps as a Device for Learning Database Concepts. In: 7th Workshop on Teaching, Learning and Assessment in Databases, Birmingham, UK, HEA (2009)
10. Novak, J.D., Canas, A.J.: The Theory Underlying Concept Maps and How to Construct and Use Them, Technical Report IHMC CMapTools 2006-01, Rev 01-2008, `http://cmap.ihmc.us/Publications/ReserachPapers/theoryUnderlyingConceptMaps`.pdf
11. James, A., Bussell, S., Robbins, P.: New Real-world Application to enhance Database Teaching. In: 7th Workshop on Teaching, Learning and Assessment in Databases, Birmingham, UK, HEA (2009) (to be published)
12. Iqbal, R., James, A.: Scenario based Assessment for Database Course. In: 6th Workshop on Teaching, Learning and Assessment in Databases, Cardiff, UK, HEA (2008)
13. Iqbal, R., James, A., Payne, L., Odetayo, M., Arochena, H.: Moving to Activity-led Learning in Computer Science. In: iPED 2008 3rd International Conference, pp. 125–131 (2008)
14. Monger, A., Baron, S., Lu, J.: More on Oracle APEX for Teaching and Learning. In: 7th Workshop on Teaching, Learning and Assessment in Databases, Birmingham, UK, HEA (2009) (to be published)
15. Proceedings of the 6th Workshop on Teaching, Learning and Assessment in Databases (TLAD 2008), Cardiff, UK, HEA (2008)
16. Tomlinson, A., Gardner, K.: Teaching Enterprise Database Application Development Using Oracle Application Express. In: 6th Workshop on Teaching, Learning and Assessment in Databases, Cardiff, UK, HEA (2008)
17. Oracle, Oracle Application Express, `http://apex.oracle.com/i/index`.html
18. Davis, K.: Enhancing Accountability in a Database Design Team Project. In: 3rd Workshop on Teaching, Learning and Assessment in Databases, Sunderland, UK, HEA (2005)

19. Davis, K., Arndt, A.: Assessing the Effectiveness of Interactive Learning in Database Design. In: 7th Workshop on Teaching, Learning and Assessment in Databases, Birmingham, UK, HEA (2009) (to be published)
20. Richlin, L.: Blueprint for Learning Sterling. Stylus Publishing, VA (2006)
21. Anderson, L.W., Krathwohl, D.R. (eds.): A Taxonomy for Learning, Teaching and Assessing: A Revision of Bloom's Taxonomy of Educational Objectives. Addison-Wesley, Longman (2001)
22. Esendal, T., Dean, M.: A Model to Make the Learning of Internet Programming with Databases Easy. In: 7th Workshop on Teaching, Learning and Assessment in Databases, Birmingham, UK, HEA (2009) (to be published)
23. Crooks, T.: The validity of formative assessments. In: Annual Meeting of the British Educational Research Association (BERA), Leeds, England (2001)
24. Teague, D.: Learning to Program: going Pair-shaped. ITALICS e-Journal 6(4) (2007)
25. Cooper, R.: Coursework Design for Teaching Distributed Data Intensive Internet Application Design. In: 7th Workshop on Teaching, Learning and Assessment in Databases, Birmingham, UK, HEA (2009) (to be published)
26. Wilson, J., Hall, F.: A learning environment for XPath. In: 7th Workshop on Teaching, Learning and Assessment in Databases, Birmingham, UK, HEA (2009)
27. Grissom, S., McNally, M.F., Naps, T.: Algorithm Visualization in CS education: Comparing Levels of Student Engagement. In: 2003 ACM Symposium on Software Visualization, pp. 87–94 (2003)
28. Moreno, A., Myller, N., Sutinen, E., Ben-Ari, M.: Visualizing Programs with Jeliot 3. In: Working Conference on Advanced Visual Interfaces, pp. 373–376 (2004)
29. Ma, L., Ferguson, J., Roper, M., Wood, M.: Investigating the Viability of Mental Models held by Novice Programmers. In: 38th SIGCSE Technical Symposium on Computer Science Education, pp. 499–503 (2007)
30. Kenny, C., Pahl, C.: Automated Tutoring for a Database Skills Training Environment. SIGCSE Bull. 37(1), 58–62 (2005)
31. Kleiner, C.: A Concept for the Automated Grading of Exercises in Introductory Database Courses. In: 7th Workshop on Teaching, Learning and Assessment in Databases, Birmingham, UK, HEA (2009) (to be published)
32. Dekeyser, S., de Raadt, M., Lee, T.Y.: Computer Assisted Assessment of SQL Query Skills. In: 18th Conference on Australasian Database, 63, Ballarat, Victoria, Australia, pp. 53–62 (2007)
33. Agarwal, R., Edwards, S.H., Perez-Quinones, M.A.: Designing an Adaptive Learning Module to Teach Software Testing. In: 37th SIGCSE Technical Symposium on Computer Science Education, CSIGCSE 2006, Houston, Texas, USA, pp. 259–263. ACM, New York (2006)
34. Goldberg, C.: Do you know SQL? About Semantic Errors in Database Queries. In: 7th Workshop on Teaching, Learning and Assessment in Databases, Birmingham, UK, HEA (2009) (to be published)
35. Brass, S., Goldberg, C.: Semantic Errors in SQL Queries: A Quite Complete List. Journal of Systems and Software 79(5) (2006)
36. Brass, S., Goldberg, C.: Proving the Saftey of SQL Queries. In: 5th Intl. Conf. on Quality Software (QSIC 2005). IEEE Computer Society Press, Los Alamitos (2005)
37. Soler, J., Boada, I., Prados, F., Poch, J., Fabregat, R.: A web-based e-learning tool for the continuous assessment and skills improvement of main database course topics. In: 7th Workshop on Teaching, Learning and Assessment in Databases, Birmingham, UK, HEA (2009)

38. Prados, F., Boada, I., Soler, J., Poch, J.: A Web-Based Tool for Entity-Relationship Modeling. In: Gavrilova, M.L., Gervasi, O., Kumar, V., Tan, C.J.K., Taniar, D., Laganá, A., Mun, Y., Choo, H. (eds.) ICCSA 2006. LNCS, vol. 3980, pp. 364–372. Springer, Heidelberg (2006)
39. Prados, F.I., Boada, I., Soler, J., Poch, J.: An Automatic Correction Tool for Relational Database Schemas. In: Information Technology based Higher Education and Training (ITHET 2005), pp. 9–14 (2005)
40. Soler, J., Boada, I., Prados, F., Poch, J., Fabregat, R.: An Automatic Correction Tool for Relational Algebra Queries. In: Gervasi, O., Gavrilova, M.L. (eds.) ICCSA 2007, Part II. LNCS, vol. 4706, pp. 861–872. Springer, Heidelberg (2007)

Research Directions in Database Architectures for the Internet of Things: A Communication of the First International Workshop on Database Architectures for the Internet of Things (DAIT 2009)

Anne James[1], Joshua Cooper[2], Keith Jeffery[3], and Gunter Saake[4]

[1] Department of Computing and the Digital Environment, Coventry University, UK
a.james@coventry.ac.uk
[2] Hildebrand, London, UK
jcooper@hildebrand.co.uk
[3] Science and Technology Facilities Council, Didcot, UK
keith.jeffery@stfc.ac.uk
[4] Department of Computer Science, University of Magdeburg, Germany
saake@iti.cs.uni-magdeburg.de

Abstract. This paper is a record of the First International Workshop on Database Architectures for the Internet of Things. The Internet of Things refers to the future internet which will contain trillions of nodes representing various objects from small ubiquitous sensor devices and handhelds to large web servers and supercomputer clusters. The workshop investigated a number of areas appertaining to data management in the Internet of Things from storage structures, through database management methods, to service-oriented architectures and new approaches to information search. Running orthogonal to these layers the matter of security was also considered. Taking a philosophical viewpoint our whole current framework for understanding data management may be ill-equipped to meet the challenges of the Internet of Things. The workshop gave participants the chance to discuss these matters, exchange ideas and explore future collaborations.

Keywords: Benchmarking, Database Architectures, Database Management, Internet of Things, Privacy, Storage Structures, Services Architecture, Searching, Security, Sensors, State, Storage Structures, Time-series, Transactions, Trust.

1 Introduction

This paper is a record of the First International Workshop on Database Architectures for the Internet of Things, which was held on 6th July 2009 at the University of Birmingham, UK in conjunction with the 26th British National Conference on Databases. The aim of the workshop was to bring together academics working in this exiting new area to share ideas and build partnerships for future work.

It is widely predicted that the next generation of the Internet will be comprised of trillions of IP connected nodes. Furthermore, it is predicted that these nodes will be specialized devices that produce and consume content in a different way to today's

A.P. Sexton (Ed.): BNCOD 2009, LNCS 5588, pp. 225–233, 2009.

predominant web server/browser combination. One analogy could be that this new Internet will resemble a very large sensor and actuator network (VLSANET), collecting, storing and organizing the data for retrieval by man and machine. Very large numbers of nodes are recording, transmitting, storing data and similarly large numbers of nodes are retrieving, analyzing and consuming the information with intermediate nodes transforming and integrating data to information. The volumes are vast, the speed is fast and the data/information space is global – indeed with space data it is universal! Given the scale of the task, existing storage, retrieval and computational models fall short.

A useful subset of this problem is framed by the use of the data in a control and signalling application. Conventional record-based, meta-data are not part of the system. Four dimensional attributes become interesting, say location in three dimensions and time with the physical reading, perhaps temperature, at that point. This can be simplified to a two dimensional storage problem if the devices are assumed to be stationary. This is a sampled reality and therefore can be represented by models of discrete mathematics and it follows well known principles of control, signal processing and statistical analysis. Problems are bounded by sampling rates, precision of the reading and any other systems anomalies introduced by the information system such as time lags.

This workshop posits that a new class of database system is required to meet the emerging application needs of the next generation Internet. The characteristics of the new class demonstrate the write speed requirements of a large transaction processing database, the relaxing of persistence guarantees, the calculation of time series statistics within ad hoc queries and read speeds guaranteed by a quality of service. Information in an Internet device environment is both supply and demand flexible, specifically devices may go on and off line without warning. Common schema for supply and demand will not exist, some standards may emerge, but the more flexible the logical data model, the more useful the system will be. Questions are not pre-determined, therefore an optimization is difficult to achieve with a priori information, such as pre-aggregation, indexing, optimal storage architectures. By its very nature, data does not remain in the same location, in some cases the data may even be mobile. Furthermore there is no central concept of state in an internet of devices. How can complex multi-level transactions be handled? Standard transaction mechanisms are meaningless. New methods need to be introduced which do not presume a global state.

Many of the trillions of devices in the Internet of Things will need to be cheap and easily maintainable. Database management systems for these, human maintenance or even frequent automatic software updates will not be feasible. We will need a new software engineering discipline for small, tailor-made, self-maintaining data management components on such devices. Creating this new discipline from what we already know about data management and combining it with traditional methods which might be supported on larger nodes will be a great challenge.

Prof. Keith Jeffery from STFC (Science and Technology Facilities Council), Didcot, UK, gave a guest lecture at the workshop [1]. Jeffery, who has been considering the idea of the Internet of billions of devices for some years, shared his findings with us. In particular Jeffery feels the matter of state is something that is key in the new Internet. The Internet of Things does not have a state and this makes transaction handling, as we traditionally know it, a near impossibility. Our traditional theoretical models will no longer work. There are many questions to be answered. What does the concept of state mean when the information map of the real world of interest is represented across millions of nodes, many of which are updating in real-time? What does integrity mean when a node in America is trying to consume information from nodes in the Far-East collecting data in real-time (and is therefore always out of date)? What does a transaction look like when the data being updated is spread across hundreds or thousands of nodes with differing update policies? A repeatable read would need to lock multiple segments of databases on multiple nodes implying continued real-time update continuing in a shadow database and a lack of integrity. Worse, how does one roll-back or compensate a transaction? The problem is that the world is changing fast, the data representing the world is on multiple nodes and database technology cannot manage. We have to rethink our notions of transaction and database management and look for new solutions. Techniques developed for streamed and real-time data may provide some direction.

Management of state is not the only challenge. Jeffery also raised the matters of security, trust and privacy. The vast amount of information being generated raises social questions concerning these important requirements. The Internet of Things opens up more opportunities for security compromises because of the greater number of interconnections and the need for trust between systems on heterogeneous distributed nodes. How do we develop trust based security techniques across multiple policies? Similarly the development of sensor webs, detecting continuously in real-time, including audio and video streams as well as conventional detector measurements, compromise privacy yet may provide security in the homeland security sense. Where is the balance? How do we prevent unauthorised use of private information yet permit authorised use? We need dynamic trust, security and privacy management. Normally one may wish medical information to be shared only by one's general practitioner and anyone authorized by her. However, a pedestrian/vehicle accident on the streets of Tokyo may well change the personal trust, security, privacy preferences. It may well be that we need a new theoretical framework.

The workshop sought papers describing ideas and current work in the general area of new generation database systems for a sensor and object rich internet. Interesting areas included: applications; artificial intelligence; handling four dimensionality of data objects; information retrieval and decision making; mathematical support models; mobile data; optimisation; locating and interconnecting; transaction management; querying data streams; sampling; security and reliability; semantics; statistical methods; and test beds.

The submissions received ranged suitably over the traditional software stack model. Contributions could be classified into the following categories: storage architectures and methods; new database management; service-oriented architecture; search applications and approaches; and security. The rest of the paper is organized as follows. Section 2 covers storage architectures and new types of database management system. Section 3 describes accepted work in service-oriented architecture. Section 4 considers the application level and methods of searching and querying. Section 5 presents an approach for improving security across the device-rich internet. Section 6 offers a conclusion.

2 Storage Architectures and Database Management Systems

Craciun, Fortis, Pungila and Aritoni of the Research Institute eAustria Timisoara, Romania, have been working as part of an FP7 project called Dehems [2]. A target of Dehems is to change the mindset of home owners towards sustainable energy consumption. The Dehems project concerns collecting sensor data on energy consumption from houses and presenting this to users in ways which will encourage good energy habits.

The contributors discussed the project and its relevance. They explained that amount of data that is collected from sensors is vast, even with a low sample rate. The collection and storage of these massive data sets needs careful consideration. The storage technology must allow for functionality like data mining and forecasting energy consumption and costs. It should allow usage patterns to be detected so that support can be given for changing consumer behaviour.

The solution adopted by this project was to use a simple data model [3]. A case study was used to test collection and aggregation speed for a stream of data generating around 100 million rows. Various database systems with different storage organizations were tested. The systems included relational, key value stores, column store, self-tuning and time-series enabled database systems. The results showed that column and key value stores performed better than relational database. Time series databases outperformed all the others. As a conclusion, taking into consideration detailed results, the authors presented a combined approach consisting of key value stores, column stores and time series databases.

Siegmund, Rosenmueller and Saake of the University of Magdeburg, Germany, together with Moritz and Timmerman of the University of Rostock, Germany, argued that robust data management is gaining in importance as device-rich Internet applications grow [4]. Reliable data management raises the requirement of continuous data availability. To achieve this in traditional systems, proxies are often used. Data is recorded and transmitted to proxies, which are connected to mains power supplies. Such solutions are not suitable for the Internet of Things because of scalability and hardware costs. The contributors proposed a redundant and robust data storage layer for wireless sensor networks based on the concept of RAID storage [5]. The diversity of hardware device in the Internet of Things means that data management systems often have to be tailored to the particular device. The contributors propose a customisable data management infrastructure, with a redundant and robust RAID based data storage layer for wireless sensor networks. A prototypical DBMS for sensor nodes has been implemented which is can be tailored according to the needs of nodes in wireless sensor networks [6]. The DBMS is used on different nodes and provides a comprehensive query interface with a tailor-made SQL dialect [7].

3 Service-Oriented Architecture

Koeppen, Siegmund, Soffner and Saake of the University of Magdeburg, Germany advocated the use of a service-oriented architecture for embedded systems created in a virtual environment [8]. For development purposes it is frequently necessary to create objects in a virtual environment. Furthermore, sometimes one may wish to combine virtual devices with real ones. The use of the service layer allows for this sort of interoperability. The authors have developed a suitable service-oriented architecture and plan to implement it in a virtual reality platform to allow different users to develop complex systems simultaneously. The virtual platform allows representation of the actual environment and brings non-visible properties such as security and safety issues, into focus. Data might be collected from different data sources and can be merged into a unique schema stored as a data collection service. Meta-data management as well as transformation rules have to be considered in this collecting process.

The authors intend to implement the interoperability platform using reuse-supporting software product line technology. This will allow for the automatic generation of a tailor-made service for a system if hardware, communication protocol, operating system and desired functionality are known. Non-functional requirements such as reliability, maintainability and software footprint can also be considered in the configuration process [9]. A solution based on software product technology to derive a tailor-made, integrated data schema, which is needed for data and schema integration of multiple sensors, has already been implemented [10].

4 Searching and Querying

According to Bizarro and Marques of the University of Coimbra, Portugal, the Internet contains thousands of Frequently Updated Timestamped Structured (FUTS) data sources [11]. These data sources are not like the semi-structured textual data one typically gets on web pages. They have a record-like structure and may appear on or form part of a web page. Examples of FUTS data sources are sports scores, stock exchange information, weather updates, real-time flight details, auction prices and traffic reports. FUTS data sources represent states and updates of real-world things. They are frequently updated and thus could be viewed as a stream. Bizarro and Marques introduced the interesting topic of FUTS data sources to the workshop and posited that FUTS data sources are not being properly handled by current technology. They identified three types of weakness of current technology when dealing with FUTS data sources: (i) there is no data management system that easily displays FUTS past data; (ii) there is no current efficient crawler or storage engine to collect and store FUTS data; and (iii) querying and delivering FUTS data is very limited. The researchers concluded that the internet contains thousands of FUTS data sources that are not being stored, parsed, aggregated or queried. New data management systems are needed with new user interfaces, parsers, storage engines and delivery mechanisms [12]. Such a system is currently being developed at the University of Coimbra.

Lansdale, Bloodsworth, Anjum and McClatchey of the University of West of England, UK, stated that the future Internet is likely to be much larger and more complex than the current Internet [13]. Data and devices are ever increasing and thus information overload could become a bigger problem than it is today. Data will be increasingly distributed and mobile. As well as effective storage and database technologies, new efficient, querying mechanisms will be necessary to be able aggregate and make sense of the massive and continuously changing information. The participants proposed that semantic meta-data can play a greater role in the structuring of knowledge contained in the Internet and could be used to facilitate efficient querying, increased personalisation and data integration. They argued that personalisation is important as it will lead to customized and more efficient querying. Ontologies can be used to structure information regarding access methods to Internet resources [14]. This can support personalisation of access approach and in turn more efficient querying. Agents can be used to support the personalisation process in a distributed environment [15]. The contributors recommended that semantic meta-data is used to represent and describe how information can be accessed and linked together. The use of a rule-engine to harness this semantic data during the processing of queries and then drive personalised agents was discussed. A prototype architecture for the querying of heterogeneous distributed data in the medical domain was presented. The contributors argue that the same approach can be applied to the future Internet. The architecture consists of an intelligent querying service which has access to ontological domain data to process, enrich and personalise queries. It features a layer of abstraction allowing it to be connected to the kinds of distributed heterogeneous data that will be found in the future Internet [16].

Podnar Zarko and Pripuzic of the University of Zagreb, Croatia, argued that the Internet of Things requires novel approaches to information search which will differ significantly from the ways we perceive the quest for information today [17]. Centralised data stores will not be suitable for a highly distributed, data intensive environment. Decentralised search engines offer a viable alternative to centralised solutions [18]. The investigators explained that users should be perceived not only as active information searchers but also as information consumers who have standard information requirements. Information consumers expect to be notified when information of interest to them is published on the global Internet. This type of continuous passive search is closely related to the publish/subscribe paradigm of the current internet.

In previous work the contributors have designed algorithms for controlling the quantity of information items delivered to users [19] and implemented systems which take into account mobility and intermittent connections of information sources and destinations in distributed publish/subscribe systems [20]. Podnar Zarko and Pripuzic believe these solutions represent the basic building blocks for the future search engines. They went on to elucidate on three factors which will be of great importance to future Internet search. These were: wireless sensor networks and data stream processing methods; mobile and telecom networks convergence where various devices can become universal interfaces to personal services; and social networks which may prove useful in honing the search space leading to more accurate personalisation.

5 Security

Wu and Hisada from NST Inc., Japan and Ranaweer of the University of Aizu, Japan commented that web security has been disastrous in the last decade due to poor identification of vulnerabilities. In the Internet of Things security will be even more challenging because of the vastly increased amounts of data, devices, distribution and heterogeneity. Wu, Hisada and Ranaweer aim to improve security matters by moving web security from black box to white box based on static analysis [21]. This involves a number of steps: (i) analysis of the syntax structure; (ii) code block identification; (iii) reverse engineering between source code and abstract syntax structure; (iv) notation of meta-data for node and node-node interference; (v) interoperation between event and node. Wu, Hisada and Ranaweer concentrated their presentation on the first three of these steps. They described their successful implementation of a universal adapter which can convert most common programming languages into a generic form of abstract syntax structure. Based on this foundation, a meta-data messaging plug-in had been applied to identify nodes and their interferences in terms of event and state. A footprint of user interaction from log-in to log-out can be provided using plug-in messaging. The researchers felt they had achieved gray box and their next step is to move this to white box by incorporating stronger input validation and meta-data strategy. They believe this approach can lead to improved web security governance which may be able to be applied to the Internet of Things.

6 Conclusion

The 1st international Workshop on Database Architectures for the Internet of Things saw some most interesting contributions from colleagues from the UK, Europe and further afield. Interesting ideas were discussed for many aspects of data management in the Internet of Things. Such areas included: suitability of storage structures; new data management systems; service-oriented architectures; and new types of information search which may involve personalisation, archive querying and information consumption as opposed to active searching. The orthogonal areas of security, trust, privacy, performance, quality of service and ease of use were also the subjects of discussion. Set against these aspects was a consideration of the suitability and feasibility of different types of architecture such as centralized, distributed with coordination and distributed with zero-knowledge of environment. The prevailing message emerging from the workshop was that we cannot rely on traditional database methods for the future Internet and that new methods and theories of database architectures for the Internet of Things are urgently required. There is certainly a rich research opportunity before us.

Acknowledgement

The DAIT 2009 Workshop and Programme Chairs, Joshua Cooper and Anne James, would like to thank all who have contributed directly or indirectly to the production of this communication and in particular the Programme Committee members for their considerate work in reviewing abstracts and full submissions to tight deadlines. Membership of the DAIT 2009 Programme Committee is as shown below:

Stuart Allen	Cardiff University,UK
Rachid Anane	Coventry University,UK
Sharma Chakravathy	University of Texas at Arlington, USA
Kuo-ming Chao	Coventry University,UK
Graham Cooper	University of Salford, UK
Joshua Cooper	Hildebrand, UK
Fortis Florin	West University of Timisoara, Romania
Mary Garvey	University of Wolverhampton, UK
Mike Jackson	Birmingham City University,UK
Anne James	Coventry University,UK
Keith Jeffery	STFC, UK
Milko Marinov	University of Rousse, Bulgaria
Robert Newman	University of Wolverhampton, UK
Dana Petcu	IeAT, Romania
Gunter Saake	University of Magdeburg, Germany
Alan Sexton	University of Birmingham, UK
Jianhua Shao	University of Cardiff, UK)
Andy Sloane	University of Wolverhampton, UK
Peter Smith	University of Sunderland, UK

References

1. Jeffery, K.G.: The Internet of Things: The Death of Traditional Database? In: 1st International Workshop on Database Architectures for the Internet of Things, Birmingham (2009) (to be published)
2. Dehems Project, `http://www.dehems.eu`
3. Craciun, C.-D., Fortis, T.-F., Pungila, C., Aritoni, O.: Benchmarking Database Systems for the Requirements of Sensor Readings. In: 1st International Workshop on Database Architectures for the Internet of Things, Birmingham (2009) (to be published)
4. Siegmund, N., Rosenmueller, M., Moritz, G., Saake, G., Timmermann, D.: Towards Robust Data Storage in Wireless System Networks. In: 1st International Workshop on Database Architectures for the Internet of Things, Birmingham (2009) (to be published)
5. Moritz, G., Cornelius, C., Golatowsksi, F., Timmermann, D., Stoll, R.: Differences and Commonalities of Service-oriented Device Architectures, Wireless Sensor Networks and Networks-On-Chip. In: International IEEE Workshop on Service Oriented Architectures. IEEE Computer Society Press, Los Alamitos (2009)
6. Rosenmueller, M., Siegmund, N., Schirmeier, H., Sincero, J., Apel, S., Leich, T., Spinczyk, O., Saake, G.: FAME-DBMS: Tailor-made Data Management Solutions for Embedded Systems. In: EDBT 2008 Workshop on Software Engineering for Tailor-made Data Management (SETMDM), pp. 1–6 (2008)
7. Rosenmueller, M., Kaestner, C., Siegmund, N., Sunkle, S., Apel, S., Leich, T., Saake, G.: SQL a la Carte – Toward Tailor-made Data Management. In: 13th GI-Conference on Database Systems for Business, Technology and Web (BTW), pp. 117–136 (2009)
8. Koeppen, V., Siegmund, N., Soffner, M., Saake, G.: An Architecture for Interoperability of Embedded Systems in the Virtual Reality. In: 1st International Workshop on Database Architectures for the Internet of Things, Birmingham (2009) (to be published)

9. Siegmund, N., Rosenmueller, M., Kuhlemann, M., Kaestner, C., Saake, G.: Measuring Non-functional Properties in Software Product Lines for Product Derivation. In: 15th International Asia-Pacific Software Engineering Conference (APSEC), pp. 187–194. IEEE Computer Society, Los Alamitos (2008)
10. Siegmund, N., Kaestner, C., Rosenmueller, M., Heidenreich, F., Apel, S., Saake, G.: Bridging the Gap between Variability in Client Application and Database Schema. In: 13th GI-Conference on Database Systems for Business, Technology and Web (BTW), pp. 297–306 (2009)
11. Bizarro, P., Marques, P.: The Internet contains Thousands of Poorly Explored FUTS Data Sources. In: 1st International Workshop on Database Architectures for the Internet of Things, Birmingham (2009) (to be published)
12. Adar, E., Dontcheva, M., Fogarty, J., Weld, D.: Zoetrope: Interacting with the Ephemeral Web. In: UIST 2008, Monterey (2008)
13. Lansdale, T., Bloodsworth, P., Anjum, A., McClatchey, R.: Querying the Future Internet. In: 1st International Workshop on Database Architectures for the Internet of Things, Birmingham (2009) (to be published)
14. Maedche, M., Staab, S.: Ontology Learning for the Semantic Web. Intelligent Systems 16(2), 72–79 (2001)
15. Huhns, M.N.: From DPS to MAS: Continuing the Trends. In: AAMAS, Budapest (2009)
16. Bloodsworth, P., Greenwood, S., Nealon, J.: A Generic Model for Distributed Real-time Scheduling based on Dynamic Heterogeneous Data. In: Lee, J.-H., Barley, M.W. (eds.) PRIMA 2003. LNCS, vol. 2891, pp. 110–121. Springer, Heidelberg (2003)
17. Podnar Zarko, I., Pripuzic, K.: Information Search in the Internet of Things. In: 1st International Workshop on Database Architectures for the Internet of Things, Birmingham (2009) (to be published)
18. Skobeltsyn, G., Luu, T., Podnar Zarko, I., Rajman, M., Aberer, K.: Web Text Retrieval with a P2P Query-driven Index. In: 30th Annual International ACM SIGIR Conference, pp. 679–686. ACM Press, New York (2007)
19. Pripuzic, K., Podnar Zarko, I., Aberer, K.: Top -k/w Publish/Subscribe: Finding k most relevant Publications in Sliding Time Window. In: 2nd International Conference on Distributed Event-based Systems (DEBS 2008), pp. 127–138. ACM Press, New York (2008)
20. Podnar, I., Lovrek, I.: Supporting Mobility with Persistent Notifications in Publish/Subscribe Systems. In: 3rd International Workshop on Distributed Event-based Systems (DEBS 2004), pp. 80–85 (2004)
21. Wu, R., Hisada, M., Ranaweer, R.: Static Analysis for Web Security in Abstract Syntax Format. In: 1st International Workshop on Database Architectures for the Internet of Things, Birmingham (2009) (to be published)

Design Challenges and Solutions: Review of the 4th International Workshop on Ubiquitous Computing (iUBICOM 2009)

John Halloran[1], Rahat Iqbal[1], Dzmitry Aliakseyeu[2], Martinez Fernando[3], Richard Cooper[4], Adam Grzywaczewski[5], Ratvinder Grewal[6], Anne James[1], and Chris Greenhalgh[3]

[1] Faculty of Engineering and Computing, Coventry University, UK
{j.halloran,r.iqbal,a.james}@coventry.ac.uk
[2] Philips Research Europe, Eindhoven, The Netherlands
dzmitry.aliakseyeu@philips.com
[3] The Mixed-Reality Lab, University of Nottingham, UK
{fxm,cmg}@cs.nott.ac.uk
[4] Department of Computing Science, University of Glasgow, UK
rich@dcs.gla.ac.uk
[5] Trinity Expert Systems Ltd, UK
adamg@tesl.com
[6] Department of Mathematics and Computer Science,
Laurentian University, Canada
rsgrewal@cs.laurentian.ca

Abstract. This paper provides an overview of several approaches, methods and techniques, of ubiquitous and collaborative computing, discussed in the papers submitted to the International Workshop on Ubiquitous Computing (iUBICOM-09). In this workshop, we aimed to balance discussion of technological factors with human aspects in order to explore implications for better design. The theme was information retrieval, decision making processes, and user needs in the context of ubiquitous computing. This paper includes work carried out on different dimensions focusing on technological as well as social aspects of ubiquitous and collaborative computing.

Keywords: ubiquitous computing, design, mobile computing, device interaction, context-aware systems, information retrieval, smart homes, assistive technology, ethnography.

1 Introduction

The fourth International Workshop on Ubiquitous Computing was aimed at balancing discussion of technological factors with human aspects in order to explore implications for better design. Thus, the theme was 'information retrieval, decision making processes, and user needs in the context of ubiquitous computing'.

A particular focus was how user needs could be incorporated into the enhanced design of ubiquitous computing systems such as smart homes, assistive technologies,

A.P. Sexton (Ed.): BNCOD 2009, LNCS 5588, pp. 234–245, 2009.

and collaborative systems. Such systems capture information of various types in order to make decisions which meet human needs, depending on the context. In order to design and develop systems which can be successfully deployed, it is necessary to take user needs into account from the outset.

This workshop attracted a wide range of participants involved in designing such systems, whose contributions addressed the following topics: user-centred design approaches for ubiquitous systems (Section 2); mobile computing to support multi-device interaction (Section 3); smart homes (Section 4); assistive and supportive technology (Section 5); ethnographically-informed system design (Section 6), context aware systems (Section 7), and information retrieval (Section 8).

2 User-Centred Design Approaches for Ubiquitous Systems

In the past, many projects have failed and one of the key reasons of their failure is inadequate analysis of user requirements [1, 2]. Most importantly, social, political and cultural factors were not taken into account for the design and development of these systems.

Motivated by the failure of systems in the past, many researchers have moved to actual working practice in order to acknowledge the work taking place in the real world, including real-time phenomena.

In line with the vision of ubiquitous computing and calm technology, several recent projects such as AMI [3], Interactive Workspaces [4] and CHIL [5] are developed using user-centred approaches. We believe it is important, when designing ubiquitous computing, to understand the social characteristics of work, people, and the real context of work. A more extensive overview of methods and approaches to support design can be found in [6]. This is not a new problem: work in CSCW as well as HCI has long recognised that the situatedness and context-dependence of collaborative work warrants, and even necessitates a user-centred design approach. However, what is new with ubiquitous computing is the complexity and distributedness of systems, together with the range and possible novelty of the activities they can support. The importance of this issue was reflected in a wide range of papers which appeared at the workshop, both explicitly and implicitly, as the following examples show.

3 Mobile Computing and Multi-device Interaction

Aliakseyeu and colleagues at Philips Research Europe (Eindhoven) have developed a technique called 'Sketch Radar', for representing computing devices in the immediate environment.

Thanks to the rapidly reducing cost of display and network technologies, situations in which many different devices with heterogeneous display sizes interact together are becoming commonplace. Often these environments present a mixture of personal devices such as Personal Digital Assistants (PDAs), tablet and laptop PCs, and shared devices such as large displays.

One way of representing these spaces is map-based techniques such as Radar View [7]. These have the potential to support intuitive system identification and interaction without necessarily requiring physical proximity to the system they interact with. The success of map-based techniques relies on being able to associate a physical device with its representation on the map. However, how this association is accomplished and maintained appears never to have been studied in detail.

The common implementation of Radar Views is based on the physical positions of interacting devices. This raises several questions/issues: (1) how is the relevant information needed to construct the radar map acquired from the interacting devices; (2) which devices should be presented on the map and should all devices be equally prominent; (3) how should devices be represented; (4) what are the boundaries of the map; and (5) how to deal with the fact that the map needs to be presented on a screen with limited size and resolution? To address these questions a Sketch Radar technique (see Fig. 1) has been proposed.

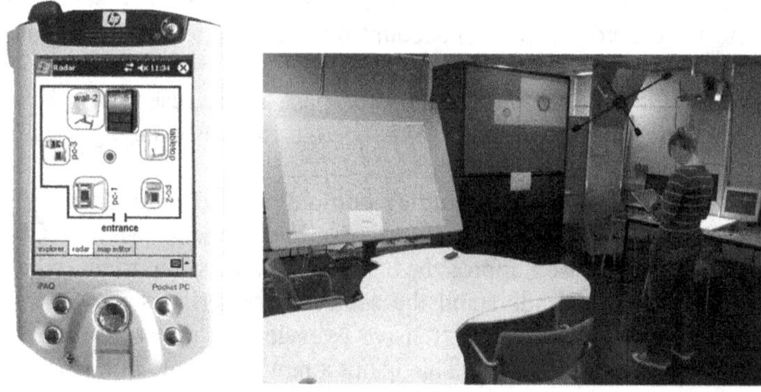

Fig. 1. Sketch Radar (left) and environment used in both experiments (right)

Sketch Radar is a novel interaction technique that is based on the radar metaphor. With it, a user is able to control how and what information is presented on the radar at any time. The representation of a device on the radar map can be acquired in a direct and explicit way. In the current prototype this is accomplished by means of a barcode reader that reads a barcode of the device. This way, the user can decide which devices to place on the map. The action of scanning the devices also enforces the mental link between the real device and its representation on the map. Only devices that are explicitly scanned with the barcode reader are presented on the map, and users are free to adjust them at any time. In addition, agglomerations of devices, such as all devices present in a single room, can also be obtained by scanning strategically placed barcodes (such as near the door entrance). One distinctive feature of the Sketch Radar is that it does not force the user to stick to a radar map that is a one-to-one reflection of the geometrical organization of the physical devices. The users are free to adjust the map in such a way that it fits better to a particular task or preference.

In a new environment it is usually wise to start with a map that is based on the physical position and size of the devices. After some time, an environment becomes more familiar and tasks become clearer. This may lead the user to readjust the positions, sizes and representations of the devices.. For example, frequently-used devices may be increased in size and placed closer to the centre of the map. Also, by allowing the user to add 'sketches' (lines, text) to the map, s/he can add elements that further strengthen the association between a specific map and a particular task. This flexibility makes the Sketch Radar useful in a range of situations, ranging from interaction in an unfamiliar space, where a close correspondence with the physical arrangement is needed to identify individual devices, to frequent and long-term usage, where the physical space is well known and users can profit from a map that is specifically tailored to their purpose.

Two user studies were performed that explore whether or not users would appreciate the possibility of adapting radar maps to particular tasks and personal preferences and if so, which criteria would be used to motivate changes. A modified version of the Sketch Radar prototype, that provides an easy and quick way to manage one or more maps of available devices, was used for implementing the experiment. The first study was done using a PDA and the task consisted of moving several files from one computer to another while being inside the room. The second study was done using TabletPC and tasks were given in a form of a game.

Both studies showed that representations of linked computers can be organised not just as maps of their real locations, but according to task demands. The studies confirmed that users modify the map for different reasons, namely type of computers, relation between computers defined by the task, visibility of the computers, spatial relation, and order of computers. Since no explicit performance measures are available it is still unknown if an altered representation is more efficient than a representation purely based on the physical locations.

4 Smart Homes

Fernando and colleagues at the University of Nottingham have developed a ubiquitous computing system called the 'activity-aware room'.

Location- and context-aware services are considered key factors in deploying collaborative resources in the home. People with special needs can be supported by smart technologies. The smart context typically explored for collaboration takes for granted that proactive collaboration will always be accepted by homeowners. The activity-aware room offers unobtrusive collaboration to parents to help monitor their children's activity.

The activity-aware room, which was set up in a real home, collects data about the children's activities, and applies a heuristic to decide whether parents should be warned about the child's whereabouts. The system identifies entries and exits into a room, activity in the room space, and activity close to artefacts within the room. Sensors installed in the activity-aware room prototype include motion, proximity and beam-break sensors. Proximity sensors collect activity near artefacts; motion sensors monitor activity at the centre of the room; and beam-break sensors identify whether a parent has entered or exited the room.

A test was carried out to classify entry/exit events from the door. Two sensors were installed on each of the frame doors – kitchen and hallway. A classification algorithm that identifies user events was explored. The implemented heuristic was able to classify entry/exit events from the parent's activity with over 90% accuracy. This is particularly important because the system must be able to identify episodes in which the parent is directly nurturing the child. However, although there was 90% success in the classification of events relating to parents, 10% uncertainty makes it likely that the system can interrupt parental nurturing. So an extra effort will be necessary to improve the classification algorithm.

Regarding the space of the room, to identify if the child is in the room, activity from the room sensors needs to be explored. The system can be aware of, for instance, changes with the motion sensor. Then, if the child is in the middle of the room and no more activity is detected from the other sensors, then the system might be able to infer that the child is in a 'safe' place. If on the contrary, there is activity close to the fireplace and the fireplace is on, then the system would decide to interrupt the parent with a report.

In terms of activity in relation to artefacts, sensors attached to artefacts should help to monitor activity around other home spaces. It could be identified whether the child is close to the fireplace, for instance. Information collected indicates that the sensor's performance is affected by environmental changes with the room illumination. The management of uncertainty from infrared technology might require an exhaustive exploration of machine learning algorithms. Consider that changes in the illumination level of the room can be due to the weather, the use of curtains, the use of interior lights and paint on walls to name some. See Fig. 2.

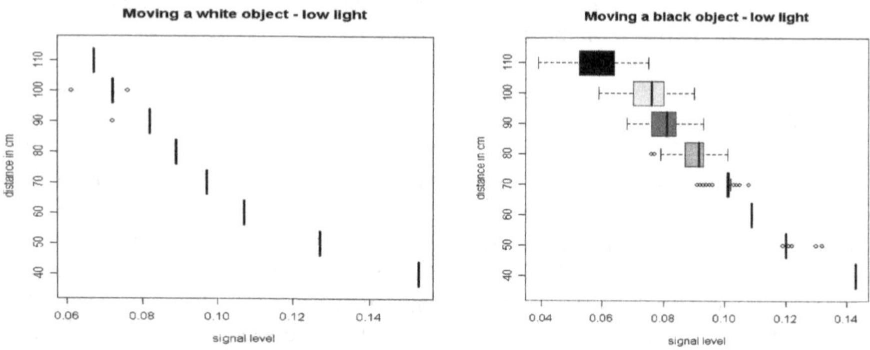

Fig. 2. Proximity sensor activity that shows how illumination levels within the room can affect its behaviour. Left side represents the sensor performance with white objects. Right side represents the sensor performance with dark objects. Both tests were done under the same light conditions.

In summary, the current exploration of the sensing information collected by the activity-aware room prototype shows that the system is not ready to derive smart collaboration. In its current stage, it is likely for the system to affect the parenting context with a high rate of ambiguous collaboration. The next section describes the way the parent can help the system to reduce uncertain collaboration. To address this a rule-based algorithm was implemented in which the user defines the level of awareness that might meet the parent's current needs. See Fig. 3.

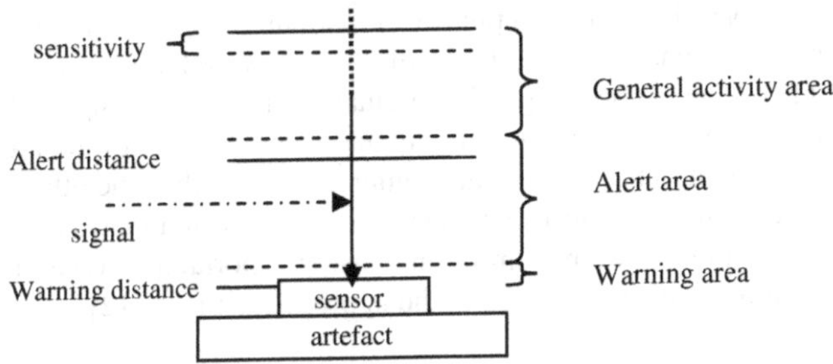

Fig. 3. The activity-aware system considers the user criterion to manage obtrusive collaboration. The user participates with the sensitivity and alert distance parameters.

The two important elements from Figure 3 are the alert distance and sensitivity parameters. The former tells the system what the individual preference is regarding the proximity of the child to the artefact. The latter represents the extent to which the parent wishes to be aware of the child's movements.

So the user's participation with the system to control ambiguous collaboration seems to be a useful resource to be included within the runtime of the system, as suggested by [8]. This degree of user's collaboration as part of the final implementation of the system's awareness, is not yet definitive. Thus, the study demonstrates the importance not only of detecting user activities but engaging with users to understand more about their activities, including user definition of when they require action.

5 Assistive and Supportive Technology

Iqbal and colleagues at Coventry University have been working in the area of assistive and supportive technology.

Assistive technology is used to support patients in carrying out their day-to-day activities. Several such systems have recently been developed to support patients at home. On the other hand supportive systems might be used to support the health staff in carrying out their tasks such as diagnosing the dementia patient [9]. A supportive system is proposed to monitor and assess dementia patients in their homes in order to support GPs in their examination of onset dementia by providing the analysis of video footage of patients' behaviour in their home for early identification of onset dementia. The system uses facial recognition techniques and artificial intelligence methods [9].

Grewal and colleagues, working on persuasive technologies at Laurentian University, Canada, show that these technologies target a variety of health issues related to diet, exercise, smoking, drug abuse and mental health [10]. Given the acute shortage of healthcare professionals, persuasive technologies are bound to catch the attention of healthcare providers', insurance companies and the pharmaceutical industry; the technology will drive these sectors to invest in this area.

In the course of Grewal and colleagues' research, a genetic algorithm has been developed that allows efficient integration of diverse biological data [11]; with development of embedded technology, this algorithm can be used in medical devices to infer decisions based on data input by a patient using the device. Persuasive messages regarding the patient's health can then be delivered at the right time through these devices [12]. The embedding of computational power along with interactive displays encourages self care among patients and they take on a much active role in self health management. The patients are better aware of their health condition and also feel comfortable to open up to persuasive devices which can be used in utmost privacy; the devices in turn are becoming increasingly portable and can be carried by the user to provide persuasive feedback at the appropriate place [12].

6 Ethnographically-Informed System Design

Halloran and colleagues at the Universities of Sussex and Nottingham developed a novel interactional workspace to support collaboration.

The eSPACE system is an interactive travel planner that allows travel consultants and customers, working face-to-face, to effectively plan complex tailor-made holidays. It is an example of how ubiquitous computing is situated in specific contexts for particular purposes, rather than being general purpose machinery as with the 'desktop' paradigm. Because the design of the system involved understanding the physical setting of face-to-face travel sales and how technology works in that setting, an ethnographically-informed observational approach was taken, integrated with user -centred design (UCD). Further details of the project can be found in [13].

A six-month ethnography study at a London travel agency showed that when planning a complex holiday involving round-the-world trips, many kinds of information are needed, in various forms and places (e.g., online booking systems, brochures, external databases, websites). However, it was also noted that the interaction between travel consultants and customers is often dominated by the consultant, working on his/her desktop PC, with the customer seated at the other side of the desk, contributing relatively little to the planning process. Such an asymmetry was found to hinder progress and customer satisfaction with the transaction process. In particular, customers find it hard to remember all the information given by consultants, and feel confused about how far they have got in terms of achieving their goal. At the same time, consultants find it hard to sell holidays, with many customers not returning after the initial consultation. A big issue here is that the planning process is a technical problem and consultants, rather than being able to spend time evoking and selling a holiday experience, have to deal with tables of product codes and their relationships. They are often drawn into attempting to 'translate' these for customers, discussing logistical issues in arranging flight sequences or hotel availability.

Thus, the ethnography led to the counterintuitive conclusion that technologies for face-to-face transactions tend to hinder rather than help travel planning conceived as a collaborative activity between customers and consultants. The project aimed to improve this situation by (i) empowering customers by making it possible for them to take a bigger part in the planning and decision-making process; and (ii) enabling the consultant to more effectively manage the transaction through the development and introduction of new ways of showing and explaining complex information. A key change was to provide better links between the different information representations needed for planning and product development to allow for more effective coordination of information. In particular, to improve joint planning, it was felt necessary to change the way information was displayed such that it could be more readily accessed, interacted with and shared.

To this end, an innovative shared workspace and accompanying software were built. This multi-screen system provided interlinked dynamic visualizations, contextually-relevant information to support decision-making activities, and user-centric interactive planning tools. The system is shown in Fig. 4.

Fig. 4. The eSPACE System

The bottom left screen shows a part of the world and the holiday is planned by dragging various objects - flights, hotels, tours - onto it. Backend databases work out whether the planned route - which can be edited and saved - is possible. The screen, bottom right, shows what the different products cost as a proportion of the budget, plus what the customer will be doing when. When a product (for example a hotel) is dragged onto a city, this activates a drop-down menu. Selecting an item brings up a brochure page, shown on the rear screen. The itinerary can be printed out and presented to the customer in a custom folder (shown back right).

Recognising that ubiquitous computing supports specific activities in given physical environments, a major commitment of the research was to develop a system in the context of use. Thus, it was used at the London travel agency for a period of months. It was possible to achieve a fairly complex itinerary at the initial consultation, and, as later interviews with customers showed, they came away with a much clearer idea of their holiday. At the same time, consultants were able to explore a number of options with customers in a much less technical way. For example, one consultant commented that the eTable enabled her to 'draw the customer in' by providing an attractive vicarious experience of the holiday: "It's much easier to sell when the client can see everything. They get excited when it starts taking shape."

The value of the eSPACE project is in showing how analysis and design of a system for collaborative work can be carried out in the context of use, employing an ethnographically-informed approach. This approach is already gaining importance in the design of ubiquitous computing, following the recognition of the radical situatedness of many systems not just physically but also in terms of the human activities ubiquitous computing supports. Such situatedness warrants - and indeed requires - approaches dedicated to close observation of those activities, in order to understand and effectively design for them.

7 Context-Aware Systems

Cooper and colleagues at the University of Glasgow are working in the area of context-aware systems, in particular the development of methods to support the design of such systems.

Ubiquitous applications like those described in this paper involve the use of a wide variety of devices connected in various ways and used by people operating in a variety of contexts. Programming such an application tends to be very complex, involving multiple programming technologies some of which, such as those for server programming, are well known and have alternatives for different platforms, while others, for instance the programming of sensors, are specific and unique to the particular device used. If every application has to be programmed separately, then keeping track of all of these options will make them very expensive to produce. Essentially good software engineering practice has tended to disappear as so many of the decisions to be made are low level. The need is to provide a modelling environment so that the whole of the application and the environment it is intended for can be described and implemented.

Model driven design is an approach which shows promise in this area. The application is described in a platform independent way at first and then transformations are described to create a deployable implementation. The work of Koch and her colleagues attempts to extend the use of UML to the descriptions of dynamic web sites [14]. However, the scope of ubiquitous applications is much greater. Firstly, applications which started out as web sites have evolved into applications intended to be available on a variety of devices in a wide range of contexts. And the situation becomes even more complex with systems involving dynamically configured mobile applications which merge more seamlessly with the physical environment – such as Smart Homes applications.

For a ubiquitous application, the model involves describing not just the functionality, data structure and user requirements, but also the delivery context and its features, as well as the implementation mechanism. The description of the context starts from an account of the devices involved and their capabilities together with the topology of the network which connects them, but must be elaborated with other contextual information such as user preferences, location and time. There follows a need to map application components to different devices and contexts.

Context aware systems [15] attempt to do this by describing application components in such a way that they behave differently in different environments. It is assumed that each function supported by an application might be accessed in a variety of situations each requiring a different behaviour or, rather, a different way of interacting to achieve that behaviour. Thus, at some point it must investigate the context and provide a version of itself which is suitable. This may be achieved using a mixture of two basic methods: providing a selection of versions of the component or a parameterised function which generates a version tailored to the context.

Take for example the user interface, the most widely studied way in which the provision of the function might change. If the requirement is to display an image to devices of different capability, the application might be supplied with a number of images each at a different resolution (plus a textual version for contexts in which a screen is unavailable) or a function which scales the image. One important research area attempts to model user interface *plasticity.* [16] describes the Cameleon framework for plastic interfaces and [17] describes USiXML a markup language to use this. MATCH [18] is one project attempting to bring all of these ideas into the context of ubiquitous applications – in this case Home Care Systems.

Cooper and colleagues describe a model and prototype environment for the creation of ubiquitous applications. The model comprises descriptions of devices, networks, tasks, user interaction and implementation languages and the prototype allows all of these to be managed in a coherent framework. This work is based on three main ideas. Firstly, everything (data, functions, context, etc.) is described in a *uniform component-based structure* – any design tool maps down to this structure [19]. Thus images, prompts, queries, tables, pages, devices, user context and whole applications take part in the same structure. Secondly, the *information schema* constructed to describe the data being managed is used to describe everything – i.e. when the underlying database structure is described, so are the structures available for describing users, messages, sessions and delivery context. Thirdly, support for *single sourcing* is attempted, not in the sense that there is only one value for each data object, but in the sense of having a single data object with multiple versions, each appropriate for different contexts. It is hoped that this can be extended to encompass version generators at a later time.

The prototype includes an editor of devices and their capabilities and a graphical network design tool with which the expected device topology is described. The application is described in terms of a task structure based on components which manipulate data objects described in terms of the application schema. The methodology for using the system allows each aspect of an application to be captured separately. A start point could be designing the physical environment by describing device capabilities and networks of devices. Applications are described in task structures involving multi-version components. Target languages are structured into a hierarchy, so that code can be generated from the designs captured by the interface. Using all of this, a consistent and coherent mechanism for multi-device ubiquitous applications can be established.

8 Ubiquitous Information Retrieval Systems

Grzywaczewski and colleagues at Trinity Expert Systems Ltd are working in the area of ubiquitous information retrieval.

With the decreasing price of mobile devices and mobile broadband connections, mobile information retrieval becomes an everyday task. However, different input/output characteristics as well as different types of search (including whether at home or at work) may change the way people estimate the relevance of results. Users' requirements for information clarity, presentation and ease of interaction may differ depending on the context. This can be addressed by context-aware personalized information retrieval systems. In order to develop such systems, there is a need to investigate what information retrieval strategies people develop through their information interaction and how they apply different search strategies in contexts that vary.

As part of this project, a user study is conducted where the search behaviour of internet users is observed across varying places (e.g., at home or at work) and devices (personal computers, laptops, or mobiles). User perception of the task is also relevant. Observations are used as a basis for finding correlations between devices, places, user perception and search strategies. Such correlations would enable automated support for searching. The aim of the research is to establish whether such correlations exist and if so to determine methods of exploiting them to provide personalised search support.

9 Conclusion

This year's workshop attracted a wide range of papers exploring challenges and solutions in the design of ubiquitous computing. A range of important issues are discussed. These include technical challenges in designing complex distributed systems; the importance of user needs; the need for user-centred design; and the improvement of human-human communication. There was repeated emphasis on applied (rather than theoretical) development of ubiquitous computing systems, and a range of applications were presented. A major theme that emerged at this workshop was thus the confluence of technical and user issues in the effective design, deployment, and evaluation of ubiquitous computing systems. We look forward to deepening our appreciation of this theme in future workshops.

Acknowledgements. All authors would like to thank colleagues who have contributed either directly or indirectly to this paper. Richard Cooper would like to thank Tony McBryan. Halloran, Iqbal and James would like to thank the programme committee of iUBICOM 2009 for supporting this workshop.

References

1. Goguen, J.A., Linde, C.: Techniques for Requirements Elicitation. IEEE Computer Requirements Engineering, 152–164 (1993)
2. Grudin, J.: Groupware and social dynamics: Eight challenges for developers. Communications of the ACM 37(1), 92–105 (1994)
3. AMI Consortium: http://www.amiproject.org
4. Fraunhofer Institute, http://www.ipsi.fraunhofer.de
5. CHIL, http://chil.server.de
6. Iqbal, R., Sturm, J., Kulyk, O., Wang, C., Terken, J.: User-Centred Design and Evaluation of Ubiquitous Services. In: Design of Communication: Documenting and Designing for Pervasive Information, ACM SIGDOC, pp. 138–145 (2005)
7. Nacenta, M.A., Aliakseyeu, D., Subramanian, S., Gutwin, C.: A comparison of techniques for Multi-Display Reaching. In: CHI 2002, pp. 371–380 (2002)
8. Huebscher, M.C., McCann, J.A.: Adaptive middleware for context-aware applications in smart-homes. In: Middleware For Pervasive and Ad-Hoc Computing (2004)

9. Iqbal, R., Naguib, R.: Real-Time Identification of Onset of Dementia Using Intelligent Techniques, Technical Report, Coventry University, UK (2009)
10. Strowes, S., Badr, N., Heeps, S., Lupu, E., Sloman, M.: An Event Service Supporting Autonomic Management of Ubiquitous Systems for e-Health. In: Distributed Computing Systems Workshops (2006)
11. Singh, P., Passi, K.: Incremental Maintenance of Ontologies based on Bipartite Graph Matching. In: Web Information Systems and Technologies (2009)
12. Zhu, J., Gao, L., Zhang, X.: Preliminary Research on Wearable Healthcare in Ubiquitous Computing Age. In: Computer Science and Software Engineering (2008)
13. Rodden, T., Rogers, Y., Halloran, J., Taylor, I.: Designing novel interactional workspaces to support face-to-face consultations. In: CHI 2003, pp. 57–64 (2003)
14. Kraus, A., Knapp, A., Koch, N.: Model-Driven Generation of Web Applications in UWE. In: Model Driven Web Engineering (2007)
15. Schilit, B.N., Theimer, M.M.: Disseminating active map information to mobile hosts. IEEE Network, 22–32 (1995)
16. Balme, L., Demeure, A., Barralon, N., Coutaz, J., Calvary, G.: CAMELEON-RT: A software architecture reference model for distributed, migratable, and plastic user interfaces. In: Markopoulos, P., Eggen, B., Aarts, E., Crowley, J.L. (eds.) EUSAI 2004. LNCS, vol. 3295, pp. 291–302. Springer, Heidelberg (2004)
17. Limbourg, Q., Vanderdonckt, J.: A User Interface Description Language Supporting Multiple Levels of Independence. In: ICWE Workshops, pp. 325–338 (2004)
18. McBryan, T., McGee-Lennon, M., Gray, P.: An Integrated Approach to Supporting Interaction Evolution in Home Care Systems. In: Pervasive Technologies Related to Assistive Environments (2008)
19. Cooper, R., Wu, X.: A Generic Model for Integrated Multi-Channel Information Systems. In: CAiSE Workshops, pp. 731–744 (2005)

Author Index